NATURAL RESOURCE MANAGEMENT AND POLICY

For further volumes:
http://www.springer.com/series/6360

EDITORIAL STATEMENT

There is a growing awareness of the role that natural resources, such as water, land, and forests, as well as environmental amenities play in our lives. There are many competing uses for natural resources, and society is challenged to manage them to improve social well-being. Furthermore, there may be dire consequences to natural resources because of mismanagement. Renewable resources, such as water, land, and the environment, are linked, and decisions made with regard to one may affect the others. Policy and management of natural resources now require an interdisciplinary approach, including natural and social sciences to correctly address our societal preferences.

This series provides a collection of works containing the most recent findings on economics, management, and policies of renewable biological resources, such as water, land, crop protection, sustainable agriculture, technology, and environmental health. It incorporates modern thinking and techniques of economics and management. Books in this series will combine knowledge and models of natural phenomena with economics and managerial decision frameworks to assess alternative options for managing natural resources and the environment.

The Series Editors

David Zilberman • Joachim Otte
David Roland-Holst • Dirk Pfeiffer
Editors

Health and Animal Agriculture in Developing Countries

Published by the

Food and Agriculture Organization of the United Nations
and
Springer Science+Business Media, LLC

Editors
David Zilberman
Department of Agricultural
and Resource Economics
College of Natural Resources
University of California, Berkeley
Berkeley, CA 94720-3310, USA
zilber11@berkeley.edu

David Roland-Holst
Department of Agricultural
and Resource Economics
College of Natural Resources
University of California, Berkeley
Berkeley, CA 94720-3310, USA
dwrh@are.berkeley.edu

Joachim Otte
Food and Agriculture Organization
of the United Nations
Rome, Italy
joachim.otte@fao.org

Dirk Pfeiffer
Department of Veterinary Clinical
Sciences
The Royal Veterinary College
Hawkshead Lane
Hatfield, Hertfordshire, UK
pfeiffer@rvc.ac.uk

ISBN 978-1-4419-7076-3 e-ISBN 978-1-4419-7077-0
DOI 10.1007/978-1-4419-7077-0
Springer New York Dordrecht Heidelberg London

Library of Congress Control Number: 2011940030

Printed on acid-free paper

Acknowledgments

This volume aims to assemble a growing literature on the economics and health of animal agriculture, mainly in developing countries. Much of the research was initiated in response to the concern about Avian Flu. The Editors would like to thank the Food and Agriculture Organization for supporting a significant part of the research as well as book preparation and the U.K. Department for International Development, which supported the multidisciplinary project on Avian Flu in Southeast Asia and Africa. We would also like to thank the Giannini Foundation for supporting a conference, where results were presented, and all of the authors and their institutions for their diligence and cooperation in the lengthy process leading to this book. Finally, we would like to thank Springer's staff for all of their assistance and technical editing and, especially, Angie Erickson for her dedication and insights in managing the book preparation, overseeing technical editing, and bringing the book to fruition.

Contents

Contributors

Drew Behnke
Department of Economics, University of California, Santa Barbara, CA, USA

Robert Burden
Serecon Management Consulting Inc., Edmonton, AB, Canada

Anaspree Chaiwan
Chiang Mai University, Chiang Mai, Thailand

Eli P. Fenichel
School of Life Science and ecoServices Group, Arizona State University, USA, AZ, USA

Guillaume Fournié
Veterinary Epidemiology and Public Health Group, Department of Veterinary Clinical Sciences, Royal Veterinary College, University of London, London, UK

Will de Glanville
Veterinary Epidemiology and Public Health Group, Department of Veterinary Clinical Sciences, Royal Veterinary College, University of London, London, UK

Samuel Heft-Neal
Department of Agricultural and Resource Economics, University of California, Berkeley, CA, USA

David A. Hennessy
Department of Economics and Center for Agricultural and Rural Development, Iowa State University, Ames, IA, USA

Jan Hinrichs
Food and Agriculture Organization of the United Nations, Rome, Italy

Amir Heiman
Department of Agricultural Economics and Management,
Hebrew University, Rehovot, Israel

Jennifer Ifft
Department of Agricultural and Resource Economics,
University of California, Berkeley, CA, USA

Yanhong Jin
Department of Agricultural, Food and Resource Economics,
Rutgers University,
New Brunswick, NJ, USA

Maria D. Van Kerkhove
MRC Centre for Outbreak Analysis and Modelling, Imperial College London,
London, UK

Mimako Kobayashi
University of Nevada, Reno, NV, USA

Raphaëlle Métras
Veterinary Epidemiology and Public Health Group,
Department of Veterinary Clinical Sciences, Royal Veterinary College,
University of London, London, UK

Bruce McWilliams
Autonomous Institute of Technology in Mexico (ITAM),
Mexico City, Mexico

Jianhong Mu
Department of Agricultural, Food, and Resource Economics,
Rutgers University, New Brunswick, NJ, USA

Clare Narrod
Markets, Trade, and Institutions Division, International Food Policy
Research Institute, Washington, DC, USA

Alan L. Olmstead
University of California, Davis, CA 95616, USA

John Omiti
Kenya Institute for Public Policy Research and Analysis, Nairobi, Kenya

Joachim Otte
Food and Agriculture Organization of the United Nations, Rome, Italy

Charles A. Perrings
Arizona State University, Tempe, AZ, USA

Brian D. Perry
Nuffield Department of Clinical Medicine, University of Oxford, Oxford, UK
College of Medicine and Veterinary Medicine, University of Edinburgh, Edinburgh, UK
Department of Veterinary Tropical Diseases, University of Pretoria, Pretoria, South Africa

Dirk Pfeiffer
Royal Veterinary College, University of London, Hertfordshire, UK

Thibaud Porphyre
Veterinary Epidemiology and Public Health Group,
Department of Veterinary Clinical Sciences, Royal Veterinary College, University of London, London, UK

Songsak Sriboonchitta
Department of Economics, Chiang Mai University, Chiang Mai, Thailand

Marites Tiongco
Markets, Trade, and Institutions Division,
International Food Policy Research Institute, Washington, DC, USA

Patrick Walker
Royal Veterinary College, University of London, Hertfordshire, UK

Qiong Wang
School of Sustainability, Arizona State University, Tempe, AZ, USA

Paul W. Rhode
University of Michigan, Ann Arbor, MI, USA

Karl M. Rich
Department of International Economics,
Norwegian Institute of International Affairs (NUPI), Oslo, Norway
International Livestock Research Institute, Nairobi, Kenya

David Roland-Holst
Department of Agricultural and Resource Economics,
University of California, Berkeley, CA, USA

Rosemarie Scott
International Food Policy Research Institute, Washington, DC, USA

Steven Sexton
Department of Agricultural and Resource Economics,
University of California, Berkeley, CA, USA

Thomas W. Sproul
Department of Agricultural and Resource Economics,
University of California, Berkeley, CA, USA

Tong Wang
Iowa State University, Ames, IA, USA

David Zilberman
Department of Agricultural and Resource Economics,
University of California, Berkeley, CA, USA

Part I
Introduction and Background

Chapter 1
Introduction

David Zilberman, Joachim Otte, David Roland-Holst, and Dirk Pfeiffer

Throughout history, animal husbandry has been a central component of agriculture and livestock has been central to agrofood systems. Animals have provided rural societies with a broad spectrum of products and services, including food, energy, fertilizers, traction and transport, pest control, security, etc. Despite the immense benefits enjoyed by humans from this symbiotic relationship, coexistence with domestic animals also poses serious risks. Most important among these are infectious diseases of animal origin that can affect humans (zoonoses). These have been prominent among the many pandemics that have wiped out millions of people and communities since the earliest human settlements. To cite a relatively recent example, the 1918–1920 Spanish Flu Pandemic, caused by a virus with Avian origins, was responsible for tens of millions of deaths worldwide (Murray et al. 2006). Recent research on the human genome suggests that we have acquired resistance to such diseases over a much longer history of recurrent viral threats.

While humans are gradually shifting away from reliance on animal traction and related services, rising incomes and population have built momentum for drastic changes in livestock consumption and production patterns, heralding a new "food revolution" (Delgado et al. 1999) driven by the emergent middle classes around the world. In response to such demanding forces, the livestock sector is changing at every level. Industrial livestock production systems are being introduced or expanded throughout the world while hundreds of millions of individual farmers

D. Zilberman (✉) • D. Roland-Holst
Department of Agricultural and Resource Economics, University of California, Berkeley, CA, USA
e-mail: zilber11@berkeley.edu

J. Otte
Food and Agriculture Organization of the United Nations, Rome, Italy

D. Pfeiffer
Veterinary Epidemiology and Public Health Group, Department of Veterinary Clinical Sciences, Royal Veterinary College, University of London, Hartfield, Hertfordshine, UK

D. Zilberman et al. (eds.), *Health and Animal Agriculture in Developing Countries*, Natural Resource Management and Policy 36, DOI 10.1007/978-1-4419-7077-0_1,

continue to raise livestock in traditional production systems. These two archetypes and all production systems in between are evolving under rapidly changing and intensifying forces, including expanding markets, global competition, increased supply chain complexity, food safety and other product standards, disease risk management, etc. Global food trade has risen more than 400% over the last generation and, with this increased product mobility, has come rapid diffusion of disease risks.[1]

As this process has accelerated, several large-scale outbreaks, such as Highly Pathogen Avian Influenza (HPAI type H5N1) or Avian Flu, Swine Flu (H1N1), Severe Acute Respiratory Syndrome (SARS), and Bovine Spongiform Encephalopathy (BSE or "Mad Cow" disease), have asserted themselves in less than a decade. Effective responses to these new threats have to be based on better public and private understanding of zoonotic-risk generation and transmission, and these insights must be integrated into a new generation of policies and practices governing animal production, processing, supply chain management, and public health. Given the complexity and socioeconomic extent of agrofood systems, meeting these challenges successfully requires multidisciplinary research, insights, and policy guidance.

This book introduces such an approach, applied to one of the leading modern pandemic threats, HPAI. While the work reported, here, is of general relevance to past, present, and future zoonotic diseases, HPAI, itself, remains a very serious threat to animal and human populations. The present research was part of a leading international effort to understand and address this disease's emergence and the larger implications of policy responses to it. The HPAI is extremely contagious and deadly to poultry, much less contagious but very deadly to people, and is still undergoing rapid mutation and reassortment in an established reservoir of tens of billions of domestic animals/poultry. To better understand this threat and devise more socially effective defenses against it, the UK Department for International Development supported our multidisciplinary HPAI research project for Southeast Asia and Africa.

The results of this research are reported in five thematic sections below. Part I provides an historical perspective on zoonotic-disease control, reviewing experience with one of the most significant zoonotic diseases—Bovine Tuberculosis. This section also presents a conceptual framework to characterize the evolution of livestock production and supply chains using Viet Nam's poultry sector as a case study. Part II presents different perspectives on managing animal disease. Zilberman et al. present a farm-level model that reveals how incentives, including penalties for selling sick animals, will affect production level, monitoring, and culling choices. While this model can be applied to traditional as well as modern livestock operations, Hennessy and Wang investigate how adoption of industrial livestock methods affects the economics of livestock production. Modern systems

[1] Indeed, some of the most deadly modern zoonotic diseases, such as HIV and the Ebola virus, were apparently long endemics in local animal populations but have only recently become larger human health threats.

are closed systems that are designed to control infectious diseases and reduce production costs. Hennessy and Wang investigate how the adoption of such system configurations and technologies affects the scale of production and the size of the industry.

One of the main challenges in developing models of disease control is integrating epidemiology and economics, i.e., the biological and behavioral realities that combine to manifest livestock disease and hold the keys to managing that risk. Horan, Wolf, and Fenichel present a dynamic framework that demonstrates how different epidemiological features affect management choice and optimality of outcomes. Wang, Fenichel, and Perrings develop a framework to deal with animal diseases in the context of international trade and movement of livestock, showing how different arrangements will affect both the magnitude and distribution of risk among different locations and producers. The control of animal disease is technology dependent, which, in turn, is dependent on public research. Sproul, Zilberman, and Roland-Holst develop a rule of thumb to determine global investment in control of zoonotic diseases based on the notion of the value of life. The investment in disease control and livestock production depends on consumers' willingness to pay for the various livestock product characteristics. Heiman, McWilliams, and Zilberman develop a hedonic price approach to assess how different characteristics of meat products affect the value perceived by consumers and thus establish the earning capacity of the livestock industry.

While Part II emphasizes economic principles, Part III is devoted to epidemiological modeling and alternative strategies to control animal disease, exemplified, here, by HPAI. Avian Flu attracted considerable media attention, because this virus has been shown to produce severe disease in humans with a case-fatality rate of approximately 50%. While there is fear that the virus could evolve into a strain capable of sustained human-to-human transmission, the virus's greatest impact to date has afflicted highly diverse poultry industries in several affected countries. Thus, HPAI control measures have, until now, focused on implementing prevention and eradication measures in poultry populations, with more than 175 million birds culled in Southeast Asia alone.

Fournié, Glanville, and Pfeiffer present an overview of HPAI type H5N1, providing a history that includes major outbreaks in Asia as well as its spread westward, among species, and within regions. Fournié et al. develop a general framework to model the spatial spread of a livestock disease over time and apply it to Avian Flu. In addition, they model several control strategies, including vaccinations and movement restrictions. Hinrichs and Otte analyze some of the economic and technological issues that are associated with alternative vaccination strategies to control Avian Flu, taking into account factors that can affect the performance of such strategies. One of the key factors affecting the spread of HPAI is the movement of poultry over space. Van Kerkhove presents a framework to analyze the impact of bird movement and interaction on the spread of Avian Flu in bird and human populations in Cambodia.

Part IV is devoted to special case studies that emphasize the diverse institutional realities associated with the HPAI control at different locations. Burden gives an overview of a disease-control programs that includes disease-prevention steps, various responses to outbreaks (including culling), as well as compensation. He emphasizes the role of alternative financial arrangements, such as insurance, in triggering these responses throughout the supply chain. Tiongco et al. develop a quantitative framework to analyze producer responses to Avian Flu with respect to information, size, and perception, applying it to case studies in Kenya. Rich and Perry present an overview of the major livestock disease threats in Africa and the different strategies taken to control them. Three other studies investigate alternative strategies to promote food safety and sustainability among small poultry producers in Southeast Asia. Heft-Neal et al. investigate the use of poultry contracting in Thailand. Behnke, Roland-Holst, and Otte investigate the potential of microcontracting to develop a sustainable supply chain in Lao PDR. And Heft-Neal, Roland-Host, and Otte analyze the changes in the structure of the poultry sector in Thailand and its capacity to deal with HPAI in Thailand. Finally, Jin and Mu report on consumer livestock valuation—willingness to pay for chicken products in China. Part V offers general conclusions and discussion of new research directions.

References

Delgado, Christopher, Mark Rosegrant, Henning Steinfeld, Simeon Ehui, and Claude Courbois. *Livestock to 2020: The Next Food Revolution*. Nairobi, Kenya: International Food Policy Research Institute, May, 1999.

Murray, Christopher, Alan D. Lopez, Brian Chin, Dennis Feehan, and Kenneth H. Hill. "Estimation of Potential Global Pandemic Influenza Mortality on the Basis of Vital Registry Data from the 1918-20 Pandemic: A Quantitative Analysis." *The Lancet*, Vol. 368, No. 9554 (December, 2006), pp. 2211–2218.

Chapter 2
The Eradication of Bovine Tuberculosis in the United States in a Comparative Perspective

Alan L. Olmstead and Paul W. Rhode

Introduction

At the dawn of the twentieth century, tuberculosis (TB) was the leading cause of death in the industrialized world. In 1900, TB caused about 1 out of every 9 deaths in the United States.[1] Death represented only a fraction of the disease's cost because, besides those that succumbed, countless others were permanently crippled and wasted away in pain. It is probable that 10% or more of U.S. TB sufferers had contracted the bovine form of the disease. Infected milk products were the main conduit to humans; however, other cattle products, direct contact with cattle, and swine products all posed a danger. Bovine-type infections were far more common in nonpulmonary cases and in children, especially infants.[2] The mysteries of this classic zoonotic disease needed to be understood before effective action could be taken.

Following the discovery of diagnostic technologies in the 1891, public health officials across the industrialized world proposed aggressive measures to stamp out the disease in the animal population. In 1917, the United States government took the first bold steps in what officials hoped would become a national campaign to eradicate bovine tuberculosis (BTB) in livestock. The program aimed to reduce the danger of human infection by attacking the disease in cattle, and in the process to increase the productivity of the livestock sector. The chief architects of this crusade

[1] U.S. Bureau of the Census, "Tuberculosis," Table 17, p. 516, and *Mortality*, pp. 16, 27. In addition to the official death toll, many who died of other causes harbored tuberculosis.

[2] Olmstead and Rhode, "Impossible Undertaking," pp. 740–42.

A.L. Olmstead (✉)
Department of Economics, University of California, Davis, CA 95616, USA
e-mail: alolmstead@ucdavis.edu

P.W. Rhode
Department of Economics, University of Michigan, Ann Arbor, MI 48109, USA

D. Zilberman et al. (eds.), *Health and Animal Agriculture in Developing Countries*, Natural Resource Management and Policy 36, DOI 10.1007/978-1-4419-7077-0_2,

were the leaders of the Bureau of Animal Industry (BAI), an agency that had been created in 1884 within the United States Department of Agriculture. By 1917, the BAI had already distinguished itself in combating other animal diseases, including contagious bovine pleuropneumonia and foot-and-mouth disease (FMD).

As a result of these experiences, the United States had already made important strides in building an integrated system of disease control that involved surveillance, preparedness and response planning, and world-class scientific research facilities. BAI leaders understood the importance of early detection and a rapid response to new outbreaks. The growing network of private and state veterinarians, supplemented by federal agents stationed throughout the country, helped identify outbreaks early. A literate farm population, farm journals and government publications, and public outreach programs added to the nation's stock of relevant social capital. The BAI learned to concentrate its monitoring efforts at high-risk locations that might serve as disease pathways such as ports, stockyards, and staging areas along railroad lines. A primitive but effective trace-back system was in operation. This had proved crucial in several earlier eradication campaigns. The preparedness and response planning programs included protocols for immediately notifying the BAI headquarters of suspected outbreaks, and for rushing state and federal experts to hotspots to handle suspect animals. While waiting for test results, state officials often quarantined specific farms. If the disease was confirmed, officials expanded quarantines and intensified monitoring. Some states had passed legislation that helped coordinate responses between the two levels of government should trouble appear; these laws set precedents for granting federal agents police powers normally reserved for state officials. Officials had also gained valuable experience in designing incentive compatible compensation schemes that delicately balanced the need to encourage farmers (and their political representatives) to cooperate with the danger that compensation would create moral hazard problems and spread diseases.

This infrastructure was far from perfect, and, when new disease outbreaks did hit, bickering and confusion often delayed the responses. Moreover, procedures developed for one disease were not directly transferable to all diseases. The BAI had been most successful in dealing with epizootic crises such as a sudden eruption of FMD.[3] It was easier to galvanize support in a dire emergency. The campaign against BTB would be different. It was unprecedented in its complexity and scale – it was so ambitious that even Henry C. Wallace, one of America's most respected agricultural spokesmen, questioned whether the proposed campaign might not be an "an impossible undertaking."[4] This campaign would require tackling a disease that was present in nearly every county in the United States. Success would require scientific advances to create a better understanding of the BTB's etiology and of the dangers it posed; this information had to be conveyed to the public to help counter

---

[3] This was also the case in Europe where many countries had succeeded in stamping out epizootics of FMD, rinderpest, and contagious bovine pleuropneumonia. Waddington, "To Stamp Out," p. 32.

[4] Smith, *Conquest*, p. 12.

the activities of hostile special interests. Tests to identify infectious, but asymptomatic animals needed to be improved; grass roots support needed to be built; working arrangements needed to be perfected with state and local governments; monitoring needed to be intensified; and compensation policies needed to be tailored to specific situations. Essentially, many of the technical, political, economic, educational, and social apparatuses needed to fight BTB still had to be perfected.

Between 1917 and 1940 veterinarians administered roughly 232 million tuberculin tests and ordered the destruction of about 3.8 million cattle (from a population that averaged 66.4 million animals over this period). By 1941, every county in the United States was officially accredited free of BTB. The savings to farmers and meat packers alone (resulting from increases in animal productivity and a decline in the number of condemned carcasses at slaughterhouses) exceeded the costs by at least a ratio of ten-to-one. However, the spillover effects on human health were the main story. By 1940, before effective chemotherapy was available, new cases of bovine-type TB in humans had become a rarity. The eradication program, coupled with the spread of milk pasteurization, prevented thousands of BTB-related deaths per year in the United States during the period immediately preceding World War II.

BTB in Cattle and Humans

BTB was an insidious disease because apparently healthy animals could be both infected and contagious. Indeed, most cattle infected with *M. bovis* appeared normal.[5] It could take years for cattle to develop lesions in organs, tissues, and bones. Eventually, infected animals had difficulty gaining or maintaining weight, cows experienced a 10–25% reduction in milk production and problems in breeding, and draft cattle lost strength and endurance.[6] As the disease progressed, cattle might show external signs of lesions, have coughing attacks (if the disease settled in the lungs), become lethargic, and die prematurely. The disease spread among cattle through direct contact with diseased animals and via contact with contaminated feed, milk, straw, water, sputum, feces, and even air. It spread between herds through the introduction of infected new animals and by incidental contact with other infected animals.[7]

The prevalence of the disease increased with the animal's age. Older stock obviously had a longer period of exposure to contract the bacteria and more time to develop full-blown, highly contagious cases. Rates of TB infection also tended to be far higher among closely confined cattle than in free-range–raised animals.

---

[5] National Research Council, *Livestock*, p. 13.

[6] Melvin, "Economic," p. 103, and in the recent literature including National Research Council, *Livestock*, p. 56; Faulder, "Bovine," p. 14 states "the producing life of a dairy animal infected with tuberculosis is often cut in half."

[7] Kiernan and Wight, "Tuberculosis," pp. 1–18; Russell and Hoffman, "Three Year Campaign," pp. 11–12; Russell, "Spread," pp. 3–5.

As a result, BTB was much more common near cities, in dairy herds and purebred stock, and more generally in the "advanced" agricultural regions of the country. Imported purebred cattle often spread the disease. In part for this reason, agriculturally backward states had relatively low BTB rates.[8]

In 1890, Robert Koch's search for a vaccine led to his development of tuberculin. Although this proved ineffective as a vaccine, it became an important diagnostic tool, making it possible to detect TB in animals without visible symptoms.[9] Early applications of the tuberculin test showed that the extent of the infection was far more widespread than had been suspected. In 1892, Leonard Pearson, who had studied under Koch, introduced tuberculin testing to the United States.

Technological advances in BTB detection greatly facilitated the eradication program. Early tuberculin testing was a very costly and time-consuming process. In order to obtain an accurate result, the veterinarian had to take an animal's temperature 3–4 times at 2-h intervals, to establish a baseline reading prior to injecting the animal. Then, beginning about 6 h after the injection, the veterinarian again took temperature readings every 2–3 h, continuing the process for up to 24 h after the injection was given. To test a herd of about a dozen animals, the veterinarian had to be on site for about 36 h, inspecting each animal from 10 to 15 times. A far simpler and cheaper approach, suited to mass testing, became the standard practice in 1920. Employing the new (intradermic) process, the tester injected a few drops of tuberculin into the skin of the subject animal – usually in a flap in the tail – and then returned to visually check for swelling 2–3 days later. One veterinarian could now do the work of five or six using the older method. Under either procedure, detecting a "reaction" was clearly a judgment call, as enraged cattle owners would at times protest. Both false positives and false negatives occurred. Recent exposure to tuberculin produced a tolerance suppressing the reaction.[10] Thus the standard testing protocols required that the subject animal had not been tested in the previous 60–90 days.

Tuberculin tests and slaughterhouse inspections suggest that about 10% of the nation's dairy cattle and 1–2% of all range cattle were infected with BTB around 1915.[11] In addition, the national infection rate was rising at an alarming pace. The proportion of cattle showing TB lesions in federal meat inspections at slaughter houses rose from 1% in 1908 to about 2.5% in 1917.[12] Nevertheless, the U.S. cattle

---

[8] Myers, *Man's Greatest*, pp. 264, 267–68, 309, 323.

[9] Myers, *Man's Greatest*, p. 115; and U.S. BAI, *Special Report* 1916, pp. 416–17.

[10] U.S. BAI, *Special Report 1916*, pp. 417–18; Smith, *Conquest*, pp. 7–9; Houck, *Bureau*, pp. 364–66; Myers, *Man's Greatest*, p. 125. Note that the proportion of false positives (or more precisely the probability that a positive test has identified an uninfected animal) depends on the prevalence of the disease. As the disease becomes less common, a positive result under a given test procedure is more likely to be false. National Research Council, *Livestock*, pp. 17–19. According to this source, the probability that an uninfected animal would test positive was less than 2%.

[11] Mitchell, "Animal," p. 168; and U.S. BAI, *Special Report 1916*, p. 409.

[12] Melvin, "Economic," pp. 101–02. As a crude indication of how fast the disease could spread, bovine tuberculosis probably did not enter Sweden and Finland until the 1840s and by the end of the nineteenth century it is likely that 25% of their cattle were infected. Myers and Steele, *Bovine*, pp. 256–57, 280–81.

infection rate was significantly below those prevailing in Europe, where between 25 and 80% of the cattle were tuberculous.[13] The European situation offers a stark picture of how the incidence of the disease might have increased in the United States if left unchecked.

Advances in public policy and science went hand-in-hand. The eradication program would not have been technically or politically feasible without a series of nineteenth-century scientific and technological breakthroughs that greatly advanced the understanding of tuberculosis and of diseases more generally. Among the important steps were Koch's discovery of the tubercle bacillus in 1882, and his subsequent development of tuberculin. Observers had long speculated that humans contracted tuberculosis from cattle. In 1898, the American scientist Theobald Smith confirmed this view when he identified differences between the bovine and human strains. Koch seriously misinterpreted Smith's findings. In 1901, he proclaimed that BTB posed little threat to humans, and might even benefit them by providing immunity against the human form. In his December 1905 Nobel laureate address, Koch asserted that "Bovine tuberculosis is not transmissible to man."[14] Despite mounting scientific evidence of the dangers of the bovine bacilli to humans, Koch adhered to his erroneous beliefs for more than a decade.[15] The acrimonious scientific debate was front-page news. Koch's declarations helped embolden special interests and galvanize the political opposition to BTB eradication efforts into the 1930s, long after the arguments were largely settled within the scientific community.[16]

Control Measures

Based on the new scientific understanding, the eminent Danish veterinary scientist Bernhard Bang proposed to test all animals using tuberculin and then to slaughter the reactors as a means to eliminate the disease from the population. This test-and-slaughter approach was aggressive, indeed radically so. Many European experts argued that the disease was so prevalent that eradicating all infected animals would cause unacceptable food shortages – better impure food than insufficient food. Even Bang admitted this in 1896: "In most European states

[13] U.S. BAI, *Special Report* 1912, p. 417; Orland, "Cow's Milk," p. 11; and Myers, *Man's Greatest*, p. 222.

[14] Miller, "Tuberculous Cattle," p. 35; *New York Times* (1 June 1904), p. 1.

[15] Some basic terminology needs clarification. The formal name of the bovine strain of tuberculosis is *Mycobacterium bovis*, which is often summarized as *M. bovis*. The corresponding terminology for the human strain of tuberculosis is *Mycobacterium tuberculosis* and *M. tuberculosis*.

[16] Myers, *Man's Greatest*, pp. 106–09, 200, 211–19, 226; Myers and Steele, *Bovine*, p. 57; and Dankner et al., "*Mycobacterium*," pp. 20–24.

a compulsory and quick butchering of all these animals is out of the question, the number of the reacting animals is so very large."[17] Bang accepted a less radical approach involving eliminating the sick animals, periodically testing the entire herd, and segregating the reactors from the nonreactors (and raising the calves of the reactors with the nonreactors). He also advocated thoroughly disinfecting the stalls.

Robert Von Ostertag, a German veterinarian, advocated an even less aggressive approach, which did not employ the tuberculin test at all. Only animals with open lesions were destroyed, and visibly infected animals were tracked. An English approach, known as the Manchester plan, called for periodic testing of milk for tubercle bacilli; when discovered, authorities traced the milk back to the dairy and tested the individual animals.[18] The French put more faith in the use of the *Bacillus* Calmette-Guérin (BCG) vaccine, developed at the Pasteur Institute over the 1906–1921 period. This offered the hope of avoiding costly test in slaughter programs. The BCG vaccine originally used a live attenuated strain of *M. bovis* to promote immunity in both humans and cattle. Over the years, various formulations of BCG vaccines (and most subsequent vaccines) have been ineffective and inefficient in combating TB in livestock. BCG vaccines did gain widespread popularity with many nations and with international agencies in fighting TB in humans. The approach offered high levels of protection in some human populations, but provided negligible protection in others. Vaccines did have the serious drawback of causing positive reactions to the tuberculin skin tests, thereby making it difficult to identify active cases of TB in patients.[19]

In addressing the BTB problem, countries differed in the approaches they employed, with a decided tendency to adopt "national" models following path dependency. That is, if a country relied heavily on vaccination in fighting human TB, it tended to rely on vaccination for BTB. Countries also differed over whether BTB was considered primarily as a public health problem affecting the human population or an animal production problem affecting the livestock sector. This helped determine which authorities were considered responsible – urban-based health or rural-based agricultural bureaucracies. Countries such as the United States, which were early to label BTB as a problem for both humans and animals, often adopted the most aggressive policies.

[17] Cited in Myers, *Man's Greatest Victory*, pp. 245–46.

[18] Myers, *Man's Greatest Victory*, p. 265.

[19] National Research Council, *Livestock Disease Eradication*, p. 32; Buddle, pp. 126–32; Meyers and Steele, *Bovine*, pp. 260–62; Myers, *Man's Greatest Victory*, pp. 188–97; Fanning and Fitzgerald, "BCG Vaccines," pp. 541–54.

Problems with Decentralized Regulations in the United States

Prior to the U.S. federal government's commitment to eradication, state and local governments experimented with a variety of schemes to slow the spread of BTB in livestock and humans. The designers of the 1917 campaign carefully studied these many experiments. In 1894, Massachusetts enacted a strict, compulsory BTB program with quarantines, comprehensive testing, and full compensation for infected cattle (up to a limit of $60 per head). The state spent liberally and slaughtered over 11,000 reactors (roughly 50% of the animals tested). High costs and the fierce opposition from cattle interests forced the state to retreat. In 1897–1898, it shifted to a voluntary program emphasizing visual inspections. The problems in Massachusetts provided valuable lessons for others. Moral-hazard drove up costs because farmers readily rid themselves of sick or unproductive bovines.[20] The experience also pointed to the danger of public officials caving in to the pressure of powerful special interests. As early as 1899, Ohio officials singled out the Massachusetts experience as providing a striking object lesson, "showing how not to do it. . .."[21]

Pennsylvania adopted a far less aggressive plan in 1896. Visibly ill animals were destroyed, with the owner keeping the salvage value and receiving a partial indemnity with lower maximum limits. The state also provided voluntary, free tuberculin testing. Owners of reactors had a choice: they could either slaughter their reactors and receive partial compensation, or they could isolate the animals and agree to heat all milk to 185°F for 10 min before sending it to market. The latter choice mirrored approach developed in Europe by Bang. Most Pennsylvania cattle owners opted for the slaughter with indemnity approach. By 1900, many states in the Northeast and Midwest had programs employing a variety of policy options.[22]

The story of BTB in Wisconsin and Illinois illustrates how uncoordinated policy responses led to serious unintended economic consequences. In 1901, Wisconsin created a Livestock Sanitary Board that established regulations concerning tuberculous cattle. State authorities also provided tuberculin and training for its use very broadly in its dairy community. After 1909, it pursued a more aggressive approach to ridding the state of diseased stock.[23] By way of contrast, Illinois made almost no lasting effort to combat BTB before 1914. In 1899, Governor John R.

[20] Teller, *Tuberculosis*, pp. 19–20; Myers, *Man's Greatest*, pp. 272–74, 283; and Reynolds, "Problem," pp. 454–56.

[21] C. E. Thorne, "*Bovine Tuberculosis*," Bulletin of the Ohio Agricultural Experiment Station, No. 108 (June 1899), p. 369.

[22] Pearson and Ravenel, "Tuberculosis," pp. 167–200; Reynolds, "Problem," pp. 451–58; Lampard, *Rise*, pp. 188–89; Myers, *Man's Greatest*, pp. 278–79; and Salmon, *Legislation*.

[23] "State and Territorial," pp. 70–72; Lampard, *Rise*, pp. 188–89; Wisconsin. Department of Agriculture, *Biennial Report 1915–1916*, pp. 83–95, and *Biennial Report 1919–1920*, pp. 41–47; Reynolds, "Problem," pp. 451–54, *Breeder's Gazette*, 30 Nov. 1910, pp. 1169–70.

Tanner issued a proclamation requiring tuberculin testing for imported dairy and breeding cattle.[24] This invoked vigorous protests from the state's powerful dairy interests. Their legal challenges led Illinois courts to declare Tanner's proclamation unconstitutional.[25] And when Chicago required tuberculin testing for its milk supply, the dairy lobby convinced the state legislature to overrule all such municipal initiatives. Thereafter (until 1914), Illinois stockowners operated in a virtually unregulated environment.

The close proximity of Illinois and Wisconsin, and the different regimes, created an ideal setting for regulatory arbitrage. The problem came to the public's attention in the fall of 1914, when the federal government and the state of Illinois cracked down on a criminal conspiracy known as the "Tuberculous Cattle Trust." The press had a field day with such banner front-page headlines as: "Elgin Clearing House For Tubercular Cows – Government Orders Quarantine of Five Illinois Counties From Which Entire West Has Flooded With Diseased Dairy Cattle for Last Ten Years – Prosecutions are Expected."[26] The five-county Elgin district neighbored Chicago and directly bordered on Wisconsin. The government charged that a number of cattle dealers in this area had defrauded legitimate farmers and endangered the public by knowingly selling diseased animals, often with falsified bills of health. By 1914, the tubercular dairy cows had been widely dispersed among the "herds supplying Western cities with milk and have sown the 'seeds of death' in thousands of homes using this milk."[27]

The charges singled out James Dorsey, who "was for many years probably the leading dealer in dairy cows in the United States."[28] At the height of his business, Dorsey was buying and selling some 20,000 animals annually; about one-half were tuberculous. His market extended across the United States and into Canada and Mexico. Beginning as a small-scale dealer in 1904, Dorsey achieved his rapid ascent by arbitraging between state regulatory regimes. He traded in "animals that had reacted to the tuberculin test or that the dairyman had reason to believe were tuberculous and wished to dispose of before the test was applied to his herd. . . ."[29] As a cover for his business, Dorsey operated a number of large, modern farms, and he advertised in the leading farm journals.

Dorsey was not acting alone. In addition to roughly 10 other unscrupulous Elgin area dealers, thousands of dairymen spread across several eastern and Midwestern

---

[24] "Bars Diseased Cattle: Governor Tanner Issues Prohibitive Proclamation," *Chicago Tribune*, 14 June 1899, p. 7.

[25] *Charles A Pierce* et al. *and State Board of Live Stock Commissioners v. E. B. Dillingham*, 96 Ill. App. 300; 203 Ill. 148.

[26] *St. Louis Republic*, 1 September 1914, pp. 1–2 and 20 September 1914, p. 1.

[27] *St. Louis Republic*, 1 September 1914, p. 1.

[28] U.S. BAI, Chief of the Bureau to Fitts, 9 July 1920. See also "U.S. Grand Jury Hears Evidence Against Dorsey," *Elgin Daily News*, 29 September 1915, p. 1; Olmstead and Rhode, "Tuberculous," pp. 929–63.

[29] U.S. BAI, Chief of the Bureau to Fitts, 9 July 1920.

states knowingly supplied the diseased stock. These cattlemen understood that if tuberculosis were to break out in a herd all they had to do was contact one of Dorsey's many buyers. After receiving infected animals in Illinois, Dorsey paid corrupt veterinarians to sign certificates of health that fraudulently claimed the cattle had passed a tuberculin test. In some cases, his veterinarians injected the animals with tuberculin, so that they would not react should the potential owner hire an independent veterinarian to test the cattle.[30] He also employed honest veterinarians to perform tests and issue certificates of health to animals in his home herd. Dorsey then used these valid certificates to ship diseased animals of similar appearances to those in his own herd.[31] In many cases, he created phony paper trails and used surrogates to market the cattle to unsuspecting buyers. The damage was immense. According to the BAI, Dorsey had created at least 10,000 foci of tuberculosis in the United States.[32] This likely led to many thousands of human infections.

In the early teens, government officials and private citizens from across the country barraged BAI administrators in Washington, D.C. with complaints about Dorsey and other Elgin cattle dealers. Major farm journals also joined the anti-Dorsey effort by warning their customers and refusing to carry his advertisements.[33] Dorsey countered these private sanctions by hiring accomplices to place ads in their names so as to hide his involvement in the transaction. When other states quarantined his cattle and even cattle from the entire state of Illinois, he drove his animals overland and shipped them out of Wisconsin under an alias. By 1914, at least a dozen states were imposing quarantines on Illinois. The actions of Dorsey and his ilk, and more generally the failure of decentralized actions to slow the spread of the BTB galvanized the demand for federal action.

Building a National Program

For all their problems, the early state and municipal programs did provide a rigorous proving ground. Many features of the post-1917 cooperative state-federal program, most specifically the partial compensation scheme, evolved directly out of the experiences of the state programs. The state-level experiments were valuable because, as the head of the BAI's BTB eradication effort J. A. Kiernan observed, the program designers could not simply "figure this out on paper."[34] These state

---

[30] *St. Louis Republic*, 1 September 1914, p. 1.

[31] "Imprisonment for Dealing in Tuberculous Cattle," *American Journal of Veterinary Medicine* 13:5 (May 1918), 236-37.

[32] U.S. BAI, Chief of the Bureau to Fitts, 9 July 1920.

[33] *Hoard's Dairyman*, 20 June 1913; *St. Louis Republic*, 20 September 1914, p. 2, and correspondence in U.S. BAI, Dorsey file of the National Archives.

[34] U.S. House. Committee on Agriculture, *Tuberculosis in Livestock*, p. 10.

initiatives also helped solidify the support of interested groups of farmers, livestock packers, and health officials to lobby for a national solution. The BAI's successful campaign (1906–1917) to clean up the milk supply in its home territory of Washington, D.C., gave authorities a better sense of what to expect.

The start of the national cooperative state-federal program dates to March 1917, when Congress appropriated $75,000 for BTB eradication. In December 1917, the BAI approved a plan advanced by veterinary and cattle interests to provide voluntary testing for cattle herds. The initial focus was on purebred animals, which composed the foundation of the industry's breeding stock. The herds that were tested and culled and which then proved BTB free in retests were certified as "Tuberculosis-Free Accredited Herds." This accreditation allowed owners to ship their animals across state boundaries for 1 year without further testing.[35] The 1917 regulations also envisioned "tuberculosis-free areas" where all of the herds in a given area were free of reactors, and "modified accredited areas," which met the less restrictive requirement that reactors made up less than 0.5% of cattle population.

Reimbursing farmers for sick animals was still highly controversial, and Congress refused to allow indemnity payments in the 1917 act. At the BAI's urging, this was rectified when legislation increased anti-BTB appropriations to $500,000.[36] Beginning in fiscal year 1919, the federal government would match state indemnities, up to one-third of the difference between the animal's appraised value and salvage value. The federal payments were initially capped at $50 per head for registered purebreds and $25 per head for grade cattle. To provide an example, for a purebred cow appraised at $200 and with a salvage value of $50, the federal government would match a state's payment up to the $50 limit. The state could pay more than $50 if it wanted, and many did. The BAI changed the federal limits as livestock prices increased and declined, and to help counter political opposition. The federal limits were raised to $70 for purebreds and $35 for grade animals in 1929, and then returned to their former levels in 1932.

Legislation passed in 1919 tripled federal funding to $1.5 million, with $1 million earmarked for indemnities. This voluntary program proved highly popular, and by 1922 all but six states were participating. Over 16,000 herds with 364,000 animals had achieved accredited herd status, and 162,000 herds with 1.5 million animals had passed one test. As of August 1922, almost 65,000 farmers, with 500,000 cattle, were on the waiting list for testing.[37] In this period, the program focused on cleaning up the breeding stock to provide healthy replacement animals when the eradication campaign became more general.

There were intense debates in state legislatures across the country over the precise details of the state programs. By 1927, state spending had risen to $13 million. Except during a brief period in the mid-1930s, the states and counties

---

[35] Kiernan, "Accredited-Herd," pp. 215–20.

[36] Myers, *Man's Greatest*, p. 295.

[37] Larson et al., "Dairy," p. 341.

largely carried the ball. In the late 1920s and early 1930s, state and local governments were spending more than twice the federal appropriation of about $6 million. By the mid-1930s, the state expenditures decreased due to the financial crisis, and emergency federal funds came online with the passage of the Jones-Connally Cattle Act of 1934. As a result, the contribution ratio flipped. At the peak in fiscal year (FY) 1935, the federal government contributed about $18 million to the program, whereas states furnished only $9 million.

Officials shifted their focus from cleaning up individual herds to treating entire areas (such as counties) during a relatively short period of time. The systematic eradication of the disease in an area was advantageous because it reduced the chances of re-infection and was more economical than testing under the accredited plan.[38] Although state policies differed, it was common to require the approval of at least a majority of the cattle owners in an area before embarking.[39]

Local participation was an important feature of federal policy, and states retained considerable flexibility in fine-tuning their eradication policies and indemnity plans. Most states adopted schedules paralleling the federal plan, with similar payment shares and limits.[40] In eastern states, where livestock prices were higher, state indemnity limits tended to be higher. States also differed over whether and how much money counties contributed to the expenses. By 1928, only three states (Alabama, Arkansas, and California) chose not to participate.

Figure 2.1 charts the progress of the testing program between 1917 and 1953 by tracking the number and percentage of all cattle tuberculin tested annually. Between FYs 1919 and 1929, the number of tests increased at an annual rate of over 35%. The Jones-Connally Act boosted the effort. At the campaign's peak in 1935/1936, the BAI and state agents tested nearly 25 million cattle per year, or roughly one-third of the nation's 68 million cattle. To accomplish this task, the BAI hired an additional 900 veterinarians and 500 assistants as temporary employees. After this peak effort, tuberculin testing slowed to the 8–12 million a year, or about one-tenth of the nation's bovines. Figure 2.1 also illustrates the increasing reliance on the area plan (as reflected by the difference between the total number of cattle tested and the number tested in the accredited herd plan). In its first year of operation (FY1922), the area plan was about four-tenths of the size of the accredited herd plan. By FY1923, the area plan tested more cattle than the herd plan, and continued to grow much faster. By the mid-1930s, it was 10 times larger. This change represented the more aggressive and better organized status of the eradication program.

As Fig. 2.2 shows, the number of U.S. counties engaged in the program and the number certified as modified accredited areas also increased rapidly. In July 1923,

[38] U.S. House, *Agricultural,* 1925, p. 136, and *Agricultural*, 1927, p. 147.

[39] Smith, *Conquest,* p. 28; and Kiernan, "Bovine," p. 182. Some states such as New York and California required the signatures of at least 90 percent of cattle owners in an area to initiate the program.

[40] U.S. House, *Agricultural,* 1930, pp. 107–08; U.S. Congress, *Congressional Record,* p. 5505.

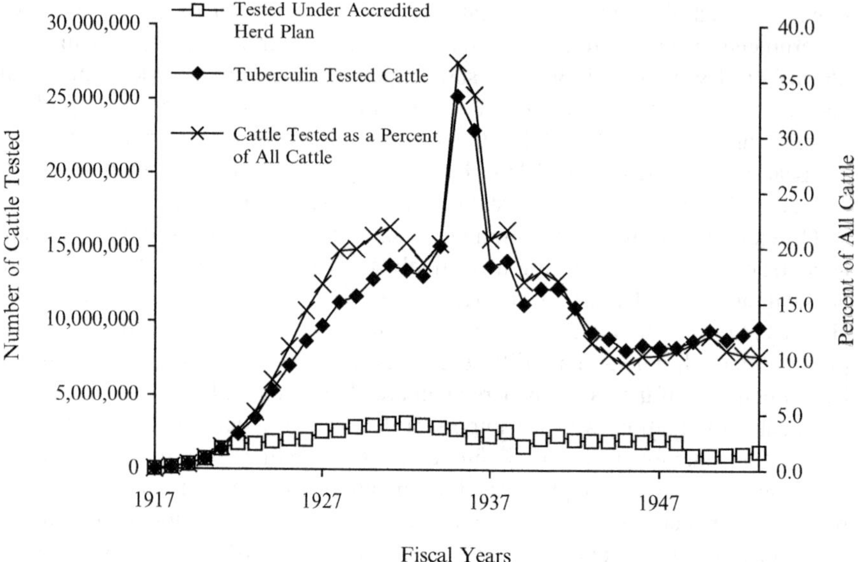

Fig. 2.1 Tuberculin test given annually in the United States, 1917–1953. *Sources*: U.S. BAI, *Annual Reports*, 1918–1954

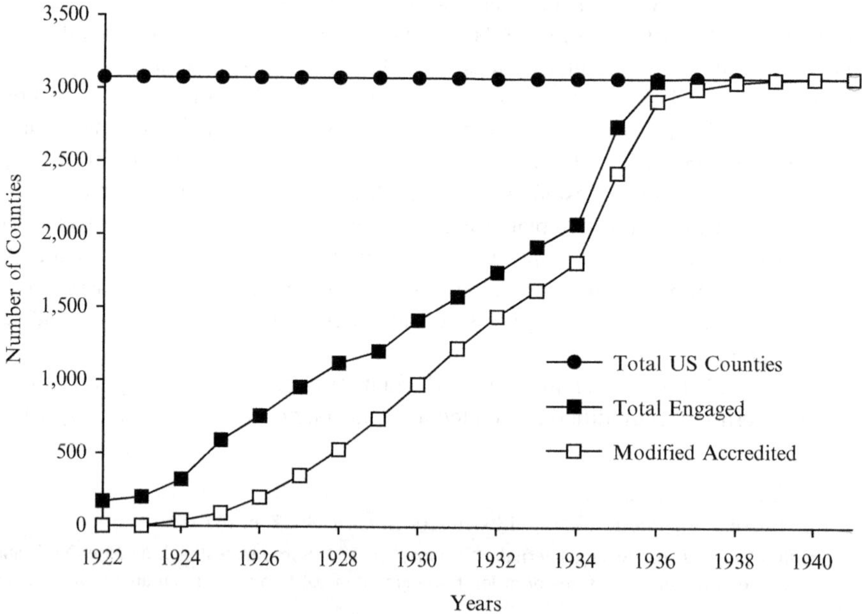

Fig. 2.2 U.S. Counties involved in TB eradication area plan. *Sources*: U.S. BAI, *Annual Reports*, 1922–1942

only 17 counties spread across Indiana, Michigan, North Carolina, and Tennessee had achieved "modified accredited areas." By July 1927, 347 counties (or 11% of all counties in the 48 states) qualified. By 1933, the share exceeded 50%. The increased federal funding that came with the Jones-Connally Act triggered a major jump in participation. When California's Merced and Kings counties were finally certified in 1940, BTB had been brought under control across the continental United States.[41] The entire campaign required only 24 years, and it required less than a decade to progress from 10% county accreditation to 90% accreditation.

BTB was not truly eradicated in the United States. Given the precision of the tests, pushing the reaction rate below 0.5% would have been increasingly expensive and therefore inefficient. (The proportion of positive results that were false rose as the actual prevalence of the disease fell.) Enforcement efforts flagged after the beginning of the Second World War, and localized outbreaks of the disease cropped up periodically over the postwar years.[42] By the early 1960s, bovine TB was sufficiently controlled that the USDA decided to suspend the continuous tuberculin testing of cattle to rely on a more targeted effort. During the testing era, the USDA had developed a tracking system that allowed it to identify the shippers of all cattle processed at federally inspected slaughterhouses. When an inspector found an infected bovine or swine carcass, a veterinarian was dispatched to control the infection on the originating farm. As of 1965, the slaughter, surveillance, and trace-back system became the main line of defense against bovine TB. In addition, officials focused resources on Mexican cattle, which were the main source of re-infection.[43]

The BTB eradication program was highly controversial, with many powerful supporters and opponents. Prominent meat packers, the veterinary and public health communities, the Farm Bureau, and the Grange enthusiastically promoted the campaign. But support was far from universal. The Farmers' Union, the American Medical Liberty League, and numerous farmers mounted stiff opposition. The most publicized grassroots protest occurred in March 1931, when up to 1,000 farmers harassed state veterinarians and their armed guards on a farm in Cedar County, Iowa. This and other threats led Governor Daniel Webster Turner to declare martial law and mobilize the National Guard. But the "Iowa Cow War," with its machine gun toting soldiers, was an exception to a generally orderly pattern of cooperation.[44]

---

[41] As Smith (*Conquest*, p. 29) notes, the program's progress nationally created market pressures for its adoption throughout the Midwest. After a large fraction of their stock proved to be reactors, eastern dairymen began to demand replacements from the more western states. Given eastern regulations aimed at preventing the re-introduction of the disease, the purchases were concentrated in clean areas, leading to premiums for dairy cows from accredited counties.

[42] U.S Agricultural Research Service, "Why Tuberculosis," pp. 1–3; and Smith, *Conquest*, p. 48.

[43] National Research Council, *Livestock*, pp. 36–39; Essey and Koller, "Status of Bovine Tuberculosis in North America," *Veterinary Microbiology* 40 (1994), 15–22. Today, when meat inspectors discover tuberculous animals, the authorities "depopulate" the entire herd.

[44] Olmstead and Rhode, "Not on My Farm," pp. 768–809.

Assessing the Returns to Eradication

At the onset of the federal eradication program, the USDA estimated that the annual cost of BTB to the livestock sector was at least $40 million.[45] This estimate only reckoned the static losses to animal production, and thus failed to capture the affects on human health and the dynamic implications of the contagion. BAI leaders noted that if the disease continued to spread as it had in the decade preceding 1917, the cost to the livestock sector would soon exceed $100 million.[46]

To obtain an assessment of the program's costs, we estimate the net losses to farmers, deflate the annual combined public and private costs to reflect changes in the price level, discount using a constant 3% real interest rate, and cumulate back to a specific reference year – 1918.[47] This exercise indicates that the discounted cost of the program over the 1917–1962 period was $258 million in real 1918 dollars (adjusting for inflation, this would amount to about $3.7 billion in 2009 dollars). The federal government contributed about 31% of the costs, the state and local governments 54%, and farmers 15%. At the assumed 3% real interest rate, the $258 million investment in the program was equivalent to borrowing with the promise to repay 7.7 million 1918 dollars (110 million in 2009 dollars) each year in perpetuity.

Employing a similar procedure and assuming an annual BTB diffusion rate of 5% in cattle in the absence of the program suggests that the BTB eradication campaign saved roughly $3.2 billion over the 1918–1962 period, or the equivalent to a perpetuity paying $98.7 million per year (in 1918 dollars). Thus, for the livestock sector alone, the annual benefits were approximately 12 times the annual costs.

The benefits from alleviating human suffering and saving lives were even larger. To evaluate the effects of the eradication campaign on human health, we must also include the effects of milk pasteurization, because both led to the decline to the bovine form of TB in humans. It is difficult to evaluate their effects separately, but we know that jointly they almost eliminated new cases of *M. bovis* in humans by 1940.[48] Assuming that 10% of TB cases in 1900 were due to the bovine form of the disease suggests, that in the absence of the pasteurization and eradication efforts, roughly 25,600 more deaths would have occurred in 1940. Value of life estimates for the early twentieth century imply that the value of the lives saved in just 1 year (1940) when converted to 1918 dollars would have been several times larger than the annual savings to the livestock industry. We realize that such estimates are

---

[45] Kiernan and Wight, "Tuberculosis," p. 2. Our evaluation of the USDA estimates indicated that the USDA figures appear reasonably well grounded (except for an over statement of the loss to pigs) and are likely lower bound estimates of the costs. For a more detailed treatment of this issue see, Olmstead and Rhode, "Impossible Undertaking," pp. 761–64, 768.

[46] Kiernan and Ernest, "Toll," pp. 280–81.

[47] To estimate the net losses to farmers, we use the difference between the appraised value and the sum of the salvage value and the government indemnities.

[48] National Research Council, *Livestock*, p. 9; Mohler, "Infectious," p. 376.

inherently problematic, but they serve the valuable purpose of highlighting that the benefits of the clean milk campaign were huge and primarily outside the agricultural sector.

International Comparisons

It is fruitful to put this experience into an appropriate international context. The U.S. was not completely alone, because Canada moved largely in parallel. There, voluntary testing began in 1897, and national meat inspection was instituted in 1907. A Canada-wide anti-BTB campaign started in 1919 with a voluntary accredited herd plan. This plan followed the example set in the United States with the hope that animals from accredited herds could bypass quarantine when shipped to the U.S. market. The voluntary plan evolved into a mandatory area plan beginning in 1923. The area plan, which required approval of 60% of cattle owners, was completed by 1961. Since 1923, program officials had administered 100 million tests, and ordered 800,000 cattle slaughtered.[49] But, with few exceptions, other developed nations did not follow close on the heels of the U.S. and Canada.

Despite sharing many approaches regarding the control of animal diseases, Great Britain lagged well behind its two North American offshoots. Following the advent of the tuberculin test, the British received a shock when 35 out of the 40 dairy cows in Queen Victoria's prized herd reacted positively to the new test.[50] A few leaders in the British veterinary and medical communities, as well as some in the meat trade, began to advocate testing with the elimination of reactors, but there was little headway. However, powerful interests, including many farmers and their representatives, were adamantly opposed. A Royal Commission on Tuberculosis established in 1890 studied the problem, held hearings, and issued a number of reports; but nothing of much significance happened. The disease was so widespread that any serious attempt at eradication would heavily burden the treasury and reduce the food supply. In addition, continued controversy over whether tuberculin posed a threat to cattle, the efficacy of the tuberculin in local use, and bureaucratic infighting between different ministries slowed progress for decades. A 1913 Tuberculosis Order to deal with "open" cases of the disease along the line of the Ostertag method was suspended after the outbreak of the Great War, and little more was done until the mid-1920s. A 1923 Milk (Special Designation) Order provided incentives for voluntary testing by creating a new milk classification, Grade A (TT), for the product of clean herds. Testing was to be conducted twice a year, all reactors removed, and all additions to the herd were required to pass a tuberculin test. The milk sold for premium prices to a limited market. The 1925 Tuberculosis

[49] Myers and Steele, *Bovine*, p. 241 and Thoen and Steele, *Mycobacterium*, p. 195.

[50] Dormandy, *White Death*, pp. 330–31; and Dubos and Dubos, *White Plague*, p. 260.

Order and the 1926 Milk and Dairy Order provided partial compensation for the removal of the most diseased livestock. To this juncture, the anti-BTB legislation and enforcement were ineffective. Even tuberculin tests were suspect because many owners of valuable, but suspect, cows evidently made a practice of injecting their animals before the official tester arrived, thereby ensuring the animals would not react. Doctoring animals to reduce reactions was a problem in many countries, and contributed to the spread of the disease in international trade. Some U.S. states began to regulate the distribution of tuberculin as early as 1909, but British authorities failed to act. In fact, tuberculin was removed without explanation from the list of regulated products before the passage of the Therapeutic Substances Act of 1925. Such was the power of the farm lobby.[51]

As late as 1934, inspections at abattoirs found at least 40% of British cattle to be diseased. Many contemporary reports in the first half of the nineteenth century maintained that BTB accounted for about 6% of British TB deaths taking between 2,000 and 3,000 lives annually.[52] In 1935, Britain instituted its first general control measures including an "attested herd scheme," with voluntary testing and compulsory slaughter of reactors. After making little progress, in 1937, the Ministry of Agriculture and Fisheries received substantially enhanced funding and powers to combat the disease. The onset of WWII slowed the campaign, but conditions also led stockowners to slaughter dairy cows made unproductive by the disease. By 1946, the estimated incidence of BTB in the national herd was still 18%. In October 1950, Britain began a compulsory eradication national program based on an area plan. The campaign started in areas with low infection rates, and all of Great Britain received attested area status in October 1960. Between 1935 and 1960, the control effort led to the slaughter of 2.2 million cattle. Maintaining low infection rates has proven difficult because local wildlife, especially badgers and deer, provided a reservoir for infection.[53] BTB in cattle and controversies involving its control have been on the resurgence in Britain since the 1990s.

In Ireland, where farmers had been particularly aggressive in opposing efforts to police milk sanitation, little was done until efforts were spurred by the infusion of $33 million in aid from the United States in the mid-1950s. A national campaign

[51] MacRae, "Eradication," pp. 81–88; Great Britain Ministry of Agriculture, Fisheries and Food, *Animal Health*, pp. 214–228; Waddington, "To Stamp Out," pp. 29–43; Dormandy, *White Death*, pp. 332–338; Proud, "Some Lessons," pp. 11–18.

[52] This range of estimates may have understated the problem. Cobbett's 1917 estimate that the bovine form accounted for about 6% of all TB mortality appears to have influenced others, but by the author's admission this was little more than "guesswork." In particular, important assumptions about the incidence of the bovine form in adults would later be shown to be an understatement. Cobbett, *Causes*, pp. 658–659. Griffith was probably the most careful student of BTB incidence in Britain in the interwar years, thought the estimate of 2,000 understatements but did not venture his own estimate. Dalling, "Tuberculosis," pp. 51–52; and Ritchie, "Bovine Tuberculosis," pp. 503–508.

[53] Myers and Steele, *Bovine*, pp. 188, 268; Goodchild and Clifton-Hadley, "Fall and Rise," pp. 100–116; http://www.defra.gov.uk/foodfarm/farmanimal/diseases/atoz/tb/abouttb/index.htm.

commenced in 1954 brought the infection rate in cattle to 0.5% in 1965. The program was initially voluntary but became compulsory within specific areas beginning in 1957 and everywhere by 1962. The Irish campaign, which provided liberal indemnities and cost $100 million over the 11 years, involved the slaughter of 831,000 cattle.[54] As in Britain, "complete" eradication has proved problematic in Ireland.

The experience was more varied on the European continent, but some general patterns stand out. The Scandinavian countries were most aggressive in addressing the problem. At the end of the nineteenth century, Finland, Sweden, Norway, and Denmark all faced infection rates higher than what existed in the United States and Canada – as an example, 25% of cows tested in Finland in 1894 were tuberculous. All four countries had experimented with voluntary BTB testing and slaughter programs (with partial indemnities) by 1900. Step by step, the campaigns were extended to include more animals and larger areas, made more mandatory, and coordinated with other measures such as milk pasteurization requirements. Finland had the greatest success, reducing reactor rates to less than 3% by 1913. By the outbreak of WWII, BTB was largely under control, giving Finland an informal status roughly on par with those of the United States and Canada.[55]

In Sweden and Denmark, the problem persisted, and in 1930s, both reported visible lesions in roughly 30% of the mature cattle slaughtered. Circa 1935, when Denmark finally adopted compulsory eradication, about 80% of the country's herds were still infected. In 1941, Swedish authorities instituted a national eradication campaign, initially focused on areas with low incidence. After 1948, progress accelerated, and by the mid-1950s, the disease was largely eliminated. Compulsory TB testing ended in 1970.[56]

BTB infection rates in cattle probably fell in Denmark during WWII because the Danes selectively exported infected animals to satisfy German requisitions. During the postwar occupation of Germany and Austria, the U.S. Army found that the Danish dairy sector was one of the few acceptable European sources of products for U.S. troops and their dependents. In 1952, the last known reactor was slaughtered, and the country was declared BTB free.[57]

In the first half of the twentieth centuries, a number of national models emerged. Most important were the aforementioned German model associated with Ostertag and the French model based on the BCG vaccine. The Netherlands, Austria, and Switzerland were mainly in the German camp until the postwar era; and Belgium,

[54] Myers and Steele, *Bovine*, pp. 273–275; Thoen and Steele, *Mycobacterium*, pp. 224–236; Good, "Ireland," pp. 154–155.

[55] Myers and Steele, *Bovine*, pp. 256–60, 276, and 280–286; Thoen and Steele, *Mycobacterium*, pp. 213, 242, and 248–249; http://www.museumsnett.no/gamlehvammuseum/vet_utstilling/html/artikler/tuberculosis.htm.

[56] Myers and Steele, *Bovine*, pp. 280–286; Thoen and Steele, *Mycobacterium*, pp. 248–249.

[57] Myers and Steele, *Bovine*, p. 249, 253, 256–260 and 280–286; Thoen and Steele, *Mycobacterium*, pp. 206, 213, and 248–49.

reflecting its cultural makeup, borrowed from both the German and French models. Nowhere was there much success with partial and largely voluntary schemes. The usual problems – political pressure groups blocking legislation and enforcement, re-infection due to the small areas addressed along with the introduction of infected breeding stock, and farmers ignoring quarantine lines, and the fraudulent use of tuberculin – all hampered progress and dampened farmer support. The hope that the BCG vaccine might provide a cheap solution to BTB delayed eradication in some countries. By 1955, even French officials recognized the vaccine's disappointing results in cattle and embarked on a nation-wide BTB control program, but a compulsory test and slaughter regime did not become universal until 1965. Eventually, the BCG failures and growing external pressures, first from the U.S. military and then from the threat of trade problems, led to the change in policy.

Central European nations and countries on Europe's southern fringe generally had even poorer control and eradication records. As examples, Italy and Spain only adopted limited measures in the 1950s and 1960s – 40–50 years after Denmark had experimented with similar policies. Neither southern nation even embarked on general compulsory programs until the late 1970s (excepting a 1954 Italian effort aimed at animals with clinical cases). The sobering fact is that circa 1950 BTB rates in dairy animals across most of Europe were typically in double digits and often approached 50%. TB-infected milk was commonplace, and pasteurization was far from universal. The milk supply in many of the areas noted for prized cheeses and expensive chocolates was among the most infected.[58]

Different cooking practices and dietary cultural patterns also affected the relative impact of the bovine form of TB on human populations. Most notably, observers credit the French practice of boiling milk, foremost to preserve it, with reducing the incidence of BTB in humans; whereas the German taste for raw and undercooked meat contributed to the spread of the disease.

It is telling that in a recent analysis of the Italian postwar experience, Guiliana Moda emphasized the "nontechnical" constraints on eradication. The science was for the most part known. The supreme problems that Italy faced were educating farmers, creating reliable epidemiological information on the extent of BTB in animals and humans, developing consistent and efficient program management along with a trained and dedicated cadre of veterinarians, creating incentive-compatible compensation schemes, and combating blatant fraud. Even after roughly 100 years of experience in other countries with farmers (and their local veterinarians) privately using tuberculin to hide reactors, the Italians were still confronting this problem. In fact, all of the nontechnical issues Moda analyzed were problems that officials in the United States, Canada, and Finland had to confront de novo in an era when the science was still rapidly evolving.[59]

---

[58] Myers and Steele, *Bovine*, pp. 244, 260, 262–268, 275–276, and 285; Thoen and Steele, *Mycobacterium,* pp. 215, 217-222, 241–242, and 249.

[59] Moda, "Non-technical Constraints," pp. 253–258.

Outside forces, first the United States and then European Economic Community (EEC) and its successor, the European Union (EU), played an important part in BTB eradication in postwar Europe. First, European officials learned much from the American experiences with both herd and area eradication, compensation schemes, etc. But more directly, in nearly every western European country, the United States kick-started the first truly compulsory area-wide programs with dedicated Marshal Plan aid and via the active involvement of U.S. Army veterinarians, many of whom had been active in the American eradication program. In a sense, American involvement in Europe paralleled the role of the federal government vis-à-vis the several states. By providing part of the funding, but only if a given state or country provided matching funds, U.S. leaders created a political dynamic to encourage local government officials to take bold steps – otherwise they would "lose" U.S. support. This process also helped local politicians stand up against farmers and other potential critics. As in the United States, once some areas were cleansed, farmers in those areas had an incentive to urge their neighbors to clean up their herds or face the consequence of trade embargos.

For decades, Europeans had claimed – with some justification – that it would be too disruptive to try to test and slaughter their herds. But the period of the first real wave of such radical programs came at a time when the economies were still recovering from the destruction of WWII, and when the agricultural sectors had yet to fully recover. Even with American aid, the culling herds circa 1950 imposed costs (as well as benefits) on consumers. Clearly, the political incentives, leadership, and technical skills provided by American veterinarians also mattered.

After their formation, the EEC and then the EU became major vehicles for eradication. Much like the American influence, these super-national bodies provided political cover for local politicians, and could credibly threaten countries that did not participate. The creation of the EEC under the 1957 Treaty of Rome was designed to reduce trade barriers. But Article 36 of the Treaty allowed for the continuation of import restrictions on health grounds, and individual Member States generally prohibited the free movement of potentially diseased animals. When EEC-wide rules were formulated in the 1964 Council Directive 64/432/EEC, movement was limited to cattle satisfying a strict officially bovine tuberculosis-free (TBOF) designation. A herd earned TBOF status if all cattle were free of clinical symptoms and passed a battery of tuberculin tests. An area earned this status if its herd infection rates were below 0.1% for the previous 6 years. This was a significantly higher standard than that required in the United States in 1940. Trade between EEC states remained low and national anti-BTB campaigns were directed to meeting domestic standards. The accession of Denmark, Ireland, and the United Kingdom in 1973 led to a re-exanimation of the 1964 regime. The end result was a 1977 Directive providing the EEC financial aid for Member States to create eradication programs aimed at meeting TBOF standards. Italy, France, Belgium, and Ireland initially received the first wave of EEC subsidies; Greece, Spain, and Portugal were part of a second wave. By 1994, the EU provided up to 50% of the funds for testing and indemnities in the national eradication campaigns. As of 2005, Austria, Belgium, Czech Republic, Denmark, France, Finland,

Germany, Luxembourg, the Netherlands, Slovakia, Sweden, and parts of Italy had been granted TBOF status.[60] Gaining access to the larger Europe-wide market proved an additional incentive for countries to participate in the EU eradication effort. External matching funds and accessing larger markets played similar roles in Europe as in the United States and Canada decades before.

Conclusion

From the start, the BTB program in the United States was built on several principles that remain central to livestock control efforts. First, federal officials laid the groundwork by meeting with and listening to a number of powerful farmer, meat packing, transportation, and veterinary interests. Not everyone was on board – far from it. But the program had the appearance of responding to citizen initiatives – even if those overtures were in part orchestrated by federal officials. Second, many important state governments were now anxious to cooperate. The Dorsey mess and similar failures of divided jurisdictions and regulatory arbitrage helped insure this turn of events. The BAI designed the program with shared financial and operational responsibility to help insure continued state participation. Third, the BAI had built up its moral capital in other, less controversial campaigns. Considerable trust and experience had been gained, and the legal foundations for eradication were largely in place. Fourth, the BAI started with a limited voluntary program, which focused on the most valuable cattle. These animals were often owned by relatively sophisticated cattlemen. Officials recognized there would be a political dynamic: farmers (and areas) with clean herds would have an interest in enticing farmers (and areas) with infected herds to participate, in order to help prevent re-infection. As more and more herds were certified BTB free, the political support for harsher measures would grow. Fifth, officials knew that they needed the carrot of indemnity payments. They were also wise enough to adjust parameters, including indemnity levels and state-federal payment ratios, to meet changing conditions. All of the above criteria were essentially political, economic, and social in nature. The final truly crucial ingredient was that the science was in place to make the attempt feasible. Further scientific and technical advances would aid the campaign as it developed.

It is paradoxical that the United States was so much more aggressive than most countries in Europe. The American philosophical bent toward pragmatism and compromise was at variance with the idealistic goal of the complete eradication of the disease. The compulsory nature of the area plan concept also ran against American voluntaristic traditions. Moreover, one might expect that the relatively centralized states of Europe, with their less democratic political systems,

---

[60] Caffrey, "Status," pp. 1–4; Reviriego Gordejo and Vermeersch, "Towards Eradication;" de la Rua-Domenech, "Bovine Tuberculosis," pp. 19–45.

established bureaucracies, and centralized research establishments might have responded more rapidly to the evolving state of scientific knowledge. The greater severity of the problem in Europe surely was an important reason for inaction. It would have been far more costly for European nations to reduce the incidence of BTB to background levels (below 0.5%) than it was in the United States. Europeans would have had to have slaughtered of a far larger fraction of their livestock, resulting in a period of shortage. Once most western European countries commenced an eradication program (typically after World War II), they did move rapidly, but this was in large part because they had the benefit of the successful American model. Additionally, U.S. financial aid and the leadership of U.S. Army veterinarians were common ingredients in most of the post-WWII campaigns.

The American model only emerged after more than 20 years of experimentation by numerous states and municipalities. This collective experience pointed to the need for a federal program to set common rules and to limit regulatory arbitrage. Private initiatives and state and local regulations to prevent the spread of BTB were largely ineffective and often counterproductive. Paradoxically, decentralized efforts in the United States to control BTB contributed to a wider geographic dispersion and more rapid increase in the overall incidence of the disease.

Success in the United States required the unflinching use of the state's police power and enormous costs. Perhaps most importantly, success required a great deal of common sense to develop incentive-compatible rules of the game and an effective bureaucracy that could gain the respect and confidence of local farmers and officials. While most of the world waited, the United States (along with Canada and Finland) led the way. The results were undeniable. The state-federal program dramatically slashed the bovine infection rate with an immediate spillover to human health. The aggressive U.S. campaign to improve milk sanitation spared hundreds of thousands of Americans from contracting the bovine form of tuberculosis over the period of 1917–1950.

References

"Bars Diseased Cattle: Governor Tanner Issues Prohibitive Proclamation." *Chicago Tribune*, 14 June 1899, 7.

Breeder's Gazette. 10 November 1910, 1169–70.

Buddle, B. M. "Vaccination of Cattle against *Mycobaterium Bovis*." *Tuberculosis* 81, iss.1 (February 2001), 126–32.

Caffrey, J. P. "Status of Bovine Tuberculosis Eradication Programmes in Europe." *Veterinary Microbiology* 40 (1994): 1–4.

Charles A Pierce et al. and State Board of Live Stock Commissioners v. E. B. Dillingham, 96 Ill. App. 300; 203 Ill. 148.

Cobbett, Louis. *The Causes of Tuberculosis*. Cambridge: The University Press, 1917.

Dalling, T. "Tuberculosis in Cattle in Great Britain." *Postgraduate Medical Journal* 26, no. 292 (1950): 51–52.

Dankner, Wayne M., Norman J. Waecker, Mitchell A. Essey, Kathleen Moser, Muriel Thompson, and Charles E. Davis. "*Mycobacterium Bovis* Infections in San Diego: A Clinicoepidemiologic

Study of 73 Patients and a Historical Review of a Forgotten Pathogen." *Medicine* 72, no. 1 (1993): 11–37.

de la Rua-Domenech, Ricardo. "Bovine Tuberculosis in the European Union and Other Countries: Current Status, Control Programmes and Constraints to Eradication." *Government Veterinary Journal* 16, no. 1 (2006): 19–45.

Department for Environment, Food, and Rural Affairs. "Bovine TB: What Is Bovine Tuberculosis?" http://www.defra.gov.uk/foodfarm/farmanimal/diseases/atoz/tb/abouttb/index.htm.

Dormandy, Thomas. *The White Death: A History of Tuberculosis*. London: Hambledon, 1999.

Dubos, René and Jean Dubos. *The White Plague: Tuberculosis, Man, and Society*. Boston:Little, Brown, 1952.

Essey, M.A. and M.A, Koller, "Status of Bovine Tuberculosis in North America." *Veterinary Microbiology* 40 (1994): 15–22.

Fanning, Anne and Mark Fitzgerald. "BCG Vaccines: History, Efficiency, and Policies." In *Reichman and Hershfield's Tuberculosis: A Comprehensive, International Approach*, third ed., Part B, edited by Mario C. Raviglione, 541–54. Geneva: WHO, 2006.

Faulder, E. T. "Bovine Tuberculosis: Its History, Control, and Eradication." *New York State Department of Agriculture and Markets Bulletin* 218 (1928).

Frøslie, A., A. C. Munthe, and K. Ingebrigtsen. "The Eradication of Bovine Tuberculosis in Norway." *The Norwegian School of Veterinary Science*, http://www.museumsnett.no/gamlehvammuseum/vet_utstilling/html/artikler/tuberculosis.htm.

Good, Margaret. "Bovine tuberculosis eradication in Ireland." *Irish Veterinary Journal* 59, no. 3 (2006): 154–161.

Goodchild, T. and R. Clifton-Hadley. "The Fall and Rise of Bovine Tuberculosis in Great Britain." In *Mycobacterium bovis infection in animals and humans,* edited by Charles O. Thoen, James H. Steele, and Michael J. Gilsdorf, 100–116. New York: Wiley-Blackwell, 2006.

Great Britain Ministry of Agriculture, Fisheries and Food, *Animal Health: A Centenary, 1862–1965; A Century of Endeavor to Control Diseases of Animal*. London: Her Majesty's Stationery Office, 1965.

Hoard's Dairyman, 20 June 1913.

Houck, Ulysses Grant. *The Bureau of Animal Industry of the United States Department of Agriculture: Its Establishment, Achievements and Current Activities*. Washington, DC: Author, 1924.

"Imprisonment for Dealing in Tuberculous Cattle." *American Journal of Veterinary Medicine* 13:5 (May 1918): 236–37.

Kiernan, John A. "Bovine Tuberculosis Being Suppressed." In *Yearbook 1926, U.S. Dept. of Agriculture*. Washington, DC: GPO, 1927.

______. "The Accredited-Herd Plan in Tuberculosis Eradication." In *Yearbook 1918, U.S. Dept. of Agriculture*. Washington, DC: GPO, 1919.

Kiernan, John A. and L.B. Ernest. "The Toll of Tuberculosis in Live Stock." In *Yearbook 1919, U.S. Dept. of Agriculture*. Washington, DC: GPO, 1920.

Kiernan, John A. and Alexander E. Wight. "Tuberculosis in Livestock: Detection, Control, and Eradication." In *Farmers Bulletin*, no. 1069. Washington, DC: GPO, 1919.

Lampard, Eric E. *The Rise of the Dairy Industry in Wisconsin*. Madison, WI: State Historical Society of Wisconsin, 1963.

Larson, C.W., I.M. Davis, C.A. Juve, O.C. Stine, A. E. Wight, A. J. Pistor, and C.F. Langworthy. "The Dairy Industry." In *Yearbook 1922, U.S. Dept. of Agriculture*. Washington, DC: GPO, 1932.

MacRae, W. D. "The Eradication of Bovine Tuberculosis in Great Britain." In *Tuberculosis in Animals*, edited by J. N. Ritchie and W. D. MacRae, 81–88. *Symposia of the Zoological Society of London*, No.4. London: The Zoological Society, 1961.

Melvin, A. D. "The Economic Importance of Tuberculosis of Food-Producing Animals." In *Twenty-Fifth Annual Report of the Bureau of Animal Industry for the Year 1908*. Washington, DC: GPO, 1910.

Miller, Everett B. "Tuberculous Cattle Problem in the United States to 1917." In *12th Formal Meeting of the American Veterinary History Society*. Orlando, Florida, 1989.

Mitchell, Edward B. "Animal Diseases and Our Food Supply." In *Yearbook 1915, U.S. Dept. of Agriculture*. Washington, DC: GPO, 1916.

Moda, Giuliana. "Non-technical Constraints to Eradication: the Italian Experience." *Veterinary Microbiology* 112 (2006): 253–258.

Mohler, John R. "Infectious Diseases of Cattle." In *Special Report on Diseases of Cattle*. rev. ed., 315–447. Washington, DC: GPO, 1942.

Myers, J. Arthur. *Man's Greatest Victory over Tuberculosis*. Springfield, IL: Charles C. Thomas, 1940.

Myers, J. Arthur and James H. Steele. *Bovine Tuberculosis Control in Man and Animals*. St. Louis, MO: W.H. Green, 1969.

New York Times, 1 June 1904, 1.

National Archives and Records Administration. Records of the Bureau of Animal Industry, Central Correspondence, 1913–1953 (BAI). RG 17. Entry 3, Box 337, 340. College Park, MD.

National Research Council (U.S.), Committee on Bovine Tuberculosis. "Livestock Disease Eradication: Evaluation of the Cooperative State-Federal Bovine Tuberculosis Eradication Program." Washington, DC: National Academy Press, 1994.

Olmstead, Alan L. and Paul W. Rhode. "An Impossible Undertaking: The Eradication of Bovine Tuberculosis in the United States." *Journal of Economic History* 64 (September 2004): 734–72.

______. " 'The Tuberculous Cattle Trust': Disease Contagion in an Era of Regulatory Uncertainty." *Journal of Economic History* 64 (December 2005): 929–63.

______. "Not on My Farm! Resistance to Bovine Tuberculosis Eradication in the United States." *Journal of Economic History* 67 (September 2007): 768–809.

Orland, Barbara. "Cow's Milk and Human Disease: Bovine Tuberculosis and the Difficulties Involved in Combating Animal Diseases." *Food & History* 1, no. 1 (2003): 179–202.

Pearson, Leonard and M. P. Ravenel. "Tuberculosis of Cattle." *Pennsylvania Department of Agriculture Bulletin*, no. 75 (1901).

Proud, Andrew J. "Some Lessons from the History of the Eradication of Bovine Tuberculosis in Great Britain." *Government Veterinary Journal* 16, no. 1 (2006): 11–18.

Reviriego Gordejo, F. J. and J. P. Vermeersch. "Towards Eradication of Bovine Tuberculosis in the European Union." *Veterinary Microbiology* 112 (2006): 101–109.

Reynolds, M. H. "The Problem of Bovine Tuberculosis Control." *American Veterinary Review* 33 (1909): 449–81.

Ritchie, J. N. "Bovine Tuberculosis." *Journal of the Royal Society for the Promotion of Health*, 68 (1948): 503–508.

Russell, H. L. "The Spread of Tuberculosis through Factory Skim Milk with Suggestions as to Its Control." *Agricultural Experiment Station of the University of Wisconsin Bulletin* 143 (1907).

Russell, H. L. and Conrad Hoffman. "A Three Year Campaign against Bovine Tuberculosis in Wisconsin." *Agricultural Experiment Station of the University of Wisconsin Bulletin* 175 (1909).

Salmon, D. E. "Legislation with Reference to Bovine Tuberculosis." *U.S. Bureau of Animal Industry Bulletin*, no. 28 (1901).

Smith, Howard R. *The Conquest of Bovine Tuberculosis in the United States*. Somerset, MI: Author, 1958.

St. Louis Republic, Various Issues.

"State and Territorial Laws Relating to Contagious and Infectious Diseases of Animals, 1901." *U. S. Bureau of Animal Industry Bulletin*, no. 43 (1902).

Teller, Michael E. *The Tuberculosis Movement: A Public Health Campaign in the Progressive Era*. New York: Greenwood Press, 1988.

Thoen, Charles and James H. Steele. *Mycobacterium Bovis Infection in Animals and Humans*. Ames: Iowa State Univ. Press, 1995.

Thorne, C.E. "Bovine Tuberculosis." *Bulletin of the Ohio Agricultural Experiment Station*, no. 108, June 1899.

U.S. Agricultural Research Service, Animal Health Division. "Why Tuberculosis in Livestock Is Increasing." Edited by ARS, 1960.

U.S. Bureau of Animal Industry. *Special Report on Diseases of Cattle, 1912*. Washington, DC: GPO, 1912.

______. *Special Report on Diseases of Cattle*, 1916. Washington, DC: GPO, 1916.

U.S. Bureau of the Census. *Mortality Statistics 1907*. Washington, DC: GPO, 1909.

———. "Tuberculosis in the United States." In *Mortality Statistics 1907*. Washington, DC: GPO, 1909.

U.S. Congress. *Congressional Record*, 70th Cong., 1st Sess. 1928.

"U.S. Grand Jury Hears Evidence against Dorsey." *Elgin Daily News*, 29 September 1915, 1.

U.S. House of Representatives. *Agricultural Appropriation Bill, 1925*. 68th Cong., 1st Sess., H. Rpt. 223.

______. *Agricultural Appropriation Bill, 1927*. 69th Cong., 1st Sess.

______. *Agricultural Appropriation Bill, 1930*. 70th Cong., 2nd. Sess., H. Rpt.1956.

______. *Committee on Agriculture*. Tuberculosis in Livestock, Hearings on H.R. 6188, a Bill Making Appropriation for the Control and Eradication of Tuberculosis in Live Stock. 65th Cong., 2nd Sess., 1918.

Waddington, Keir. "To Stamp Out 'So Terrible a Malady': Bovine Tuberculosis and Tuberculin Testing in Britain, 1890–1939." *Medical History* 48 (2004): 29–48.

Wisconsin Department of Agriculture. *Biennial Report, 1915–1916*. Wisconsin Department of Agriculture: Madison, WI.

Wisconsin Department of Agriculture. *Biennial Report, 1919–1920*. Wisconsin Department of Agriculture: Madison, WI.

Chapter 3
The Evolution of Animal Agricultural Systems and Supply Chains: Theory and Practice

Jennifer Ifft and David Zilberman

Introduction

The current efforts to find technical and policy solutions to zoonotic diseases, such as Avian Flu and Swine Flu, are part of a continuous process of evolution of production and supply chains for food and other agricultural products. This chapter provides a conceptual framework to analyze this coevolutionary process and, in particular, the role of technology, economics, and institutional factors as well as biological coevolution. This chapter will build upon the basic principles of production economics, theory of innovation, urban economics, and public finance to show how changes in relative prices, consumer preferences, and technological innovation affect the management and regulation of farming and, in particular, animal husbandry. This theoretical framework is then applied to poultry supply chains in Northern Vietnam.

Evolution of Agricultural Supply Chain

Throughout history, humans have manipulated biological systems for their own survival. Plants and animals have provided humans with food, clothing, and energy. One manifestation of human evolution was the development of agricultural and food production systems, which also contributed to the coevolution of natural systems. To compare various food systems, it is useful to recognize that they have several components that are presented in Fig. 3.1.

The essential components of the system are (1) breeding, which is still a natural process that may be assisted by humans; (2) feeding, which may include the

J. Ifft (✉) • D. Zilberman
Department of Agricultural and Resource Economics, University of California
Berkeley, CA 94720-3310, USA
e-mail: jifft@berkeley.edu; zilber11@berkeley.edu

D. Zilberman et al. (eds.), *Health and Animal Agriculture in Developing Countries*, Natural Resource Management and Policy 36, DOI 10.1007/978-1-4419-7077-0_3,

BREEDING	FEEDING	HARVESTING	PROCESSING	CONSUMING

Fig. 3.1 Agricultural and food production systems

physical feeding of livestock or providing inputs, such as fertilizer and irrigated water, for field crops; (3) harvesting, which has different forms, such as fishing, hunting, as well as using machinery to harvest fruits, vegetables, and livestock; (4) processing, which may include many steps, such as transportation, cleaning, refining, and producing various consumer products; and (5) consuming, which can be done both in households as well as in institutions, such as restaurants.

Over time, production systems become more complex. Hunter gatherers were basically harvesters and processors of food. Through learning by doing and experimentation, humans identified the species that they harvested and developed alternative procedures of cooking that resulted in more refined foods. As humans evolved, they introduced farming and animal husbandry. In these systems, humans involved themselves in breeding as well as in feeding, which increased the effort and uncertainty associated with harvesting. Carlson and Zilberman (1993) developed a model analyzing the extent of harvesting vs. farming efforts. They argue that, in the case of livestock, for example, food can be produced by farming or by hunting, and the allocation of effort is done to maximize the discounted net social benefit (which consists of benefits from food minus the cost of effort taking into account the dynamics of affected population). The resulting allocation of effort is such that the net marginal benefit from effort allocated to hunting (taking into account the cost of hunting as well as future cost of depletion of the stock) is equal to the net marginal benefit of effort allocated to farming. Thus, as the human population increases and technology improves, the relative importance of farming systems vs. hunting increases. This analysis is consistent with the results of Bosrup (1965), who analyzed the economics of slash-and-burn systems and emphasized the role of increasedpopulation density in transition to more intensive agricultural systems. Similarly, Binswanger and McIntire (1987) have analyzed the economics of the transhumance system, where grazing is rotated across location, and the evolution of this system to more permanent farming systems as animal density increases. Hoagland, Jim, and kite-Pawell (2003) investigate the equilibrium relationship between fisheries and aquaculture systems and how population growth and increases in demand and technological improvement tend to increase the share of aquaculture in total production.

Diamond (1999) traces the evolution of agricultural systems from prehistoric periods and emphasizes the role of capacity to expand technological possibilities, including animal husbandry, in enhancing productivity and facilitating economic growth and progress. His analysis emphasizes the important role of technological exchange and transfer of technologies among societies. He argues that lower cost of the transfer of knowledge will tend to enhance economic growth. For example, one reason for the relative success of societies spanning all the way from Western Europe to China is the ease of latitudinal transportation among these regions. Diamond's findings are consistent with the results of much of the literature on agricultural productivity and growth. Ruttan (2001) argues that differences in biophysical and sociopolitical conditions, economic situation, and technological capabilities are leading to the emergence of different technologies in different locations and different rates of technological change. He argues that investment in research, both by private and by public sectors, has a high rate of return and leads to transformation of agricultural systems. In countries where the relative cost of labor is increasing, agriculture becomes less labor intensive, farms become larger, and production becomes more capital and energy intensive. Kislev and Peterson (1982) suggest that the increase in farm size in the United States is largely attributed to increased opportunities for labor in the urban sector that led to migration from the rural sector and thus an increase in farm size. The economic logic and methodology can be applied to the rest of the world and the processes of change that are occurring elsewhere. Boehlje (1999) argues that technological change and increases in consumer income are leading to drastic changes in agriculture first observed in developed countries, such as the United States, but spreading throughout the world where the role of processing and value added in production is increasing. There is a transition from traditional commodity agriculture to agribusiness, and sometimes industrial farming, where the emphasis is on the differentiation of products to meet consumer demands and cater to different market niches and production systems that emphasize contracting to specification and "just in time" supply management.

Returning to Fig. 3.1 and the food supply chain that it depicts, modern agriculture has much more emphasis on human intervention of breeding through selective breeding and genetic manipulation. There is growing specialization in various tasks of farming. Increased understanding of the principles of genetics has led to investment in breeding and the establishment of units that are specialized in herd improvement. There are several strategies for feeding. They range from grazing with low-concentration to high-concentration feedlots. Furthermore, animals may be moving to different forms of feeding throughout their lives. In many cases, harvesting becomes highly concentrated and there is proximity between the final stages of feeding as well as processing. Processing meat products has intensified with increased emphasis on value added and convenience. As argued elsewhere in this book (Heiman et al. 2000), consumers appreciate value and convenience significantly and food processing has responded accordingly. Different strategies of production of animals result in different types of organizational structures. These structures are built to assure coordination, continuity, and reduced production cost but vary according to socioeconomic and biophysical characteristics of different locations. In particular, we can distinguish between traditional systems where livestock is integrated with other farming activities and where consumption is, for the most part, local. In more modern systems, the key institutions within the supply chain include cooperatives, vertical integration, and contracting.

Contract Farming, Cooperatives, Vertical Integration, and Animal Supply Chains

Livestock production systems around the world have undergone significant changes in their structure since the 1950s. Traditionally, grain and livestock production systems were integrated into one farming system. In systems such as the corn-hog systems in the Midwest, animals were fed on grain that was grown and fertilized with their manure. In other regions, farmers relied on mixtures of grain and pastures to feed their animals. The reduction in transportation costs, higher labor costs, development of automation, and differences in productivity and location resulted in specialization. Some regions specialize in producing grains and others in feeding and processing livestock. The extent to which we have specialized vs. integrated production systems varies within and between countries. While the specialized livestock systems originated in the United States and spread to Europe, it is now becoming very prominent throughout the world and has become a major manifestation of the industrialization of agriculture.

There has been significant research on contract farming in livestock production systems in the United States. While industrial livestock systems vary among countries, they have much in common. Some of the lessons of research on the industrialized livestock systems in the United States are general. Geographic specialization and a high degree of contract farming characterize the livestock sectors in the United States. The actual growers of livestock do not purchase input and sell output in traditional markets but operate under contract and receive inputs and provide outputs to intermediaries known as integrators. The integrator provides the farmers with specifications for the product and the manner in which it will be produced as well as with inputs, including genetic materials. Growers will receive payment for the output delivered, based on a prespecified financial agreement. Other growers have market contracts in which processors and growers agree on predetermined output prices. These contracts are not as prevalent as production contracts. To some extent, these contracting arrangements are similar to the relationship between a franchisee and a franchise in retail. McDonald's provides its franchisees technologies as well as inputs to produce their product. While retailers sell to the public, in contract farming, the integrator markets the product.

Contracting has been especially important in broiler production. Historically, chicken provided a joint product of eggs and meat, while specialized production of meats from poultry was rather recent (Cochrane 1993) and based on innovations in breeding as well as management that occurred in the 1940s. Broiler farming was a "new technology" that farmers were unwilling to adopt because of the unstable prices of meat. Grain marketers realized that the broiler sector was a potential source of growing demand, and they were instrumental in introducing the technology. Feed suppliers introduced production contracts as well as credits to growers. They supported research and outreach that improved broiler technologies and, in the meantime, developed proprietary varieties. Over time, growing emphasis has been placed on product quality and increasing the value added for the output. New organizations, such as Tyson and Purdue, have emerged, developing unique methods, genetic material, and processing technologies that resulted in differentiated poultry products that were shipped to different parts of the world to meet different market niches. For example, while the United States market has preferences for chicken breasts, other parts were sent to Asia and the Soviet Union. The emphasis on quality characteristics, uniformity, and precision has led to increased reliance on contracting. More than 90% of U.S. broilers are currently produced under contract arrangements, and the share of contracting is growing fast around the world. Yet, most of the production of chicken under contracts is "industrialized," but there is growing demand for free-range chicken and concern about the manner in which chickens are raised (Mitchell 2001). Nevertheless, even free-range chicken can be produced by contracting.

While livestock contracting emerged in poultry, over time, the swine sector went through a similar process of industrialization. Pork production was historically viewed as "value added" to corn production (Martin and Zering 1997). However, increasing returns to scale in the feeding and processing of swine have led to the emergence of highly concentrated livestock production farms that rely on exported purchased inputs. These operations resulted in a large volume of waste, and the disposal of this waste has become a major policy and legal challenge. The movement to contract farming in swine production is not unique to the United States. The economics of scale in the processing of hogs have resulted in an emergence of contract farming in The Netherlands, as well as Denmark (Schrader and Boehlje 1996). However, an alternative route where farmers can take advantage of the economics of scale in processing while preserving farmers' independence is to form cooperatives that process and market pork products. In these situations, farmers are more independent in their production decisions, buying and growing their own feed. Yet, their outputs have to adhere to the product specifications laid out by the cooperative.

Some of the major processors and marketers of poultry products are vertically integrated organizations with large production facilities. Even some processors that get most of their animals from contractors have integrated production in order to develop expertise and unique knowledge as well as production technologies that they can later introduce to their contractors. Obviously, vertical integration has advantages of greater control on the production process and less risk of information and technological leakages. Nevertheless, contract production is growing significantly in the poultry sector for several reasons.

As mentioned earlier, production contracting can be compared to franchise contracting. Studies by Lafontaine (1992) and Lafontaine and Slade (2000) suggest several reasons for franchising. One is that franchising requires revenue sharing, and thus revenue risk sharing, and provides the franchisee with incentive to engage in efforts even when they are unobservable to the franchisor. Another is that franchising allows the franchisor to share the cost of expanding the productive capacity in the early stages of the organization. The plausible reasons for contracting livestock include

1. Growers provide the finance for production capital faster growth for the organization. Generally, the integrator has unique technological knowledge but is facing financial constraints that limit expansion. By contracting, others invest in production facilities and real estate that are very expensive and integrators can focus investment in processing, marketing, and technological development.
2. Integrators provide market power and knowledge to the contractors. The integrators have returns to scale that would allow them to obtain volume

discounts as well as monopoly premiums associated with the brand. They also have human capital and unique knowledge that reduce the cost and effort required for management for growers.
3. Development of farming operations requires major investments in real estate as well as good access to local labor markets. These are relative advantages of the growers.
4. Contracting may provide good incentives for efficient production. The grower is sharing the risk with the integrator and, thus, is induced to make the extra effort when unobserved. Furthermore, in some cases, the payment to different growers is based on their relative performance, so the contract of individuals who continuously fail may not be renewed (Hueth and Ligon 2001).
5. Contracting is likely to emerge when an entrepreneur develops a unique production system for livestock but is lacking the resources to bring it to scale as a vertically integrated system. It is less likely to emerge if the technology is owned by a rich organization that is concerned about the control of the technology and has enough resources or access to credit to bring it to scale. Furthermore, in regions where there are well-established farming populations and significant ability to adjust and learn, as well as social capital to conduct joint ventures, one may observe the emergence of farmer cooperatives that allow producers much more economic independence yet allow the group take advantage of economies of scale through processing and marketing. Government policies may give the cooperatives rights and capacity to enforce quality standards as well as pricing and allocation rules.

Externalities, Regulation, and the Livestock Supply Chain

Environmental, human, and animal health externalities are pervasive throughout the livestock supply chain. These externalities are multidimensional. Production of livestock may result in groundwater contamination, runoff of contaminated materials, noxious smells, generation of greenhouse gases (e.g., methane), and the spread of zoonotic diseases. Bacterial, viral, and other diseases are spread through the processing and distribution stages of production. Other chapters of this book emphasize results of research on controlling these externalities. Here, we will emphasize how these externalities are related to different supply chain structures.

In traditional systems, there is a low level of animal concentration. Waste is used to fertilize feed production that may result in a sustainable system with a low level of groundwater contamination. However, an increase in animal concentration, frequently associated with modern livestock management systems, results in increasing severity of both groundwater contamination as well as random runoff. There are various policies that aim to control both groundwater contamination and random runoff (Zilberman et al. 2006). However, enforcement of these policies can encounter several challenges.

Information must be considered first. When many small units conduct production, it is difficult if not impossible for a policy maker to assign responsibility for each unit of pollution, e.g., in cases of nonpoint source pollution, problems that are not amenable to efficient policies, such as Pigouvian taxes. Another issue is the ability to pay when damages occur. In some cases, for example, storm runoff from farms or spread of disease may result in significant damage that cannot be paid by an individual producer who may be the source of the problem. Thus, the ability to introduce heavy taxation to control risks is limited (Sproul 2010). However, the capacity to regulate and the nature of regulation can vary within different organizational structures.

In cases where the livestock supply chain is vertically integrated and one organization controls all production within a distinct region, then this organization will be liable for much of the damage. The organization can then assign responsibility to different units within itself, so it may introduce penalties for activities that may generate pollution that will reflect the policies that are imposed on the organization. For example, when we have a vertically integrated poultry production, and this organization is taxed for contamination of a body of water in a region where it is active, then the organization is likely to develop monitoring activities as well as incentives that will identify units that cause pollution and penalize them for it to reduce damage.

In the case of contracting, the situation is more complex. When it comes to externalities generated in production, the ultimate control is by the grower. However, the grower production system is designed according to specifications provided by the integrator, as the integrator provides the growers with inputs as well as management guidelines. While there is likely to be some asymmetric information between the integrator and the contractor, it is likely that the contractor will have better information than the government about growers' activities. Therefore, Ogishi et al. (2003) as well as Aggarwal and Lichtenberg (2005) argue that integrators should be held liable for polluting activities in livestock production. Since the integrator cannot control some activities of the growers, the liability should be shared. For example, when integrators share liability, they will develop information systems and contribute to enforcement that would result in growers being ultimately liable for negligence on their part that results in externalities.

When individual growers form a cooperative that may market their product, this cooperative or a similar entity may be held liable for several types of externalities, e.g., nonpoint source externalities or other externalities that are difficult to assign to individuals. The cooperative then may develop the capacity to assign penalties to the perpetrator and to induce behavior that will reduce externalities and improve product safety. In other situations, for example, in cases of random storm runoff, it may be worthwhile to introduce a strict punishment (such as an ambient tax) that will serve as an inducement for individual and collective preventative activities (Segerson and Wu 2006).

When individual growers are held responsible for pollution-control activities or are liable for externality damage, they may lack the resources to pay and, in some cases, it may lead to either bankruptcies or noncompliance with environmental regulations. In some cases, government may introduce policies, such as payment for environmental services (PES), that would subsidize behavior that would reduce externalities by small producers. Given the multidimensionality of externalities generated by livestock production systems, a mixture of carrots and sticks has to be established to control externalities and, at the same time, enable economic viability. However, the design of policies should correspond with the structure of the supply chain.

The Economics of Poultry Supply Chains in Northern Vietnam

Vietnam, like many developing countries, has several poultry supply chains. Poultry is being produced at some level by most rural households and marketed over a dispersed area with different trading systems. Chicken varieties in Vietnam are highly differentiated, and market channels have become fragmented. This differentiation is related to production scale. Poultry market channels in Vietnam might be generalized into two to three groups. In one, small farms produce local chicken and sell to nearby markets or to urban areas through informal channels. In another, medium and large farms sell through formal, regulated channels, such as wholesale markets. In a few cases, large companies have built their own slaughterhouses, nearing complete vertical integration. Farms tend to be small although the number of farms raising chicken on an industrial scale is increasing.

In the small-scale traditional system, chickens are grown as part of a diversified farming system. The farmer may grow rice and other products, and chickens are being used for eggs as well as for meat and are sold in wet markets. At the same time, in the cases of larger scale production, we see the emergence of more modern industrial chicken grown in relatively large facilities and fed by exported grain or fish meal. The industrial chicken is processed to produce a product mix that includes whole chickens and chicken parts. Some of it is sold in traditional markets, but much of it is sold in supermarkets or to institutions.

The existence and characterization of these parallel supply chains resulting in different chicken products can be explained by economic conditions. In a rural setup, where farmers grow a field crop, such as rice, there are a certain number of birds that can be supported per unit of land with minimum supplementary inputs. The birds gain their calories from grain and their protein from worms and other creatures that exist naturally in the land. Continuously, farmers are taking birds out of consumption either for self-consumption or for sale and introducing new ones. The system is continuous. What enables the farmers to market a small number of birds is the availability of traders that use motorcycles or other means of transportation with small capacity. The existence of this class of traders is feasible by the relatively low wage rate of workers and the low level of capitalization in the society. This allows these traders to earn a competitive wage from buying and selling a relatively small volume of chicken using a cheap means of transportation.

However, in the commercial sector, production is conducted at chicken houses with a significant amount of scale. Generally, production is done in batches on a periodic basis. Efficient management is very important since much of the input is purchased and the scale of production is large. Thus, the birds are selected to grow within a shorter period of time – cycles of 1½ to 3 months (Gillespie and Flanders 2009). Once a cohort is ready for shipment, it will be shipped in large quantities by trucks. Because of the cost of transport, there is much more emphasis on coordination within the system to reduce idleness of equipment and, at the same time, provide a continuous supply of food to different outlets that include supermarkets, small stores, and open markets.

The major poultry supply chains serving Hanoi, the capital of Vietnam and largest city in Northern Vietnam, illustrate many themes in the conceptual framework we have developed for the evolution of animal agriculture and supply chains. In the following sections, we will discuss the structure of these supply chains and related issues. The data presented are based on the results of a resource flows study that is documented in (Ifft et al. 2008) and Roland-Holst et al. (2007). This study included chicken producers, chick (stock) producers, local traders, wholesale traders, slaughterhouses, and urban market vendors in Hanoi and surrounding poultry-producing provinces.

Production Structure

Production structure for chicken is closely related to farm size in Vietnam. A few backyard or "local" chickens are raised by most rural families in Vietnam. Local (free-range) chicken breeds cannot be caged, cannot efficiently process concentrate feed, grow slowly, and generally require few inputs. They are also known for being "hardy" or being able to handle a free-range environment with greater stress. Production is constrained by overall farm size, open space, and availability of farm/household food byproducts. Industrial (exotic) chicken breeds are fast growing under a diet of concentrate feed and can also be caged. They require more health inputs and can be grown in small spaces. Industrial breeds require a protected environment; they do not respond as well to environmental stress. The main constraints are production infrastructure (cages, protective buildings) and credit constraints for concentrate feed.

For the reasons given above, poorer farm households tend to specialize in local chicken while larger, wealthier farmers tend to specialize in industrial chicken. Crossbred chicken is almost exactly "in between" local and industrial at all levels, especially in terms of input requirements and prices. For clarity, we will refer largely to industrial and local breeds except when specifically discussing the emerging role of crossbred chicken. Tables 3.1-3.5 below refer to production-level quartiles for farmers in various major poultry production areas in Northern Vietnam: The first quartile has the smallest annual production (number of batches × average batch size) while the fourth quartile has the largest production level.

Table 3.1 Breeds produced and income levels by annual production

Farm production level	Local	Crossbred	Industrial	Household income from poultry production
First quartile (%)	76	20	4	16
Second quartile (%)	55	38	6	20
Third quartile (%)	39	50	11	23
Fourth quartile (%)	8	61	28	34

Table 3.2 Source of chicks

Farm production level	Own farm	Village farms	Commune farms	District farms	Own province farms	Other province farms	Market	Trader
First quartile (%)	38	14	6	8	2%	8	19	5
Second quartile (%)	33	15	14	16	7	10	3	1
Third quartile (%)	22	26	10	12	10	15	2	3
Fourth quartile (%)	4	11	13	12	18	33	0	9

Table 3.3 Percent of farms keeping hens and selling chicks

Farm production level	Farms keeping hen	Farms selling chicks
First quartile (%)	56	18
Second quartile (%)	54	25
Third quartile (%)	53	25
Fourth quartile (%)	23	10

Table 3.4 Location of chick sales

Farm production level	Village	Commune	District	Province	Other Province	Market
First quartile (%)	76	12	4	0	4	4
Second quartile (%)	69	16	10	0	3	2
Third quartile (%)	57	21	8	4	2	8
Fourth quartile (%)	32	13	9	3	18	25

Table 3.5 Average age of chicks purchased

Farm production level	Local	Cross	Industrial
First quartile	19.7	4.7	1.0
Second quartile	9.6	3.2	1.3
Third quartile	8.5	1.9	1.3
Fourth quartile	3.4	1.3	1.1

Table 3.1 shows the percent of each variety of chicken produced by different annual production levels and the contribution of chicken to total household income. The first quartile produces less than 150 head chicken per year, the second quartile produces 150–300 head, the third quartile produces 300–800 head, and the fourth quartile produces more than 800 head. Comparing the first and fourth quartiles, it is easy to see that the largest farmers are wealthier, more specialized in raising chicken, and largely raise crossbred and industrial breeds while the farmers with lower levels of production are poorer, less specialized, and mainly raise local breeds.

Smaller farms are more likely to maintain their own breeding stock and buy any chicks from local producers (Table 3.2). They are also more likely to sell chicks to other farms, largely farms in the same commune (Tables 3.3 and 3.4). The majority of the commercial chick producers interviewed raises crossbred chicks, which indicates that any farms raising local chicken are generally limited to sourcing from local farms for chicks. Larger farms and those raising crossbred and industrial breeds are more likely to buy chicks over larger distances. For all types of farms, purchasing chicks from traders is rare. Small farms tend to buy chicks at an older age (averaging 18 days for local breeds) while, for large farms, 1 day is the norm for industrial and about 3 days for crossbred (Table 3.5). The age at which local chicks are bought and the reliance on local sources indicates that local chick production might also be constrained by availability of breeding stock. Some reasons for the purchase of local chicks at an older age include: (1) production of local chicks is dispersed and small scale and, hence, it is more difficult to coordinate purchases and (2) farmers buying local chicks are more credit constrained and, hence, often buy them at a later date. Restocking projects have found that locating larger producers of local chicks is difficult, consistent with (1).

The timing of rice harvests and availability/affordability of concentrate feed may play a large role in constraining poultry production. Industrial chickens are largely fed concentrate feed, which can be stored more easily than byproducts but is more expensive. Over half of the industrial chicken producers sell more than four batches per year while almost no producers of local and crossbred chicken sell more than four batches (Table 3.6). Many producers of local chicken are able to sell 3–4 times per year while this is less likely for crossbred chicken producers. Crossbred chickens are fed a combination of concentrate feed and household byproducts. Therefore, production would be most likely to be constrained by harvest timings and byproduct quantity as batch size tends to be larger (Table 3.7). Households

Table 3.6 Number of batches by type of chicken produced*

Annual number of batches	Local chicken producers	Crossbred chicken producers	Industrial chicken producers
1–2 Batches (%)	54	71	26
3–4 Batches (%)	42	28	25
>4 Batches (%)	5	1	49
Observations	254	229	68

*About 12% of producers interviewed raise more than one type of chicken. For clarity, they are not included in these calculations.

Table 3.7 Batch size by type of chicken produced

	Local chicken producers	Crossbred chicken producers	Industrial chicken producers*
Average batch size	90	519	5542
Median batch size	70	250	325
Observations	254	229	68

* One industrial chicken farmer sells 300,000 birds per batch; the next largest producer sells 15,000 birds per batch.

Table 3.8 Farmers' percent increase in average sales by month*

District	Jan	Feb	Mar	Apr	May	Jun	Jul	Aug	Sep	Oct	Nov	Dec
Gia Binh	40	28	20	19	0	0	0	0	117	10	10	6
Tien Du	20	93	14	0	0	0	0	50	50	0	15	18
Ly Nhan	25	200	20	25	0	50	50	20	28	23	25	27
Kim Bang	28	82	19	0	5	0	0	20	400	30	55	0
Soc Son	23	25	15	0	0	30	60	35	30	57	45	23
Tu Liem	43	23	37	25	0	0	30	23	26	26	27	44
Hoai Duc	0	0	0	0	40	0	0	30	35	40	0	0
Phu Xuyen	30	185	13	10	0	0	0	0	0	0	30	20
Chuong Mi	0	0	0	0	0	0	0	0	0	0	0	0
Dong Anh	10	34	0	0	0	0	200	0	100	0	50	0
Yen Phong	0	0	0	0	0	0	0	0	0	0	0	0
Total	40	28	20	19	0	0	0	0	117	10	10	6

* This is only for farmers reporting an increase in their sales and thus only represents an average for those reporting.

raising crossbred chicken might also face a liquidity/credit constraint if they are only able to purchase concentrate feed after harvests. Crossbred chicken producers may also be producing for periods when demand is higher.

Farms report sales peaks in the early fall and winter, which indicates that sales are, indeed, following harvests as the two rice harvests tend to fall in June and September/October (Table 3.8). In many districts, sales increased in the early fall, experienced a month or two of lower sales, and then increased again in the winter. These patterns are not uniform across districts, which could be due to different harvest timings or marketing specialization. Different areas supply urban areas at different times. Generally, sales during spring and summer appear to be low, which would also coincide with low levels of feedstock and the feeding period for chicken directly after the rice harvest. Large farmers (mainly represented by the last three districts) do not appear to be affected by seasonal demand or seasonal feed constraints. Most likely, they are operating under contract for regular sales, while smaller farmers increase sales during seasonal demand and feed availability.

Local traders (Table 3.9) also show peaks in sales in the early fall throughout the winter with a glut in sales in the spring and summer that follows patterns reported by farmers. Because local traders mainly deal in local chicken and crossbreds,

Table 3.9 Commune traders' percent increase in average sales by month

District	Jan	Feb	Mar	Apr	May	Jun	Jul	Aug	Sep	Oct	Nov	Dec
Gia Binh	8	4	0	0	0	1	1	2	3	1	3	9
Tien Du	14	10	10	0	0	0	0	0	11	11	0	13
Ly Nhan	0	10	10	0	4	4	0	0	2	2	4	4
Kim Bang	14	8	8	0	0	0	0	0	2	3	5	7
Soc Son	17	4	5	3	0	0	0	0	6	2	5	5
Tu Liem	0	0	0	0	0	0	0	0	0	0	18	18
Hoai Duc	36	0	0	0	0	0	0	3	4	1	0	19
Phu Xuyen	24	4	4	0	0	0	0	5	6	6	0	24
Total	16	5	5	1	0	0	0	2	5	3	3	12

Table 3.10 Slaughterhouses' percent increase in average sales by month

	Jan	Feb	Mar	Apr	May	Jun	Jul	Aug	Sep	Oct	Nov	Dec
Bac Thang Long	3	0	0	0	0	0	0	3	3	0	0	18
Ha Vi	15	39	2	0	0	0	7	2	1	0	6	3
Soc Son	0	60	0	0	0	0	0	0	0	0	0	0
Tu Liem	0	0	0	0	0	0	0	0	0	0	0	0

Table 3.11 Hanoi market vendors' percent increase in average sales by month

Month	Increase in sales	Standard deviation	Max
January	52	127	700
February	21	105	1,000
March	2	7	50
April	0	2	30
May	0	2	20
June	0	1	10
July	0	4	50
August	2	18	250
September	1	4	30
October	2	7	50
November	7	31	400
December	6	12	60

they would be more subject to the date of rice harvests. Slaughterhouses (Table 3.10) appear to be responding to different supply and market conditions with variable increases in sales similar to those of farmers and local traders. The slaughterhouses largely serve Hanoi and do not appear to be receiving the chicken from increasing sales in various districts in the fall. Hanoi market vendors (Table 3.11), however, report the most significant increases in sales in January and February, in contrast to increased sales in the fall for other market-chain players. This could be caused by most of the extra chicken being produced in the fall being consumed locally or through the chicken being sold through other channels, such as increased restaurant sales.

Production Decisions and Constraints

When making production decisions, farmers must consider capital, labor, credit/liquidity, and farm/household byproduct availability. Labor is not a constraint for local chicken production, but it might be for larger farms producing industrial chicken. Credit availability mainly affects producers of crossbred and industrial chicken. Farms that are constrained by capital and credit might choose to raise local chicken. Farms that are not constrained by these things would choose to raise industrial chicken. Another constraint would be the availability of chicks whether from one's own farm or other farms. Given the existence of commercial producers of crossbred and industrial chicks, this is most likely a constraint for local chicken farmers. Prices would also factor into farm production decisions. However, farmer survey data indicate that, on average, scale does not have a large impact on farm gate chicken prices.

At all levels of the supply chain, informal contracts dominate agreements to purchase poultry (see Tables 3.12–3.17; Fig. 3.2). Instead of operating as a commodity market with perfect competition where all producers and buyers can freely enter/leave markets, poultry producers and intermediaries have locked-in informal relationships with their trading partners. This also indicates that producers rarely initiate production without predetermined buyer(s). Many producers report selling to traders on credit, and they are paid only after the trader sells to the next intermediary. If selling to an unknown trader did not entail a significant cost or risk, farmers could freely sell to anyone at "market clearing" prices. More research is necessary to better understand the nature of these informal contracts, but it is clear that supply chain players are linked through verbal or implicit agreements and that these agreements play a critical role in how supply chains function.

The fact that most producers rely almost entirely on informal agreements, while eschewing formal contracts, indicates not only that levels of trust are high among supply chain players but locked-in trading partners are necessary in the current market environment. Chickens have fairly short production cycles compared to other livestock, so producers and traders would have regular interaction. Hence, producers and traders are playing a "repeated game" where, if one player defects (i.e., does not deliver the agreed product), then that player can be excluded from future interactions or otherwise "punished." Poultry trade in Northern Vietnam, especially trade of local breeds among small players, is thus governed by a series of high-trust informal relationships that are essential for market participation. Larger players are better able to adopt formal contractual relationships.

Among commercial chick producers in several districts, formal relationships were more likely to occur with larger farms and traders but verbal contracts were the norm. In the farmer survey, verbal contracts dominated agreements for purchases except those with end users with whom farmers have no agreement. Wholesale traders working exclusively with companies (such as CP) had formal agreements for the purchase of chicken. Agreements with farms tended to be verbal except at one smaller wholesale market where traders had no contractual agreement

Table 3.12 Chick producer relationship with customers by district

	None (%)	Verbal (%)	Formal (%)
Soc Son			
Backyard farms	0	100	0
Small farms	0	100	0
Tu Liem			
Small farms	0	100	0
Medium farms	0	100	100
Large farms	0	0	100
Hoai Duc			
Small farms	33	67	0
Medium farms	33	67	0
Large farms	20	80	0
Traders	0	100	0
Phu Xuyen			
Backyard farms	10	90	0
Small farms	0	100	0
Medium farms	0	88	13
Large farms	0	50	50
Traders	0	71	0
Dong Anh			
Backyard farms	100	0	0
Small farms	0	100	0
Medium farms	0	100	100
Large farms	0	0	100
Traders	0	0	100
Yen Phong			
Backyard farms		100	
Small farms	–	100	–
Medium farms	0	100	0
Large farms	0	100	0
Traders	0	100	0

Table 3.13 Contractual standards for the farmer survey

All farms	None (%)	Verbal (%)	Formal (%)
Commune trader	4	96	0
District trader	4	95	1
Province trader	5	95	4
Other Province trader	2	98	0
Commune market	11	89	0
District market	0	100	0
Wholesale market	0	100	0
Slaughterhouse	0	100	0
End user Hanoi	33	50	17
Local end user	23	77	0

Table 3.14 Supplier contractual standards for wholesale traders

	None (%)	Verbal (%)	Formal (%)
Bac Thang Long "entering"			
Farms	45	55	0
Ha Vi "entering"			
Farms	6	71	26
Traders	0	100	0
Company	0	0	100
Ha Vi "middlemen"			
Farms	0	58	42
Traders	0	94	6
Company	0	0	100
Tien Du "middlemen"			
Farms	0	100	0

Table 3.15 Customer contractual standards for wholesale traders

	None (%)	Verbal (%)	Formal (%)
Bac Thang Long "entering"			
Traders	0	100	0
Slaughterhouses	42	58	0
Shops	0	100	0
Markets	0	100	0
Ha Vi "entering"			
Traders	9	85	6
Slaughterhouses	31	62	8
Shops	100	0	0
Consumers	67	33	0
Markets	9	91	0
Ha Vi "middlemen"			
Traders	20	80	0
Slaughterhouses	36	64	0
Shops	17	83	0
Consumers	82	18	0
Tien Du			
Traders	67	33	0
Shops	0	100	0
Consumers	100	0	0
Markets	0	100	0

Table 3.16 Supplier contractual standards for slaughterhouses

	None (%)	Verbal (%)	Formal (%)
Bac Thang Long			
Farms	83	17	0
Trader (delivered)	25	81	0
Ha Vi Market			
Farms	0	100	0
Market	2	97	0
Soc Son			
Trader (delivered)	0	100	0
Company	0	100	100
Tu Liem			
Farms	0	100	0
Company	0	0	100

Table 3.17 Customer contractual standards for slaughterhouses

	None (%)	Verbal (%)	Formal (%)
Bac Thang Long			
Trader	13	53	20
Consumer	67	33	0
Shops	8	33	67
Ha Vi Market			
Trader	0	100	0
Consumer	19	79	0
Shops	13	84	0
Retail	0	100	0
Soc Son			
Consumer	0	100	0
Shops	0	100	0
Retail	0	100	0
Tu Liem			
Consumer	0	100	0
Shops	50	50	50

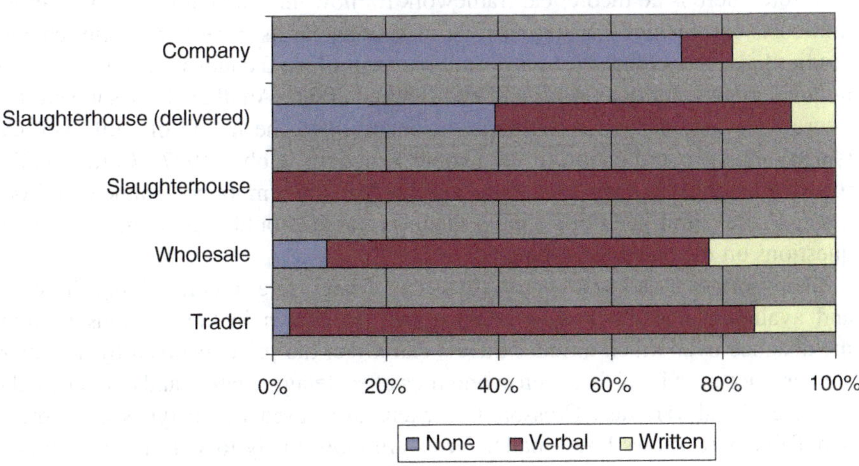

Fig. 3.2 Supplier contractual standards for HANOI market vendors

with, on average, 45% of the farmers from whom they purchased chicken. Wholesale traders also tended to have verbal agreements with other traders. Overall, wholesale traders tended to have more varied relationships with their customers although the majority relied on verbal agreements. Slaughterhouses tended to have verbal agreements with traders but formal agreements with companies. Verbal agreements were the norm with slaughterhouse customers although those operating in one market relied heavily on formal agreements with shops. Hanoi market vendors had verbal agreements with many of their suppliers although those purchasing from companies or slaughterhouses that delivered birds to them often had no contracts.

Table 3.18 Crossbred chicken customers by farm production level

Annual farm production level	Commune traders	District traders	Own Province traders	Other Province traders	Commune level end users	Other
First quartile (%)	16	11	22	7	33	11
Second quartile (%)	21	26	17	2	26	8
Third quartile (%)	31	27	19	7	13	3
Fourth quartile (%)	14	43	32	5	3	4

Contractual agreements for regular purchase might include arrangements for quantity, price, timing of purchase, or specific product characteristics. If farmers are contracting on product characteristics, our survey work indicates that they are not being rewarded for it. Farmers generally do not receive a premium for any type of safety, special breeds, or any quality characteristics (Ifft et al. 2008). Future research on contractual relations could highlight many of the constraints facing producers and intermediaries as well as the type of information failure and transaction costs that these contracts are intended to mitigate. In the economics literature, there is no theoretical framework for how information failure or transaction costs can restrict the involvement of smaller farms in certain supply chains. Studies have shown that trust plays an important role in the industrial supply chains in developing countries (Handfield and Bechtel 2002). Another study showed that transaction costs played a large role in determining the nature of cattle vertical (supply chain) coordination in the United Kingdom (Hobbs 1997). Certainly, the role of trust must be even greater in supply chains in a more constrained environment. Future farm-level and supply chain research should include more detailed questions on the nature of contracts.

Transportation and scale constraints play a large role in determining the price and availability of chicken in urban areas. As shown in the previous section, the distance over which a trader travels can affect the price received by a farmer. Traders face a trade-off between economics of scale/aggregation and delivering the preferred local varieties. Crossbred chickens are raised by all types of farmers, and Table 3.18 shows how smaller farms are more likely to sell in local markets directly to end users. These small commercial farmers (second and third quartiles) were more likely to deal with commune traders, who, in turn, largely operate from motorbikes or even bicycles (Table 3.19). Reaching more distant and lucrative urban markets is more likely if there are economics of scale related to transportation and aggregation. In the Ha Vi wholesale market, the largest wholesale market in Northern Vietnam, several large traders utilized trucks to deliver birds to slaughterhouses or other traders based on the daily number of birds traded.

Aggregation of different varieties is also a likely significant transportation constraint. To come to a certain area, a truck must be able to pick up enough market-ready birds in a specific area or route. For enough local chickens that are market ready to fill a truck, a larger geographical area would have to be covered due to the biological requirements of raising local varieties. Similarly, aggregating from

Table 3.19 Commune traders' method of chicken collection

District	Motorcycle (%)	Bike (%)	Other (%)
Gia Binh	37	63	0
Tien Du	100	0	0
Ly Nhan	100	0	0
Kim Bang	50	50	0
Soc Son	54	44	3
Tu Liem	90	10	0
Hoai Duc	56	44	0
Phu Xuyen	89	11	0
Total	0	0	0

Table 3.20 Chicken varieties traded by wholesale market traders

	Local (%)	Crossbred (%)	Industrial (%)	Egypt (%)
Bac Thang Long	55	5	30	10
Ha Vi "entering"	26	24	47	3
Ha Vi "middlemen"	13	44	43	0

several small farms with any breed would add to the costs of utilizing a truck or larger vehicle. Traders generally deliver birds in the early morning (1:00 am to 6:00 am) to wholesale markets. These birds are slaughtered immediately and then sold in Hanoi that day. Gathering birds from farms in concentrated areas and larger farms would allow traders to meet these daily time constraints.

An epidemiological study (Soares Magalhaes et al. 2007) indicated that many of the traders serving Hanoi wholesale and larger retail markets sourced from larger farms. About 80% of farms supplying poultry (duck, Muscovy, and chicken) to these traders had farms with more than 200 head of poultry. Almost two-thirds of the farm suppliers (63%) had more than 500 head of poultry. Given that household consumption of local chicken is still much higher than that of industrial chicken, and the larger wholesale market tends to receive more crossbred and industrial chicken, local chicken appears to be coming into Hanoi through less formal channels that do not operate on the scale of wholesale markets. Bac Thang Long is a smaller scale wholesale market than Ha Vi (Table 3.20).

The poultry sector in Vietnam is developing fragmented supply chains, where a more formal sector caters to urban areas and a smaller, more informal sector raises chickens that are consumed within a relatively small geographical area or sells through less formal marketing channels. Specialization is occurring based on scale and variety and, while having an important role, smallholders may not benefit from this fragmentation. This trend is influenced by the challenges that producers face: production constraints, heterogeneous regulation, scale and transportation constraints, and informal contracting. The poultry sector has also experienced upheavals due to Avian Influenza epidemics. Although not discussed in depth in this chapter, other reports detail national, regional, and farm-level impacts that occurred immediately after outbreaks. These events clearly have influenced the poultry sector structure.

Table 3.21 Local chicken customers by farm production level

Annual farm production level	Commune traders	District traders	Own province traders	Other province traders	Commune level end users	Other
First quartile (%)	17	10	7	1	59	6
Second quartile (%)	11	31	8	0	36	13
Third quartile (%)	14	39	8	3	31	4
Fourth quartile (%)	25	44	13	7	10	0

Larger farmers raise industrial breeds (or crossbreds) and sell to urban consumers through larger markets and intermediaries, with greater integration into existing regulatory systems. Larger farms tend to operate through the existing wholesale markets or large retail markets in Hanoi's outer district, which are more regulated, while some "companies" (CP, Phuc Thin, etc.) are selling directly to urban market vendors. Generally, larger farmers are better able to sell at further distances and in urban areas as indicated by the higher percentages of large farms selling to traders operating over larger distances. These trends hold when the same varieties are compared across different size producers although, for industrial chicken producers, it is more difficult to draw many conclusions due to the low sample size.

Larger farms also report that they are able to utilize more of their existing production capacity, which is likely due to more market integration with urban areas. Birds sold from the larger quartiles of producers also are more likely to be at full market weight, indicating that these farms have less difficulty selling at optimal market weight. Furthermore, large farms sell more batches per year, supporting data that these farms are better able to negotiate regular sales with traders and that they are able to overcome production constraints. Smallholders sell local varieties to traders who tend to work through smaller markets that are likely to be over shorter distances. Small farmers are also much more likely to sell to local end users. Relationships with end users tend to be without contract while formal contracts are common in the Chuong Mi district, where only large farmers were interviewed. This indicates that, in the current market environment, larger entities are more likely to have formalized, legal supply chain relationships. When birds from smallholders do enter urban areas, they are more likely to come through unregulated channels (Tables 3.21–3.25).

Commercial chick producers in several districts report specialization in crossbred and industrial chicken, which supports data that farmers source local chicken from other farmers. Chick producers in these districts tend to specialize in both varieties and areas to which they sell. Some producers operate locally while others have a large portion of their market outside of their own district or in other provinces (Tables 3.26 and 3.27). Chick producers also tend to specialize in the type of farm to which they will sell. In Phu Xuyen, there are many chick producers, and small to medium-sized farmers from other provinces will often travel there by

Table 3.22 Industrial chicken customers by farm production level

Annual farm production level	Commune traders (%)	District traders (%)	Own province traders (%)	Other province traders (%)	Commune level end users (%)	Other (%)	Observation
First quartile	0	0	61	1	38	0	8
Second quartile	3	34	42	7	15	0	12
Third quartile	2	31	40	17	5	5	21
Fourth quartile	2	41	27	25	1	4	48

Table 3.23 Farmer survey: Percent of capacity used during survey period

Farm production level	Mean (%)	Standard deviation (%)
First quartile	56	26
Second quartile	54	24
Third quartile	54	25
Fourth quartile	62	25

Table 3.24 Farmer survey: Percent of chicken sold at full market weight

Farm production level	Mean (%)	Standard deviation (%)
First quartile	74	19
Second quartile	82	13
Third quartile	84	12
Fourth quartile	89	10

Table 3.25 Number and size of batches sold per year

Farm production level	Average number of batches	Average batch size
First quartile	2.1	45.5
Second quartile	2.5	97.9
Third quartile	3.0	191.2
Fourth quartile	4.5	3048.3

Table 3.26 Varieties produced by commercial chick farmers

District	Local (%)	Crossbred (%)	Industrial (%)	Egypt* (%)
Soc Son	0	100	0	0
Tu Liem	0	100	0	0
Hoai Duc	0	79	21	0
Phu Xuyen	1	77	8	14
Dong Anh	10	60	30	0
Yen Phong	17	42	17	25

* Egypt chicken is a relatively new variety that might be compared to crossbred or industrial chicken.

Table 3.27 Location of client farms for commercial chick producers

District	Commune (%)	District (%)	Own province (%)	Neighboring province (%)	Other province (%)
Soc Son	100	0	0	0	0
Tu Liem	0	0	20	0	80
Hoai Duc	57	37	4	2	0
Phu Xuyen	3	8	39	43	6
Dong Anh	0	0	30	0	70
Yen Phong	18	45	0	23	13

Table 3.28 Size of client farms for commercial chick producers

District	Backyard: <50 head (%)	Small: 50–200 head (%)	Medium: 201–1,000 head (%)	Large: >1,000 head (%)	Size unknown (%)
Soc Son	73	28	0	0	0
Tu Liem	0	20	30	50	0
Hoai Duc	0	19	68	13	0
Phu Xuyen	26	49	15	4	5
Dong Anh	5	10	35	50	0
Yen Phong	0	60	10	30	0

motorbike to pick up chicks. In Soc Son, we see that the chick producers raise crossbred chicken for local small farms. In Tu Liem and Dong Anh, industrial and crossbred chicks are sold to farmers over long distances. Generally, the large distance over which some chick producers sell indicates that markets for crossbred chicks are well developed (Table 3.28).

Several small poultry traders operate in rural areas in Vietnam, sourcing from local farms and selling into local markets. Several traders operating in farmer survey districts were interviewed. They specialize in local chicken and source most chicken from smaller farms. For almost all traders, all customers were located within one district and customers were individual consumers (usually more than 90%). Traders and slaughterhouses at Ha Vi market, the largest wholesale market, tend to specialize in crossbred and industrial varieties while local traders and slaughterhouses at Bac Thang Long market handle more local varieties. Bac Thang Long market is much smaller with almost all traders operating from motorbikes. The chicken breed as well as the average number of birds per trader per day varies considerably for each market (Table 3.29). Table 3.30 shows the number of chickens traded per day for each of the "entering" traders, who bring birds into the Bac Thang Long and Ha Vi markets. Traders with lower volume tend to trade the higher value local chicken while larger traders are more likely to source crossbred and industrial birds (Table 3.31).

Slaughterhouse data support the trend of market fragmentation (Tables 3.32 and 3.33). The large registered slaughterhouses in Soc Son and Tu Liem source almost entirely from companies; they also slaughter mostly crossbred and

Table 3.29 Breeds traded by local traders

District	Local (%)	Crossbred (%)	Industrial (%)
Gia Binh	83	7	10
Tien Du	67	0	33
Ly Nhan	0	12	88
Kim Bang	81	4	15
Soc Son	90	4	5
Tu Liem	100	0	0
Hoai Duc	47	14	39
Phu Xuyen	70	1	29

Table 3.30 Size of farm suppliers for local traders

District	Backyard: <50 head (%)	Small: 50–200 head (%)	Medium: 201–1,000 head (%)	Large: >1,000 head (%)	Traders or market (%)
Gia Binh	62	25	11	2	0
Tien Du	26	36	29	0	10
Ly Nhan	0	28	72	0	0
Kim Bang	52	38	10	0	0
Soc Son	67	18	15	0	0
Tu Liem	100	0	0	0	0
Hoai Duc	27	20	53	0	0
Phu Xuyen	37	17	4	0	42
Total	46	25	21	0	8

Table 3.31 Type of birds traded by sales volume of traders delivering birds to wholesale markets

Birds traded per day	Local (%)	Crossbred (%)	Industrial (%)	Egypt (%)	Average birds traded daily
First quartile	61	0	33	6	81
Second quartile	31	22	39	8	150
Third quartile	17	26	50	7	305
Fourth quartile	27	28	45	0	1,185

industrial. These large-scale slaughterhouses are also more specialized while the family operations in Bac Thang Long and Ha Vi generally diversify with chicken and poultry. In Ha Vi, most slaughterhouses buy chicken from larger markets, where traders have higher volume, so most birds are crossbred and industrial varieties. Bac Thang Long slaughterhouses are supplied local chicken by the small-scale traders as previously discussed.

In urban markets, 40% of vendors sell only one variety while most (68%) vendors have one variety account for two-thirds of their total sales. Despite the fragmentation of chicken markets, it does appear that chicken producers have some market power. Farmers, on average, receive roughly two-thirds of the urban retail value of the chicken (Fig. 3.3). Urban market vendors receive less than 10% on

Table 3.32 Type of birds of slaughtered and weekly volume

Slaughterhouse location	Local (%)	Crossbred (%)	Industrial (%)	Head chicken	Head other poultry
Bac Thang Long	56	14	27	881	836
Ha Vi	14	37	48	148	103
Soc Son	20	0	80	500	0
Tu Liem	20	25	55	17,500	0

Table 3.33 Source of birds for slaughterhouses

Slaughterhouse location	Farm (%)	Market (%)	Delivered by trader (%)	Company (%)
Bac Thang Long	13	0	88	0
Ha Vi	3	97	0	0
Soc Son	0	0	20	80
Tu Liem	25	0	0	75

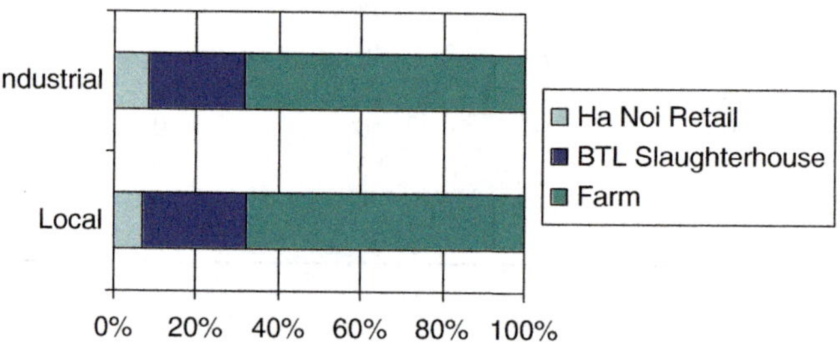

Fig. 3.3 Local and industrial chicken margins

average and, between the farm gate and slaughterhouses, a little more than 20% of the final value is added. Given that chicken often passes through several hands and often is transported by motorbike, there does not appear to be large mark-up by a market player that could be asserting significant monopoly or market power.

Conclusion

The production, transport, and regulatory constraints discussed previously limit the ability of small producers to sell to urban markets. From a broader perspective, the constraints on production of local chicken and market failure, combined with high demand (Ifft et al. 2007), have led to the high prices currently observed in urban markets. These high prices could support the continuing informality of local chicken supply chains and vending, which includes illegal activity, such as selling live chickens in Hanoi. Strengthening the regulatory system might further exacerbate these issues unless smallholder participation is taken into account. Cooperatives and development of new methods of contract farming with smallholders may be ways to overcome market fragmentation. Further research should elucidate ways to overcome these marketing problems.

References

Aggarwal, Rimjhim M., and Erik Lichtenberg. *Pigouvian taxation under double moral hazard.* Journal of Environmental Economics and Management. Volume 49, Issue 2, March 2005, pp. 301–310.

Binswanger, Hans P., and John McIntire. *Behavioral and Material Determinants of Production Relations in Land-Abundant Tropical Agriculture.* Economic Development and Cultural Change, The University of Chicago Press, Vol. 36, No. 1 (Oct., 1987), pp. 73–99.

Boehlje,Michael. *Structural Changes in the Agricultural Industries: How Do We Measure, Analyze and Understand Them?* American Journal of Agricultural Economics, Vol. 81, No. 5, Proceedings Issue (Dec., 1999), pp. 1028–1041.

Bosrup, E. *Conditions of Agricultural Growth.* Chicago: Aldine Pub. Co., 1965.

Carlson, Gerald A., and David Zilberman. *Emerging Resource Issues in World Agriculture..* Agricultural and Environmental Resource Economics. Gerald Carlson, David Zilberman, John Miranowksi (eds) Oxford University Press, New York, 1993.

Cochrane, Willard. *The Development of American Agriculture: A Historical Analysis.* Minneapolis: University of Minnesota Press, 1993.

Diamond, Jared. *Guns, Germs, and Steel: the Fates of Human Societies.* WW Norton & Company, 1999.

Gillespie, James R., and Frank B. Flanders. *Modern Livestock and Poultry Production.* Eighth Edition. Delmar Cengage Learning, Inc., 2009.

Handfield, R. and Bechtel, C. (2002) "The role of trust and relationship structure in improving chain responsiveness." *Industrial Marketing Management* 31: 367– 382.

Heiman, Amir, David Just, and David Zilberman. "The Role of Socioeconomic Factors and Lifestyle Variables in Attitude and the Demand for Genetically Modified Foods." *Journal of Agribusiness*, 18,3 (Fall 2000), pp. 249–60.

Hueth, B. and Ligon, E. "Agricultural Markets as Relative Performance Evaluation." *American Journal of Agricultural Economics*, Vol. 38, No. (May, 2001), pp. 318–328.

Hoagland, Porter, Di Jin, Hauke Kite-Powell. *The optimal allocation of ocean space: aquaculture and wild-harvest fisheries.* Marine Resource Economics, Summer 2003, Vol. 18, Issue 2, pp 129–147.

Hobbs, J. (1997) "Measuring the Importance of Transaction Costs in Cattle Marketing." *American Journal of Agricultural Economics*, 79(4): 1083–1095.

Ifft, J., Otte, J., Roland-Holst, D. and Zilberman, D. (10/07) Demand-Oriented Approaches to HPAI Risk Reduction, Research Report Number 07-14, Rural Development Research Consortium, University of California at Berkeley.

Ifft, J., Otte, J., Roland-Holst, D. and Zilberman, D. (01/08) Poultry Market Institutions and Livelihoods: Evidence from Vietnam. Research Report Number 08-02, Rural Development Research Consortium, University of California at Berkeley.

Kislev,Yoav and Willis Peterson. *Prices, Technology, and Farm Size.*The Journal of Political Economy, Vol. 90, No. 3 (Jun., 1982), pp. 578–595.

Lafontaine, F. "Agency Theory and Franchising: Some Empirical Results. " *The RAND Journal of Economics*, Vol. 23, No. 2 (Summer, 1992), pp. 263–283.

Lafontaine, F. and Slade, M. "Incentive Contracting and the Franchise Decision." *Game Theory and Business Applications,* Vol. 35, 2000, pp. 133–188.

Martin, L, and Zering, K. "Relationships Between Industrialized Agriculture and Environmental Consequences: The Case of Vertical Coordination in Broilers and Hogs." *Journal of Agricultural and Applied Economics*, 29, 1 (July 1997): 45–56.

Mitchell, L. 2001. "Impact of Consumer Demand for Animal Welfare on Global Trade." In Changing Structure of Global Food Consumption and Trade.Edited by A. Regmi. Agriculture and Trade Report WRS-01-1. Washington, D.C.: U.S. Department of Agriculture, Economic Research Service.

Ogishi, Aya, David Zilberman, Mark Metcalfe. *Integrated agribusiness and liability for animal waste*. Environmental Science & Policy, Volume 6, Issue 2, April 2003, pp. 181–188.

Roland-Holst, D., D. Chadwick, J. Ifft, and V. Reed (8/07) Livestock Surveys for IPALP, Research Update, PPLPI, FAO, Rome.

Schrader, Lee F., and MichaleBoehlje. *Cooperative Coordination in the Hog-Pork System: Examples from Europe and the U.S.* Staff Paper 96-21, Department of Agricultural Economics, Purdue University. September 4, 1996.

Segerson, Kathleen, and JunJie Wu. *Nonpoint pollution control: Inducing first-best outcomes through the use of threats*. Journal of Environmental Economics and Management, Volume 51, Issue 2, March 2006, pp. 165–184.

Soares Magalhaes, R., H.D. Quoc, and L.T. Lan (05/07). Farm gate trade patterns and trade at live poultry markets supplying Hanoi: Results of a rapid rural appraisal. PPLPI, FAO, Rome.

Sproul, T. "Accidents Happen: The Effect of Uncertainty on Environmental Policy Design" Working Paper, 2010. Available: http://ecnr.berkeley.edu/vfs/PPs/Sproul-ThoW/web/Sproul_JobMarketPaper_10302010.pdf.

Ruttan, Vernon W. *Technology, growth and development : an induced innovation perspective*. New York : Oxford University Press, 2001.

Zilberman, David, Mark Metcalfe, Aya Ogishi. *Creative Solutions to the Animal Waste Problem*. Animal Agriculture and the Environment: National Center for Manure and Animal Waste Management White Papers. J. M. Rice, D. F. Caldwell, F. J. Humenik, eds. 2006. St. Joseph, Michigan: ASABE.Pp. 161–180.

Part II
The Economics of Managing Animal Diseases

Chapter 4
The Economics of Zoonotic Diseases: An Application to Avian Flu

David Zilberman, Thomas W. Sproul, Steven Sexton, and David Roland-Holst

Introduction

There is growing concern about diseases that exist in animal populations but can jump to humans with lethal consequences. Emerging zoonotic diseases include West Nile Virus (WNV), Bovine Spongiform Encephalitis (BSE), and Avian Influenza and are responsible for hundreds of deaths in the past quarter century. The emergence in 2003 of a strain of Avian Influenza lethal to humans has raised the specter of a pandemic that could claim millions of human lives and cost hundreds of billions of dollars (McKibbin and Sidorenko 2006). Prevention of such a pandemic has become an urgent priority for epidemiologists and has engaged the medical and agricultural industries in developed and developing countries alike. Since 2003, 243 people in 10 Asian and African countries have died of the Highly Pathogenic Avian Influenza (HPAI) through direct contact with infected birds. More than 200 million birds around the world have been either killed by the H5N1 strain of Avian Influenza or destroyed in human efforts to prevent the spread of the disease. Unlike other human disease, the risk of zoonotic disease is not strictly a function of human interactions. Rather, zoonotic diseases spread through animal populations. The BSE is spread through cattle populations via contaminated feed but is not transmitted by casual contact. Avian Influenza, however, does spread through bird populations through direct and indirect contact with infected birds. Human infection follows direct or indirect contact with infected animals. Avian Influenza does not yet spread

D. Zilberman (✉) • S. Sexton • D. Roland-Holst
Department of Agricultural and Resource Economics, University of California, Berkeley, CA, USA
e-mail: zilber11@berkeley.edu

T.W. Sproul
Department of Environmental and Natural Resource Economics, University of Rhode Island, Kingston, RI, USA
e-mail: sproul@mail.uri.edu

D. Zilberman et al. (eds.), *Health and Animal Agriculture in Developing Countries*, Natural Resource Management and Policy 36, DOI 10.1007/978-1-4419-7077-0_4,

by human contact though it is feared the H5N1 strain may mutate to become transferable from human to human.

Few health developments in recent history present such intractable challenges to policymakers as the emergence of a lethal strain of bird flu and BSE, which is also known as Mad Cow Disease. The costs of an outbreak exceed the human death toll, which is 243 and 200 for the two diseases, respectively. Costs also include the expenditures on control, the loss of livestock (from disease and culling), and the associated market impacts, including losses to producers and consumers. Nearly 200,000 cattle have been infected by the lethal BSE, and more than four million have been slaughtered to control the spread of the disease. Tight controls on slaughtering have largely kept the disease from afflicting humans.

Control of zoonotic diseases, such as Avian Influenza, is difficult because individuals, firms, and states do not face incentives to achieve socially optimal outcomes. The risk of a bird-flu pandemic is the result of market failures and missing markets. There exists, therefore, a strong role for policy intervention to achieve socially optimal levels of risk. Development of policy, however, is complicated by a number of social and economic factors, including the public-good nature of disease prevention and free rider problems, externalities associated with investment in prevention and treatment, uncertainty, endogenous risk, myopia and dynamic considerations, and budget constraints in developing countries. In this chapter, following a brief background on Avian Influenza and other emerging zoonotic diseases, we describe how each of these characteristics effects the provision of disease prevention and control and explain how they cause departures from social optimality. We next summarize recent efforts to combine epidemiology and economics in modeling farm level behavior and government response. A prototype model of prevention and control activities is then presented in the context of avian flu. The model realistically captures many of the epidemiologic attributes of animal disease transmission as well as the economic decisions made by market participants. Yet, the model remains tractable. The free-market equilibrium levels of disease prevention and control are derived along with conditions for social optimality. We offer policy prescriptions in a concluding section.

Background on Avian Influenza and Other Zoonotic Diseases

Zoonotic diseases may spread with varying impacts on host animal populations. The BSE, for instance, is lethal to infected cattle, which experience brain degeneration but may not be symptomatic for several years. Some strains of avian flu, however, have only mild impacts on bird health. Much of the concern with zoonotic diseases surrounds high rates of spread among animals or humans. The rate of transmission may determine the strategic approach to disease response, such as investment in prevention or control.

Though many strains of Avian Influenza are spread throughout bird populations with minimal impacts on bird health (and no impacts on human health), some forms are highly pathogenic and can kill 90% of infected birds within two days of infection

[Centers for Disease Control and Prevention (CDC) 2008)]. The HPAI is primarily spread through contact with secretions from infected birds. It may be introduced to domesticated bird populations (i.e., poultry farms) through direct contact with wild fowl or domesticated birds at live bird markets. It may also be transferred on the clothing or equipment of people who work in proximity to infected birds. The BSE, in contrast, is not spread through casual contact. Though the transmissibility of BSE is not well understood, it is believed the pathogenic prion that carries the disease is spread through infected animal parts, especially the brain and spinal cord. An outbreak of BSE in London, beginning in 1986, is likely to have been caused by British cattle consumption of infected meat. The WNV is transmitted by arthropods, such as mosquitoes, and transmitted to animals and humans when bit by infected arthropods. It is not spread by animals or humans who may become sick and die in some cases.

The probability of an outbreak of zoonotic diseases, such as Avian Influenza or WNV, once they are introduced into vector populations (birds and mosquitoes, respectively) is a function of several factors that determine the rate of virus transfer and the amount of virus that is sufficient to infect otherwise healthy arthropods, animals, or people. First, the rate of virus transfer depends on the amount of virus produced by the hosts (infectiousness), the rate of virus transfer through contact, and the rate of contact between sick and susceptible populations. Risk of an outbreak is increasing in infectiousness, rate of transfer, and rate of contact. Transmissibility is also determined by the amount of virus that is required to make a vulnerable animal or human sick (susceptibility) and by the fraction of the animal or human population that is susceptible. Susceptibility increases as the amount of virus needed to infect a vulnerable population decreases. Transmissibility increases in susceptibility and the size of the susceptible population (Stegeman and Bouma 2006).

The rate of transmission can be controlled by a number of interventions practiced by farming operations and public-health departments. For Avian Influenza, monitoring for signs of infection in bird populations and quarantining or destroying infected birds reduces the rate of contact with sick birds. Biosecurity practices, such as sterilization of clothing and equipment and restrictions on movement and contact with bird populations, reduce the likelihood of contact with contaminated material. Susceptibility of populations can be reduced by vaccination. Risk of BSE transmission is reduced by precluding the slaughter of symptomatic cattle and by banning the use of highly contagious cow parts in the human or animal food supply. Risk of WNV can be reduced by use of protective clothing and efforts to drain stagnant bodies of water, which attract mosquitoes.

Damage from the spread of zoonotic disease can be mitigated by efforts to prevent the introduction of the disease in animal populations, regions, and countries and by efforts to control the spread of the disease once it is introduced into a population, region, or country. In this respect, responses to the risk of zoonotic diseases are similar to the risk of introductions of other biological and genetic material, including invasive species. While expenditure on ex-ante prevention may seem preferable to outlays for ex-post control, particularly over an indefinite time horizon, it may be more costly than control. The cost of reducing the risk of disease introductions to arbitrarily low levels can be prohibitive and include foregone benefits from the unimpeded movement of goods and people. Similarly, once a disease is introduced to a region, it may be preferable to control the damage from disease transmission rather than attempt to eradicate the disease. The costs of the latter can be infinite, and the probability of eradication can approach zero (Finnoff et al. 2005a, b; Jensen 2002; Mack et al. 2000).

Economics of Animal Disease and Avian Influenza

Economic theory predicts that animal-disease prevention and control will be underprovided by market participants absent policy intervention. Disease prevention and control are, to some extent, public goods as classically defined. First, some mitigation activities are nonrivalrou. That is, the benefit that an individual derives from mitigation is not decreasing in the number of other beneficiaries nor is the cost of the mitigation effort increasing in the number of beneficiaries. Second, prevention and control can be nonexcludable. Investment in risk reduction by one individual can reduce risk to other individuals, who cannot be excluded from sharing in the benefits of the investment. Adoption of quarantine and other biosecurity practices are preventative measures that exhibit public-good characteristics. Likewise, culling of infected animals is nonrivalrous (risk to humans and livestock operations is reduced) and nonexcludable (the other beneficiaries cannot be precluded from enjoying benefits). Not all responses to Avian Influenza and other zoonoses have these attributes however. Vaccinations, for instance, are often used to reduce the susceptibility of populations to infection. They are rivalrous; vaccines used on one farm cannot be reused on other farms. Indeed, in recent history, flu vaccines have been scarce from time to time, sparking intense rivalry and rationing to the most at-risk demographic groups (Connolly 2004). To some extent, vaccines are also excludable. Benefits accrue principally to the one who receives the vaccine though others may benefit from reduced risk of infection as the susceptible population declines.

Given that at least some aspects of disease prevention and control exhibit public-good characteristics, such efforts will suffer from free-rider problems in which consumers benefit from the production of a good or service but do not contribute to its production. Because risk-reducing behavior provides benefits beyond those recognized in private cost-benefit analysis, provision by the private sector will be too low relative to the social-welfare maximizing level. This creates a role for policy intervention to mandate safety practices or incent prevention and control effort through subsidies, compensation, or penalties.

With a high rate of movement of people and goods among countries, an outbreak of Avian Influenza or other zoonotic disease in one country could easily spread to neighboring countries and even to continents oceans away. Since the first known case of BSE was identified in London in 1986, BSE has spread to more than 20 countries on four continents. The lethal strain of HPAI that threatens human and bird populations today was first discovered in China in 1997. Within 10 years, it had spread to four continents and infected people in 16 countries (World Health

Organization 2008). Today, governments around the world fear an outbreak of HPAI in their countries. The ease with which zoonotic disease can spread around the world makes prevention and control efforts global public goods. The positive externalities associated with risk-reducing behavior extend beyond political boundaries and cannot be fully internalized even at the national level. Therefore, theory predicts that absent a supranational governing body, investment in disease prevention and control will be below the global social optimum.

In addition to being a global public good, mitigation of Avian Influenza and other zoonoses is a weakest link public good. Benefits to society of prevention effort are determined by the weakest member of society (Sandler 1997). Disease introduction is the result of interactions among countries, including a country that is home to infected animals and other countries connected to it through trade in goods or people. Efforts to contain or eradicate pests will fail even if all members of society do so successfully, and one member of society does not. As Perrings et al. (2002) explain, "If one quarantine facility does not contain an invasive pathogen, the fact that all the others may do so is irrelevant." This characteristic of disease mitigation suggests a role for stronger members of society to support weaker members by sharing experts, technology, and information. On the level of state actors, developing countries may lack the funds to provide effective monitoring and quarantine. Developed countries could provide funds for biosecurity investments or provide information from their disease surveillance and research agencies, such as the U.S. CDC. On the level of firms, small farmers may not be able to afford vaccines to reduce the susceptibility of a community's livestock population. Larger firms could provide the vaccines in order to reduce risk of disease transmission within the community, including to their own farm operations.

While economics can address the market failures posed by the spread of zoonotic disease, it is important to realize the role of political institutions in determining responses to the problem. Optimal responses to disease introduction risk may be to increase trade restrictions. Such policies may not be implemented because of political pressures for trade liberalization. Responses to livestock disease have typically been swift, however, with countries banning imports from regions with known cases of infection. The bans often remain in place for years until infected regions can verify that their food exports are safe beyond any doubt. Despite the risk aversion demonstrated by such trade restrictions, society is typically unwilling to support considerable expenditures on prevention when the probability of disease introduction is low and the cost of an outbreak is uncertain. Democratic institutions, therefore, can be contributing factors to underinvestment in prevention and control.

The determination of optimal disease-control policies is hindered by the endogeneity of control dynamics. Risk of HPAI outbreaks and their costs are functions of human responses to risk. Free trade can increase the risk of spreading Avian Influenza whereas adoption of biosecurity practices and vaccination can reduce risk. The probability of disease transmission, therefore, is internal to the system—a function of human decisions. Besides the endogeneity of risk, the nature of infection risk poses additional challenges. The probability of disease outbreak is typically low, but the consequences are quite high. A single outbreak can be calamitous. Consider that past flu pandemics have killed tens of thousands and that epidemiologists fear that millions could be killed by Avian Influenza. By 2005, outbreaks of the H5N1 strain were estimated to have cost more than $10 billion in losses to the Southeast Asian poultry industry. The power of expected utility is diminished with low-probability catastrophic events (Chichilinsky 1998). People treat unlikely events by either overestimating their likelihood of realization or setting their probabilities to zero. Disease outbreaks are also one-time events often independent of history, which makes estimation of probability density functions impossible (Horan et al. 2002).

As has already been mentioned, rational farmers are expected to undersupply disease control and prevention because they do not internalize all of the costs and benefits associated with their behavior. Suboptimal provision at the farm and government level can also be attributed to myopia and lack of information. First, if farmers are unaware of Avian Influenza infections among their livestock populations, they will not be able to respond with profit-maximizing mitigation, let alone social-welfare maximizing mitigation. Farmers may also be unaware of the risk of Avian Influenza outbreaks, which are a function of interactions with wild-fowl and live-bird market transactions. An outbreak at a given farm, for instance, would increase the risk to adjoining farms. But, if adjoining farms are unaware of the outbreak, they will not adopt prevention and control measures that may be profit maximizing in light of the higher risk, such as limited movements of livestock and people, heightened biosecurity protections, and vaccinations of own livestock populations.

Myopia may result in economic agents (farmers and governments) failing to consider the optimal disease response in the long run. The cost of eradication of disease may seem too substantial in the short run and yet be less costly than the net-present value of an indefinite stream of control expenditures.

Previous Modeling of Animal Disease

Given that farmer response and even national government response to Avian Influenza may be suboptimal from the standpoint of global social-welfare maximization, economists have attempted to model the behavior of market participants and compare outcomes from private response to socially desirable outcomes. Beach, Poulos, and Pattanayak (2007) combined the basic epidemiology of HPAI, including endogeneity of risk and externalities associated with farm behavior, with the economic foundations of the poultry market to determine the divergence of private prevention effort from socially desirable prevention and to examine the effects of various mechanisms to compensate farmers for destroyed birds. Unlike the model presented in this chapter, the Beach model assumes farmers may only choose prevention effort and focuses on the effect of government compensation. Farmers are assumed not to invest in control. Rather, in the event that an infection is realized at an individual farm, the poultry at that farm are all culled.

Assuming that the government does not fully indemnify farmer losses and that private investment in prevention on a given farm reduces risk to another farm in the region, Beach, Poulos, and Pattanayak find that private investment in prevention is too low from a social-welfare standpoint. When governments offer compensation for destroyed poultry, private prevention investment is declining in the degree of compensation. Full compensation for disease outbreak costs eliminates private incentive for prevention. Only if the risk of spreading the disease from a given farm is small and the output on the infected farm is large will the private optimum converge to the social optimum. If compensation is tied to investment in prevention, the disincentive for prevention from compensation is partially—but only partially—mitigated.

Sproul et al. (2007) consider a sequential decision-making process and incorporate interventions in a risk-generating function (see Lichtenberg and Zilberman 1988) to analyze trade-offs between the scope of livestock operations, prevention, and monitoring in agricultural operations affected by animal disease. The risk-generating function is akin to a production function in economic analysis. It models risk as a function of decisions made by the public and private sectors; biophysical, economic, and environmental parameters; and random variables to incorporate uncertainty. The specification of the risk-generating function is based on risk-assessment models developed by epidemiologists and public-health professionals (Wilson and Crouch 1987; Bogen 1995). Sproul et al. find that higher output prices and higher penalties for the sale of infected meat will increase prevention effort. Higher output prices, surprisingly, are shown to reduce the monitoring effort. Branding and quality certifications are considered as market-based approaches for achieving health-risk reductions.

A New Model of Avian Influenza Prevention and Control

Studies on the economics of zoonotic diseases vary in their modeling of the problem and policy alternatives and, generally, emphasize micro- or macroeconomic issues. Beach, Poulos, and Pattanayak for instance, limit analysis to prevention effort and, therefore, are unable to consider the trade-off between ex-ante prevention and ex-post control of disease. Sproul et al. emphasize microlevel choices and not regional effects. The analysis in this chapter considers both private and public intervention, where private intervention includes both preventive and responsive measures. It also models individual farm outcomes and industry outcomes. Furthermore, it models risk of animal loss, health risk at the farm region, and health risk outside of the farm region. This diverse set of possible outcomes leads to a more general policy setup.

Therefore, let us consider a region with N livestock farms, which are assumed to maximize expected profits. Let $i = 1, \ldots, N$ be a farm indicator. Define potential output as the level of output (measured in animal units) produced absent damage from disease or pests (see Lichtenberg and Zilberman 1986) and denote it by y_{Pi}. This is the planned level of output. For simplicity, we will suppress the i subscript when presenting the choices or characteristics of individual farms. The cost of production for a farm that realizes output y_P (i.e., there is no disease or pest damage) is $c(y_P)$, where $c'(y_P) > 0$, $c''(y_P) > 0$. The final output of the farm—the quantity sold at the market—is denoted by y. The total output of the industry is denoted by $Y = \sum_{i=1}^{N} y_i$. Assume that the industry faces inelastic demand given by $Y = D(p)$, with $D'(p)<0$ and output price p.

The industry faces periodic episodes of disease infestation that occur randomly. Disease may initially infect a few farms but, then, through contacts among farmers and movements of animals, spread among farms. Economists rely on epidemiological research for specification and modeling of these processes, especially for simulation and numerical analysis. These specifications are essential to make this economic decision-making framework more operational.

We will distinguish in our analysis between primary and secondary infection, where primary infection refers to the introduction of disease in a farm region. Primary infections may spread to other farms or regions and cause secondary infections. The animals initially infected at a farm or region can transmit the disease to other animals on the farm or in the region, making other animals sick. A given fraction of the sick population dies. For simplicity, assume that the initial infection strikes one of the farms randomly. For example, a wild bird carrier of Avian Influenza may come into contact with and infect livestock on farm i. Let q_{I_i} denote the probability that a particular farm i will be affected by the primary infection q_{I_i}. Assume $\sum_{i=1}^{N} q_{I_i} < 1$, namely, primary infection does not afflict every farm in a given period, and that the risk of primary infection facing an individual farm is small. Assume that protection from infection can be controlled by a public policy, such as culling wild birds. Let the investment in this policy be denoted by K. With the policy, the risk of primary infection is $q_I = q_I(K)$ with $q_I' = \frac{\partial q_I(K)}{\partial K} < 0$.

Once the primary infection occurs, disease will spread among livestock of the infected farm. The fraction of infected animals is reduced by preventive activities that may require fixed cost I (i.e., for cleanup and disinfection of facilities) and per-animal cost v (i.e., for vaccinations and culling). The risk to an animal on an infected farm, measured as the fraction of animals infected with the disease, is $r = r(v, I, \beta)$, where β is an indicator of the severity of the strain of the disease (i.e., infectiousness). This risk reflects vulnerability to the disease and farmer choices. Given v, I, and β, it is assumed that the risk is the same regardless of whether disease was introduced to the farm by primary or secondary infestation. It is further assumed that $\partial r/\partial v < 0$, $\partial r/\partial I < 0$, and $\partial r/\partial \beta > 0$.

Once the disease is introduced at one farm (primary infection), it may spread among other farms, as the result of movement of animals, people, and equipment among farms and interactions at live-animal markets. Assuming the ith farm was the source of the primary infection, we assume that the secondary infection is a cumulative process with probabilities $q_{Sj} (j \neq i)$ that farm j among the N–1 farms that were not initially infected will suffer secondary infection. Assuming that monitoring and death occur near the end of the period, the probability of secondary infection is a function of the animal risk within infected farms, r, so that $q_S = q_S(r)$, where $\partial q_S(r)/\partial r > 0$. Higher risk of animal disease within primary infected farms increases the probability of disease transmission among other farms. With this definition, the expected aggregate number of infected animals, given that the ith farm is the source of primary infection, is

$$Y_{Ii} = r_i y_{Pi} + \sum_{j \neq i} q_{Sj}(r_i) r_j y_{Pj}. \tag{4.1}$$

We assume that a fraction of the infected animals, α, dies on the farm because of disease. In addition to preventive activities, farmers may engage in responsive activities once they discover that infestation has occurred. Responsive activities may include monitoring and treatment or monitoring and culling or both. Disease response may vary between the farms that were initially infected and the farms suffering secondary infections. The monitoring effort of an infected farm is denoted by M, and the monitoring efficiency, which is the fraction of sick animals discovered by monitoring, is $m(M)$ where $m'(M) > 0$ and $m''(M) > 0$. In cases where treatment is feasible, the cost of treatment is u per animal. It is assumed that animals that receive treatment recover and produce the same output as uninfected animals. When an animal is culled, the cost of treatment is the lost output. We will present alternative specifications to address these two situations.

If a fraction of the sick animals, $m(M)$, is discovered by monitoring, and is either cured or culled, the fraction of sick animals that reach the market from an infected farm is $r_S = r(v, I_1, \beta)(1 - \alpha)(1 - m(M))$. The expected aggregate number of infected animals given that the ith farm was initially infected is

$$Y_{Si} = r_i(1 - \alpha)(1 - m(M_i)) y_{Pi} + \sum_{j \neq i} q_{Sj}(r_i) r_j (1 - \alpha)(1 - m(M_j)) y_{Pj}. \tag{4.2}$$

Concern about animal diseases is not restricted to their impact on livestock productivity and prices but also includes health effects on humans. We can distinguish between impacts on individuals on the farm and off the farm. Animal disease on infected farms may risk the health of the farmers or their families. Let $n_F(y_I)$denote the expected number of people on the farm contracting the disease, where $y_I = r y_P$ and $n'_F(y_I) > 0$. The price per disease case is c_F. The expected aggregate number of people contacting the disease given that the ith farm was initially infected is

$$Y_{Fi} = n_F(r_i y_{Pi}) + \sum_{j \neq i} q_{Sj}(r_i) n_F(r_J y_{Pj}). \tag{4.3}$$

Contact with sick animals outside of the farm gate, e.g., at the market, can also contribute to human health problems. The expected number of people outside of the farm region infected by the disease originated at a particular farm, $n_S(y_S)$, where $y_S = r_S y_P$, is a function of the sick animals sold by the farm, y_S, with $n_S'(y_S) > 0$. The cost of a sick person outside the farm is denoted by c_S. Given that the ith farm was initially infected, the expected aggregate number of people contracting the disease outside of the farm gate is

$$Y_{Si} = n_S(r_{Si}y_{Pi}) + \sum_{j \neq i} q_{Sj}(r_i)n_S(r_{Sj}y_{Pj}). \tag{4.4}$$

This setup permits determination of optimal decision rules for the farmer and the government and incorporates alternative responses that characterize real-world decisions. We will first use this model to consider a laissez-faire equilibrium (no government intervention) with identical, price-taking farms. We will then introduce policy to achieve the social-welfare maximizing outcome and use the model to indentify features that should be incorporated in government policy.

No Regulation and No Awareness of Health Effect

Consider the extreme case in which there is no regulation of farmer behavior and the farmer is not aware of adverse human health effects on and off of the farm. We assume that farmers are identical and price taking. The individual farmer maximizes profit and, therefore, optimizes

$$\underset{v,y_P,I}{\text{Max}}\, py_P - c(y_P) - py_Pr(v,I_1,\beta)\alpha q_I(K)[1+(N-1)q_S(r^*)] - vy_P - I. \tag{4.5}$$

The first two elements of the objective function are profits without losses due to disease. The third element is the expected cost of the disease, which is the probability that a farm will be infested $(q_I(K)[1+(N-1)q_S(r^*)])$ times the expected loss of revenue in case of infestation due to death of the sick animals, $py_Pr(v,I_1,\beta)\alpha$. The variable r^* is animal disease risk on other farms, which affects the likelihood of secondary infection to the farm considered here. The other elements of the optimization problems are the preventive cost of risk reduction.

The first-order conditions to this optimization problem are

$$p - c'(y_P) - pr(v,I_1,\beta)\alpha q_I(K)[1+(N-1)q_S(r^*)] - v = 0. \tag{4.6}$$

This condition suggests that optimal herd/flock size (output) is determined at a level where price is equal to the sum of the marginal cost of production plus the sum of expected marginal costs of animal-disease loss $(pr(v,I_1,\beta)\alpha q_I(K)$ $[1+(N-1)q_S(r^*)])$ and the marginal disease-prevention cost with respect to herd size. The condition, with respect to variable cost of prevention per animal, is determined where

$$-py_P\frac{\partial r}{\partial v}\alpha q_I(K)[1+(N-1)q_S(r^*)] = y_P. \tag{4.7}$$

This condition sates that the marginal value of increased prevention cost is equal to the level of output (each unit increase in variable cost affects all output). The marginal value of an increase in v is equal to the reduction in the expected loss due to death, which is a product of marginal reduction of risk of death among infected farms $(-(\partial r/\partial v)\alpha)$; the expected number of infected farms $q_I(K)[1+(N-1)q_S(r^*)]$; and revenues. This result indicates that higher output price and mortality rate on the farm (α) and lower investment in public prevention policy will increase the variable prevention efforts, v.

Comparing condition (4.6) to the familiar marginal rule for determining output and marginal costs of output, it is determined that the added marginal costs associated with the disease is $pr(v,I_1,\beta)\alpha q_I(K)[1+(N-1)q_S(r^*)]+v$. It includes the extra variable costs of prevention and the expected marginal value of mortality loss. Higher output price, disease spread, and mortality rate on the farm will increase the extra marginal costs and reduce output.

The foregoing farm-level analysis provides the foundation for industrywide analysis, recalling that the industry faces a negatively sloped demand curve. The disease makes the output and price of the industry randomly distributed. In particular, output is distributed according to

$$Y_N = Ny_P \qquad \text{with probability } 1 - N[q_I(K)],$$

$$Y_I = Ny_P[1 - r(v,I_1,\beta)\alpha(1+(N-1)q_S(r^*))] \text{with probability } N[q_I(K)],$$

where Y_N is the output level when there is no infestation and Y_I when infection occurs. The price distribution is determined from $p = D^{-1}(Y)$. Condition (4.2) suggests that the occurrence of the disease shifts the supply curve upward and produces reduced output Y and higher prices p relative to equilibrium without disease risk. This holds even in years when there is no disease infestation. When infestation occurs, output declines and price further rises. While producers must carry higher production costs due to the presence of disease risk, they may benefit from larger surplus if demand is sufficiently inelastic. This result is not surprising and is consistent with evidence that increased pest pressure may benefit farmers (Zilberman et al. 1991). Note, however, that consumers are big losers when supply declines; they consume less and pay more.

The Socially Optimal Resource Allocation

We now consider the case where overall social welfare is maximized. We will also consider the cases with inelastic demand so that, assuming additive utility with respect to chicken, we can maximize expected consumer and producer surpluses. Let the social benefits from meats, in monetary units, be denoted by $B(Y)$, where Y is aggregate output. The $B(Y)$ is the area under the inverse demand curve for meat, and $B'(Y)$ is marginal benefit from consumption of the meat and is equal to the competitive price when output is Y. Using this notation, the social-optimization problem is

$$L = \underset{v_i, y_i, I_i, K, M_i}{\text{Max}} \left\{ \begin{array}{l} B\left(\sum_{i=1}^{N} y_{Pi}\right)\left[1 - \sum_{i=1}^{N} q_{Ii}(K)\right] \\ + \sum_{i=1}^{N} B\left[y_{Pi}(1 - r_i(\alpha + (1-\alpha)m(M_i)) + \sum_{j \neq i} y_{Pj}[1 - q_{Sj}(r_i)r_j(\alpha + (1-\alpha)m(M_j))]\right] q_{Ii}(K) \\ - \sum_{i=1}^{N} \left\{ \begin{array}{l} c_F\left[n_F(r_i y_{Pi}) + \sum_{j \neq i} q_{Sj}(r_i) n_F(r_J y_{Pj})\right] \\ + c_S\left[n_S(r_{Si} y_{Pi}) + \sum_{j \neq i} q_{Sj}(r_i) n_S(r_{Sj} y_{Pj})\right] \end{array} \right\} q_{Ii}(K) \\ - \sum_{i=1}^{N}\left[c(y_{Pi}) + \left[M_i + \sum_{j \neq i} q_{Sj}(r_i) M_j\right] q_{Ii}(K) + v_i y_{Pi} + I_i\right] - K \end{array} \right\}. \tag{4.8}$$

Assume identical farms with identical probabilities, then the optimal-decision rules are

$\dfrac{dL}{dy_{pi}} = 0$ implies

$$\begin{aligned} & p_N[1 - Nq_I] + p_I(1 - [(1 + (N-1)q_S)/N]r(\alpha + (1-\alpha)m(M))Nq_I \\ & = rc_F n'_F[1 + (N-1)q_S]q_I + r_S c_S n'_S[1 + (N-1)q_S]q_I + c' + v, \\ & \textit{where } p_N = B'(Ny_P) \\ & p_I = B'[Ny_P[1 - [(1 + (N-1)q_S)/N](r(\alpha + (1-\alpha)m(M))]]. \end{aligned} \tag{4.9}$$

Condition (4.9) suggests that the optimal planned output is at a level where the expected marginal benefit of output (the benefit of producing an additional animal), which is price when there is no infestation times the probability of no infestation, $p_N[1 - Nq_I]$, plus the price corrected for expected infestation damage $p_I(1 - [(1 + (N-1)q_S)/N]r(\alpha + (1-\alpha)m)$ times the probability of an infestation Nq_I, is equal to the expected marginal health costs of output both on and off of the farm plus the marginal cost of production and prevention. Comparison of Conditions (4.9) and (4.6) suggests that government intervention may lead to monitoring, and will require an extra fee per unit of output to compensate for risk to people on and off the farm. If a farmer is aware of the own marginal health cost of production, she may take it into consideration in determining y, but she will ignore the marginal health cost of production outside her own farm. Let $t = rc_F(N-1)q_S n'_F q_I + r_S c_S n'_S [1 + (N-1)q_S]q_I$, be an extra tax on production to reduce the externalities. Alternatively, (4.9) can be used to set direct control on production at the farm level.

Equation (4.9) suggests that the consideration of human health effects will lead to higher output price and reduced output, as output per farm is associated with a health effect that must be taken into consideration. Similarly, by differentiation of (4.9) with respect to v, the socially optimal level of variable prevention cost is set where

$$\begin{aligned}\frac{\mathrm{d}L}{\mathrm{d}v_i} &= 0 \text{ implies}\\ &-P_I\left[(1+(N-1))\left(q_S+\frac{\partial q_S}{\partial r}r\right)y_P(\alpha+(1-\alpha)m(M))\frac{\partial r}{\partial v}\right]q_I\\ &+\left[c_F(n_F'y_P+(N-1))\left(q_S n_F' y_P+\frac{\partial q_S}{\partial r}n_F\right)\frac{\partial r}{\partial v}\right]q_I\\ &+\left[c_S(n_S'y_P+(N-1))\left(q_S n_S' y_P+\frac{\partial q_S}{\partial r}n_S\right)\frac{\partial r}{\partial v}(1-\alpha)(1-m(M))\right]q_I\\ &= y_P.\end{aligned} \tag{4.10}$$

Condition (4.10) states that the optimal variable-prevention cost of a farm is determined where the expected value of marginal product of preventive expenditure is equal to its price. The expected value of marginal product of v is the sum of the expected marginal benefits generated by reducing economic loss of animals [the first line of equation (4.10)] and the expected marginal benefits from reducing the impact on health on the farm (the second line) and off the farm (the third line). The marginal cost of variable preventation is the level of output.

The expected marginal benefits from reducing output loss includes direct and indirect marginal benefits to the own farm, $z_D^O = -P_I[1+(N-1)q_S]y_P\ (\alpha+(1-\alpha)m(M))\times(\partial r/\partial v)q_I$, and an indirect effect through prevention activities that reduce disease spread and loss to other farms, $z_I^O = -P_I[(N-1)(\partial q_S/\partial r)ry_P(\alpha+(1-\alpha)m(M))\partial r/\partial v]q_I$. The self-interest of the farmer will lead the farmer to consider only the direct output effect, z_D^O; extra incentives are need to induce consideration of the indirect output effect, z_I^O. The marginal benefits of variable prevention costs on health loss can also be decomposed to direct effects from reduction of health losses on the own farm, $z_D^{HF} = [c_F(n_F'+(N-1))q_S n_F')y_P(\partial r/\partial v)]q_I$, and the indirect effect from reducing the spread of the infection and the resulting health effects to other farms, $z_I^{HF} = (\partial q_S/\partial r)n_F(\partial r/\partial v)q_I$. If farmers are aware of and concerned about the health effects on the farm, they may take into account z_D^{HF}. But self-interest will not lead them to consider z_I^{HF}, so it needs to be included as a part of a policy incentive. Farmers are not likely to consider health effects outside of the farm region, when determining v, unless it affects demand (which we do not consider here) so that the marginal benefits of extra variable cost,

$$z^{HO} = \left[c_S(n_S'y_P+(N-1))\left(q_S n_S' y_P+\frac{\partial q_S}{\partial r}n_S\right)\frac{\partial r}{\partial v}(1-\alpha)(1-m(M))\right],$$

must be included as part of a policy to affect the farmer choice of v.

Thus, our analysis suggests that, given other variables, farmers will tend to underinvest in variable costs of prevention. Therefore, an intervention in the form of either a direct control that determines the level of v or a subsidy that will contribute $s_v = z_I^O + z_I^{HF} + z^{HO}$ to the variable prevention cost is needed to induce the optimal variable preventive effort. This analysis suggests a case for subsidizing vaccination and similar activities. Note from Condition (4.10), however, that the magnitude of the preventive effort depends on the initial probability of infection. Smaller infection probabilities will reduce v and, if q_I is sufficiently small, Condition (4.10) does not hold, the marginal cost of v is greater than the expected marginal benefits, and it is suboptimal to conduct preventive efforts on the farm at all.

Differentiation of (4.8) with respect to monitoring cost per farm, M, leads to the optimality condition

$$\frac{\mathrm{d}L}{\mathrm{d}M_i} = 0 \text{ implies}$$

$$\begin{aligned} & c_S y_P n_S' r(1-\alpha)m'[1+(N-1)q_S]q_I \\ & = [p_I y_P' r(1-\alpha)m' + 1][1+(N-1)q_S]q_I. \end{aligned} \tag{4.11}$$

Condition (4.11) suggests that the optimal monitoring level is where the net marginal benefits from monitoring, including reduction of health cost outside of the farm when infestation occurs, $c_S y_P n_S' r(1-\alpha)m'$, minus lost sales due to culling, $P_I y_P' r(1-\alpha)m'$, are equal to the marginal monitoring costs, which is equal to one $= 1$. This condition suggests that monitoring is not optimal unless the marginal health cost per unit of output is greater than the price of output, $c_S n_S' > p_I$. The self-interest of farmers will not lead them to monitor because they do not consider the health costs off of the farm. The government may, therefore, provide monitoring itself, may require it of farmers (direct control), or may impose a penalty of c_S if sick animals can be traced back to the originating farm. Sproul et al. (2007) analyzed the implication of this policy, which will affect other choices beside the choice of M.

The decision rule for the optimal level of regional protection from infestation, K, is determined from (4.8) to be

$$\frac{\mathrm{d}L}{\mathrm{d}K} = 0 \text{ implies}$$

$$Nq_I'(K)\left\{\begin{array}{l} B[Y_I] - B(Y_N) - M[1+(N-1)q_S] \\ - c_F[n_F + (N-1)q_S n_F] - c_S[n_S + (N-1)q_S n_S] \end{array}\right\} = 1. \tag{4.12}$$

Condition (4.12) suggests that optimal investment in regional protection against infestation, which in essence is an invasive species policy, occurs where the expected marginal gain from reduced infestations because of higher K is equal to its marginal cost, \$1.00. The expected marginal gain is the product of the marginal reduction in probability of infestation times the gain from realizing—a period without infestation as opposed to realizing a period with infestation. These gains include benefits from higher output and lower health costs and monitoring costs.

The benefits from reduced likelihood of invasion are shared by all farmers and by the potential victim of the negative health effects inside and outside of the region. Thus, reducing probabilities of infestation to the region has public-good properties and should be pursued by the government. Since the beneficiary is inside and outside of the region, the finance of activities to reduce infestation or invasion to the region may be financed by agencies inside and outside of the region. In some situations, central activities to control invasion (quarantine, culling of wild species) will be cheap relative to other activities and will result in a much higher level of output than without intervention. Consequently, the notion that environmental polices curtail output does not apply in this case.

The analysis, so far, has identified that control of animal diseases has many aspects of externalities and public goods, which warrants intervention. Policies that combine control of the source of the infestation, control of the size of output, and incentives for preventive activities and monitoring may altogether increase expected production levels compared to the laissez faire (no intervention) outcomes, especially if invasion control is a central element of the policy. If intervention reduces the level of production (through output taxes, monitoring, and culling) because of concern about health effects outside of the region, then local consumers may be major losers, whereas outsiders may win. Policy design should consider these distributional implications.

Limitations of Analysis

This analysis assumed a particular pattern for the spread of the disease. While the qualitative results obtained here may hold under broad circumstances, quantitative analysis requires more accurate modeling of the spread of the disease and its interaction with the economic system. The analysis assumes that consumers are not affected by disease considerations, i.e., they do not respond to disease outbreaks by reducing demand for meat. Future research should address issues associated with product differentiation based on quality and health consideration and how these considerations affect optimal policies. We also considered gradual, incremental policies. Many food and animal-health policies have an element of "all or nothing." For instance, if Mad Cow Disease is discovered once in a region, the region may lose many good buyers. Development of strategies to address these situations is an area for further research.

Other issues related to regulation include the impact of alternative market structures on social-welfare maximizing outcomes. Wholesalers may play an important role in upgrading food quality. Yet, at the same time, they may capture rents and reduce the well-being of farmers. Developed and developing countries need to establish a regulatory framework that takes into account various supply-chain and market-structure conditions. Similarly, technological considerations affect policy choices. Weak monitoring strategies emphasize the role of preventive measures at the regional and farm levels. Effective vaccination and cures may affect policy choices drastically.

Analysis of animal disease emphasizes conceptual modeling, especially econometric and statistical strategies. Simulations based on detailed knowledge can take advantage of empirical findings and allow better examination of new options.

Conclusion

Externalities associated with the choices of individual farmers (and even individual governments) mean that prevention and control of zoonotic disease will be undersupplied in private markets unless social planners intervene. We have presented a model that merges the epidemiology of Avian Influenza (and other animal diseases) with features of livestock production to model the decisions of farmers and determine socially optimal policy interventions. The analysis aims to reflect the dynamics of disease transmission and the process of poultry production and yet remain tractable and intuitive. The conceptual analysis, here, can be operationalized with econometric estimation of parameters to provide estimates of the magnitudes of policy interventions, such as the size of subsidy or mandated provision of biosafety.

We find that farmer investment in variable prevention, such as vaccination and culling, increases in output price and mortality rate on the farm. Private variable prevention efforts also increase as public prevention declines. This suggests that, sometimes, public prevention crowds out private effort and may save farmers from significant outlays. Much depends on the relative cost effectiveness of the different measures. Farm output declines as output price, disease transmissibility, and farm mortality rate increase. Total industry production is reduced in the presence of Avian Influenza introductions even in periods when no introductions occur. This causes higher prices than would prevail absent disease risk. Given a sufficiently inelastic demand for meat, farmers may actually benefit from the presence of animal disease (producer surplus increases). Consumers, however, will lose from reduced meat production and higher prices.

The foregoing analysis identifies socially optimal public investments in prevention effort and control. It confirms that farmers will underinvest in prevention and control because they do not consider the risk of disease transmission to animals on other farms and to people in the production region and other regions connected through trade. Where monitoring is effective, ex-post control of disease infestations (by culling, for instance) can be preferable to ex-ante prevention efforts, such as vaccination. But where monitoring is costly and ineffectively identifies sick animals, vaccinations should be pursued in the hope of preventing disease introductions.

Policies that seek to reduce production because of health risks to individuals outside the production region (such as in importing countries) will hurt local consumers and benefit those consumers outside of the production region. These distributional issues should be addressed, perhaps with direct transfers. Concern among foreign consumers has triggered demand for greater safety than that demanded by consumers in the production region, who may have better information about health risks or who may have lower willingness to pay for risk reduction. Foreign institutions and governments may pay producers for additional biosafety effort, but it is not clear if such interventions will achieve social optimality.

Responses to animal disease outbreaks, such as responses to environmental hazards, often include quarantines, bans, and boycotts. But the cost of these responses is not based on the size of the outbreak or the number of infected animals. Producers pay a near infinite price whenever one animal is identified to be sick. A challenge for policy makers is to reduce the cost of outbreak response. Draconian measures are intended as penalties to farmers for risky behavior. The punishment, often times, falls instead on consumers who face reduced supply and higher prices for meat.

References

Bogen, KT. "Methods to Approximate Joint Uncertainty and Variability in Risk." *Risk Analysis* 15 (1995): 411–419.

Beach, R, C Poulos, and SK Pattanayak. "Farm Economics of Bird Flu." *Canadian Journal of Agricultural Economics* 55(2007): 471–483.

CDC. Key facts about avian influenza (bird flu) and avian influenza A (H5N1) virus. Washington (DC); 2008. Accessed online at http://www.cdc.gov/flu/avian/gen-info/facts.htm on August 20, 2008.

Chichilinsky, G. "The Economics of Global Environmental Risks," In The International Yearbook of Environmental and Resource Economics 1998/9, eds. T. Tietenberg and H. Folmer, Cheltenham: Edward Elgar; 1998.

Connolly, C. "CDC Announces Plan to Ration Flu Vaccine." Washington Post: 2004 November 10; pA6.

Finnoff, D, J Shogren, B Leung, and D Lodge. "Risk and Nonindigenous Species Management." *Review of Agricultural Economics* 27(2005a): 475–482.

Finnoff, D, J Shogren, B Leung, and D Lodge."The Importance of Bioeconomic Feedback in Invasive Species Management." *Ecological Economics* 52(2005b): 367–381.

Horan, RD, C Perrings, F Lupi, and EH Bulte. "The Economics of Invasive Species Management: Biological Pollution Prevention Strategies under Ignorance: The Case of Invasive Species." *American Journal of Agricultural Economics* 84(2002): 1303–1310.

Jensen, R. "Economic Policy for Invasive Species." Working Paper, University of Notre Dame; 2002. Accessed at: http://www.nd.edu/~rjensen1/workingpapers/InvasiveSpecies.pdf.

Lichtenberg, E, and D Zilberman. "The Econometrics of Damage Control: Why Specification Matters." *American Journal of Agricultural Economics* 68 (1986): 261–273.

Lichtenberg, E, and D Zilberman. "Efficient Regulation of Environmental Health Risks." *Quarterly Journal of Economics* CIII (1988): 167–178.

Mack, RN, D Simberloff, WM Lonsdale, H Evans, M Clout, and FA Bazzaz. "Biotic Invasions: Causes, Epidemiology, Global Consequences and Control." Ecological Applications 10(2000): 698–710.

McKibbin, WJ and AA Sidorenko. "Global Macroeconomic Consequences of Pandemic Influenza." Sydney, Australia: Lowry Institute for International Policy; 2006.

Perrings, C, EB Barbier, M Williamson, D Delfino, S Dalmazzone, J Shogren, P Simmons, and A Watkinson. "Biological Invasion Risks and the Public Good: An Economic Perspective." *Conservation Ecology* 6(2002): 1.

Sandler, T. *Global Challenges*. Cambridge University Press, Cambridge, UK; 1997.

Sproul, TW, D Zilberman, and D Roland-Holst. "The Economics of Managing Animal Disease." UC Berkeley Department of agricultural and Resource Economics Working Paper 2007.

Stegeman, J. A. and A. Bouma. "Epidemiology and Control of Avian Influenza." In Proceedings of the 11th International Conference of the Association of Institutions for Tropical Veterinary Medicine and 16th Veterinary Association Malaysia Congress, 141–43. Patalang Jaya, August 2006.

World Health Organization. "Cumulative Number of Confirmed Human Cases of Avian Influenza A/(H5N1) Reported to WHO." Washington (DC): Accessed online August 20 2008.

Wilson, R. and EA Crouch. "Risk Assessment and Comparisons: An Introduction." *Science* 236 (1987): 267–270.

Zilberman, D, A Schmitz, G Casterline, E Lichtenberg, and JB Siebert. "The Economics of Pesticide Use and Regulation." *Science* 253(1991): 518–522.

Chapter 5
Animal Disease and the Industrialization of Agriculture

David A. Hennessy and Tong Wang

Introduction

Descartes' perspective that animals are machines, and perhaps little more, is a matter of great ethical disquiet in contemporary society (Cottingham 1978). Sweeping developments in the life sciences since about 1950 have provided technical insights on how to control life and growth in ways that have made the animal-as-machine analogy more real. The moral principles and economic tradeoffs at issue have become more clearly defined, in large part because production sciences and the systems they support demand clear definition of the production environment. Animal disease confounds control efforts, and also belies the attitude that an animal's technical performance can be abstracted from its environs.

Demand for technical performance in animal protein production is large and growing. Global meat consumption has increased at about 2.4% per annum over the period 1975–2007 due in equal measure to population growth and growth in per capita consumption, and this pattern is projected to continue through to 2017 (Trostle 2008). In part as a response to market pressures, the dominant animal production format has changed markedly in developed countries in recent times. The newer format, often referred to as industrial animal agriculture, has also made inroads beyond the developed world (Steinfeld et al. 2010; Li 2009). Although controversial in many ways, the approach has proved to be remarkably successful in providing

D.A. Hennessy (✉)
Professor of Economics, Department of Economics and Center for Agricultural and Rural Development, Iowa State University, Ames, IA, USA
e-mail: hennessy@iastate.edu

T. Wang
Doctoral Candidate in Economics, Department of Economics, Iowa State University, Ames, IA, USA
e-mail: wangtong@iastate.edu

D. Zilberman et al. (eds.), *Health and Animal Agriculture in Developing Countries*, Natural Resource Management and Policy 36, DOI 10.1007/978-1-4419-7077-0_5,

animal produce suitable for processing at low market prices (Key et al. 2008; MacDonald and McBride 2009; Mosheim and Knox Lovell 2009).

The general matter of this chapter is how industrial animal agriculture and animal health interact. This is a broad topic where detailed economic scrutiny has been largely absent. As typically considered, industrialization refers to growing focus on and specialization in defined tasks accompanied by scale expansion, market development, and structural realignment. While modern theories of industrial development encompass a wide set of phenomena, technical change remains central (e.g., Galor and Moav 2002). Although the endogeneity of technical change is a primary concern of this literature, in our case we may assume it to be exogenous. This is because animal agriculture is a comparatively small sector of the global economy. The primary sources of technical change in the sector are likely to have been spillovers from innovations in the much larger human medicine, biotechnology, and manufacturing sectors.

Industrialization in animal agriculture has involved confinement in climate-controlled buildings. So confined, animals expend less energy on foraging, defense against predators, and temperature regulation. The opportunity also exists to build physical protections against infectious and other diseases. In short, nature's influences can be at least partly controlled. Additionally, genetic innovations can seek to modify an animal's make-up so as to optimize production given the physical conditions under which production occurs.

The specific focus of this chapter is to provide a general understanding of how the closing off of animal herds affects production. We will do so by looking at three, among many possible, dimensions to how innovations intended to control infectious animal disease can affect the nature of production. Each dimension is modeled in a separate section. The first dimension we will consider pertains to how disease externalities affect the equilibrium structure of production. By structure we mean the stocking rate and the number of farms in the sector. Surprisingly little in the way of formal analysis has been conducted on the topic, perhaps because the result was deemed obvious. As infectious disease is a public bad, one might expect there to be free riding in the form of excessive stocking per farm. In a rather standard model of commons behavior, we show that this is true. But we also show that there are *too few* farms in the sector, precisely because there is excessive stocking per farm. We develop the analysis to address how efforts to close off herds to disease exposure should affect sector structure.

The second aspect looked at is the use of control technologies. On any given farm, one way to mitigate the effects of external disease is through subtherapeutic use of antibiotics.[1] Infectious disease and animal density are of course related. High animal density is a feature of confined animal production. Research has found

[1] The precise mechanisms through which sub-therapeutic antibiotics affect performance are unclear (Proctor 2010). They may inhibit bacteria as sources of sub-clinical disease, or suppress gut bacteria that prevent food absorption, or some combination thereof. Under any of these settings, non-use of antibiotics increases intra-herd performance variability.

animal density on a farm and/or farm density in a region to be factors in increased disease risk (Mortensen et al. 2002; Rose and Madec 2002; Vandekerchove et al. 2004). Infectious disease among confined animals can lead to sector-wide losses of up to 70%, as in China's shrimp sector during the early 1990s (Smil 2000, p. 178). Although scale expansion had been underway in poultry since the 1930s, the advent of antibiotics and other biosecurity innovations likely propelled the growth of confined agriculture in much of the developed world since the World War II (Finlay 2004).

The use of antibiotics has come under scrutiny in recent years because of concerns that excess use could increase resistance among animal and human infection agents (Miller et al. 2006; Graham et al. 2007). A prominent feature of the technology is that it promotes uniformity in animals, since a random disease event is less likely to differentiate animals in growth rate and product quality (Hayes and Jensen 2003). It is not alone in this trait, as feed uniformity-promoting technologies also promote animal performance uniformity (Ciftci and Ercan 2003; Madsen and Pedersen 2010).

Uniformity allows producers to avoid penalties for marketing lightweight animals and/or the capital costs of keeping some animals in half empty barns. Increased uniformity should also promote the substitution of capital for labor, especially when wages are high. In addition, slaughter animal uniformity allows processors to better automate packing lines. In short, uniformity-promoting technologies increase process control and so might be viewed as a factor contributing to an industrial approach. We provide a brief model to argue that the presence of antibiotics should increase scale of production, and also that antibiotics should lead to an increase in the capital-to-labor ratio.

The third dimension surrounds incentives to guard against infectious disease risk and how these incentives interact with scale of production. Once it enters a large feeding operation, disease can spread rapidly throughout the operation. Scale, biosecurity, and other forms of process control tend to go hand-in-hand. For example, World Bank (2006) and Beach et al. (2007) accept the Dolberg et al. (2005) classification of four stylized global poultry production systems. In it, the biosecurity level increases as production scale increases from backyard format to industrial format.

Subject to engineering constraints, the construction cost of a storage vessel scales up in proportion to the square of the scaling factor (Besanko et al. 2004). But capacity scales up in proportion to the cube of the scaling factor so that the unit capital cost of storage declines with an increase in scale. Similarly, a feedlot's perimeter length per unit production capacity declines with an increase in production capacity.[2] To the extent that biosecurity regards protecting an enterprise's perimeter, there should be scale economies in doing so.

---

[2] Consider a square animal feedlot with side of length l and perimeter $4l$. If animal stocking density is 1 per unit area then animal capacity per lot is $Q = l^2$ so that $l = Q^{0.5}$. With annual maintenance cost per unit side length as $0.25z$, total annual maintenance cost is $zl = zQ^{0.5}$ or $zQ^{-0.5}$ per animal. Maintenance cost per animal is decreasing in scale.

Our third model provides two counterintuitive results. In light of discussions to this point, one may infer that if the unit cost of a biosecurity input declines with an increase in capacity then production capacity and the level of biosecurity should go hand-in-hand. We show that the inference is not valid. One may also infer that if biosecurity risks originating off-farm are high then there will be an incentive to use biosecurity inputs more intensively on-farm, i.e., internal biosecurity substitutes for external biosecurity. To the contrary, we show that an internal biosecurity input may complement biosecurity inputs external to a farm so that efforts to improve the external biosecurity environment do not crowd out efforts to do so on-farm. We provide reasonable conditions under which this is so.

The chapter's format comprises a section for each of the three issues studied. We conclude the chapter with a summary and a brief discussion.

Model 1: The Commons, Social Efficiency, and Deficient Entry

In a region, there are N, a large number, identical livestock farms labeled as $n \in \{1, 2, \ldots, N\} \equiv \mathbf{N}$. Each farm stocks s_n animals at cost $c(s_n)$, a strictly increasing and convex function. Disease afflicts the region, with negative animal health spillovers such that output per farm is,[3,4]

$$q_n = F(s_n, \lambda S_{\setminus n}), \quad S_{\setminus n} = \sum_{i=1}^{N} s_i - s_n. \tag{5.1}$$

Here $F(s_n, \lambda S_{\setminus n})$ is strictly increasing and concave in its first argument, to reflect positive marginal product and decreasing returns to the input. It is strictly decreasing in its second argument, to reflect the adverse region-wide effects of stocking density on disease prevalence and so on productivity. Parameter $\lambda \in [0, 1]$ is an index of how open production is to disease externalities, where external disease effect $\lambda S_{\setminus n}$ has value 0 under $\lambda = 0$.

[3] Throughout, all functions are assumed to be twice continuously differentiable whenever differentiability is found to be convenient for analysis.

[4] Hardin's (1968) "Tragedy of the Commons" pastoral example notes that common grazing " ... may work reasonably satisfactorily for centuries because tribal wars, poaching, and disease keep the numbers of both man and beast well below the carrying capacity of the land." But when these problems are solved " ... the inherent logic of the commons remorselessly generates tragedy." In this, he didn't view the disease externality environment to be the same as that of an over-exploiting resource and nor do we. But both problems involve unaccounted for negative spillovers so that the formal presentation of both problems can be similar, where the technology in (5.1) is very similar to the commons analysis in, e.g., p. 27 of Gibbons (1992).

With decreasing inverse demand function $P(Q)$ and fixed cost per farm K, private profit is

$$\pi_n = P(Q)F(s_n, \lambda S_{\setminus n}) - c(s_n) - K, \quad Q = \sum_{n=1}^{N} F(s_n, \lambda S_{\setminus n}). \tag{5.2}$$

The assumptions made to this point on demand and technology are henceforth referred to as the monotonicity assumptions. We posit a two-stage game in which potential entrants make the entry decision simultaneously. Knowing the number of entrants, each entering firm then decides on stocking rate. The information structure is closed loop in that information from the stage 1 entry decision is known to all who make the stage 2 stocking rate decision.

Under simultaneous-move Nash behavior, and given that N is large, the stage 2 problem involves setting s_n at level $\hat{s}_n$ solving

$$P(Q)F_1(s_n, \lambda S_{\setminus n}) - c'(s_n) = 0, \tag{5.3}$$

where $F_i(\cdot)$ is the partial derivative with respect to the ith argument, $c'(s_n)$ is marginal cost and producers have been assumed to take price as given.

Since

$$\frac{\partial^2 \pi_n}{\partial s_n^2} = P(Q)F_{11}(s_n, \lambda S_{\setminus n}) - c''(s_n) < 0 \tag{5.4}$$

it follows that private profit is concave in its own action. Notice too that, for $i \neq n$,

$$\frac{\partial^2 \pi_n}{\partial s_n \partial s_i} = P(Q)\lambda F_{12}(s_n, \lambda S_{\setminus n}) \tag{5.5}$$

so that the actions are strategic substitutes whenever the cross-derivative of $F(\cdot)$ is nonpositive. The sign $F_{12}(\cdot) \leq 0$, which we assume, indicates that the marginal product of own stocking rate becomes less positive whenever other farms in the region stock more. With common firms, the symmetric solution to (5.3) given N region firms (labeled as $\hat{s}^N$) requires

$$P(\hat{Q}^N)F_1[\hat{s}^N, \lambda(N-1)\hat{s}^N] - c'(\hat{s}^N) = 0, \quad \hat{Q}^N = NF[\hat{s}^N, \lambda(N-1)\hat{s}^N]. \tag{5.6}$$

Here $\hat{Q}^N$ is region-wide aggregate output while it is understood that $F[\hat{s}^N, \lambda(N-1)\hat{s}^N]$ is the evaluation of $F[s_n, \lambda S_{\setminus n}]$ at point $(s_n, \lambda S_{\setminus n}) = (\hat{s}^N, \lambda(N-1)\hat{s}^N)$.

Given stocking rate $\hat{s}^N$ per firm, social welfare is

$$W(N) = \int_0^{\hat{Q}^N} P(u)\,\mathrm{d}u - Nc(\hat{s}^N) - NK, \quad \hat{Q}^N = NF[\hat{s}^N, \lambda(N-1)\hat{s}^N]. \tag{5.7}$$

Writing farm profit at Nash equilibrium as

$$\hat{\pi}^N \equiv P(\hat{Q}^N)F[\hat{s}^N, \lambda(N-1)\hat{s}^N] - c(\hat{s}^N) - K \tag{5.8}$$

the derivative of social welfare with respect to farm number is

$$\begin{aligned} W'(N) = \hat{\pi}^N &+ N(N-1)\lambda P(\hat{Q}^N)F_2[\hat{s}^N, \lambda(N-1)\hat{s}^N]\frac{\mathrm{d}\hat{s}^N}{\mathrm{d}N} \\ &+ N\{P(\hat{Q}^N)F_1[\hat{s}^N, \lambda(N-1)\hat{s}^N] - c'(\hat{s}^N)\}\frac{\mathrm{d}\hat{s}^N}{\mathrm{d}N}. \end{aligned} \tag{5.9}$$

Under free entry it follows that $\hat{\pi}^N = 0$. Furthermore, we may apply Nash optimality condition (5.3) so that derivative (5.9) reduces to

$$W'(N) = N(N-1)\lambda P(\hat{Q}^N)F_2[\hat{s}^N, \lambda(N-1)\hat{s}^N]\frac{\mathrm{d}\hat{s}^N}{\mathrm{d}N} \tag{5.10}$$

when evaluated at Nash equilibrium. Because $F_2(\cdot) \le 0$, it follows that $W'(N) \ge 0$ whenever $\mathrm{d}\hat{s}^N/\mathrm{d}N \le 0$. So as to avoid technicalities at the expense of insight we make[5]

Assumption 1. Nash equilibrium exists, is unique, and is locally stable.

In particular, we assume that the function

$$L(s;\lambda) = P(NF[s, \lambda(N-1)s])F_1[s, \lambda(N-1)s] - c'(s) \tag{5.11}$$

is decreasing in s.

Note that

$$\begin{aligned} L_1(s;\lambda) = P(\cdot)\{F_{11}[\cdot] + \lambda(N-1)F_{12}[\cdot]\} + P'(\cdot)\{F_1[\cdot] \\ + \lambda(N-1)F_2[\cdot]\}NF_1[\cdot] - c''(\hat{s}^N). \end{aligned} \tag{5.12}$$

Given $F_{11}[\cdot] < 0, F_{12}[\cdot] \le 0, P'(\cdot) < 0, F_1[\cdot] > 0$ and $c''(\cdot) > 0$ it follows that $F_1[\cdot] + \lambda(N-1)F_2[\cdot] \ge 0$ ensures $L_1(s;\lambda) < 0$. If $L_1(s;\lambda) < 0$ and an equilibrium is unique, then a differentiation of (5.11) establishes that $\mathrm{d}\hat{s}^N/\mathrm{d}N$ has the sign of

$$\begin{aligned} &P(\cdot)F_{12}[\cdot]\lambda\hat{s}^N + P'(\cdot)\{F[\cdot] + N\lambda\hat{s}^N F_2[\cdot]\}F_1[\cdot] \\ &= \left(\sigma + \frac{1}{\xi_d}\right)\frac{\lambda\hat{s}^N F_1[\cdot]F_2[\cdot]}{F[\cdot]}P(\cdot) + P'(\cdot)F[\cdot]F_1[\cdot], \end{aligned} \tag{5.13}$$

when evaluated at $s_n = \hat{s}^N\ \forall n \in \mathbf{N}$. Here, ξ_d is the own-price elasticity of demand and $\sigma = F_{12}[\cdot]F[\cdot]/(F_1[\cdot]F_2[\cdot]) \ge 0$ is the elasticity of substitution between own stocking rate and other stocking rates. So $\sigma \ge -1/\xi_d$ suffices to ensure that

[5] See pp. 47–52 in Vives (1999) for details on uniqueness and stability in non-cooperative games.

$\mathrm{d}\hat{s}^N/\mathrm{d}N \leq 0$. Condition $\sigma \geq -1/\xi_\mathrm{d}$ is likely to apply as a region's absolute demand elasticity is likely to be small.

Thus, we have shown that marginal social welfare, when evaluated at Nash equilibrium participation level $\hat{N}$, is likely positive or $W'(\hat{N}) \geq 0$. In conclusion, we have shown

Proposition 1. *Given Assumption 1, the monotonicity assumptions, $L_1(s;\lambda) < 0$ and $\sigma \geq -1/\xi_\mathrm{d}$, then the free entry number of livestock farms is below the number that is socially optimal.*

An analogy exists with the well-known excess entry result in the theory of imperfectly competitive markets, as developed by Mankiw and Whinston (1986) and others. There the private incentive is to account for own price effects by producing too little. Foreseeing this, firms enter in anticipation of high profits. In our case, the situation is quite the reverse. The private incentive is to ignore external disease effects by stocking too densely. In anticipation of low profits due to disease, too few farms enter production.

Finally, we turn to assessing how closing off the herd, or decreasing the value of λ, will impact equilibrium. The stage 2, or stocking rate, effect is established through differentiating

$$P(NF[\hat{s}^N,\lambda(N-1)\hat{s}^N])F_1[\hat{s}^N,\lambda(N-1)\hat{s}^N] - c'(\hat{s}^N) = 0 \tag{5.14}$$

with respect to $\hat{s}^N$ and λ given the value of N set in stage 1. This allows us to identify the stage 2 response, or reaction function sensitivity, as

$$\begin{aligned}\left.\frac{\partial \hat{s}^N}{\partial \lambda}\right|_{\text{stage 2 reaction}} &= -\frac{\{P(\cdot)F_{12}[\cdot] + NP'(\cdot)F_2[\cdot]F_1[\cdot]\}(N-1)\hat{s}^N}{L_1(\hat{s}^N;\lambda)} \\ &= -\left(\sigma + \frac{1}{\xi_\mathrm{d}}\right)\frac{P(\cdot)(N-1)\hat{s}^N F_1[\cdot]F_2[\cdot]}{F[\cdot]L_1(\hat{s}^N;\lambda)}.\end{aligned} \tag{5.15}$$

So a necessary and sufficient condition for $\partial\hat{s}^N/\partial\lambda|_{\text{stage 2 reaction}} \leq 0$ is that $\sigma + 1/\xi_\mathrm{d} \geq 0$.

Proposition 2. *Given Assumption 1, the monotonicity assumptions and $L_1(s;\lambda) < 0$, then the participation-conditioned stocking rate $\hat{s}^N$ increases (decreases) with a closing off of production to infectious disease whenever $\sigma + 1/\xi_\mathrm{d} \geq (\leq)\ 0$.*

While the presence of the externality may encourage overstocking, the marginal effect of a more open system is to decrease stocking. The stage 1 effect is then obtained from differentiating free-entry condition $\hat{\pi}^N = 0$ with respect to N and λ when recognizing the implications for $\hat{s}^N$. To do so, use (5.3) to write the effect of a change in stocking rate only on farm profit under symmetric actions as

$$\begin{aligned}\frac{\partial \hat{\pi}^N}{\partial s} &= \{P(\cdot) + P'(\cdot)NF[\cdot]\}\{F_1[\cdot] + \lambda(N-1)F_2[\cdot]\} - c'(\cdot) \\ &= P'(\cdot)NF[\cdot]F_1[\cdot] + \{P(\cdot) + P'(\cdot)NF[\cdot]\}\lambda(N-1)F_2[\cdot]\end{aligned} \tag{5.16}$$

when evaluated at Nash equilibrium.

The total differential of firm profit with respect to participation and the disease openness parameter is

$$\begin{aligned}&\{P(\cdot)F_2[\cdot]\lambda\hat{s}^N + P'(\cdot)NF_2[\cdot]\lambda\hat{s}^N F[\cdot] + P'(\cdot)(F[\cdot]))^2\}\,\mathrm{d}N\\ &+\frac{\partial\hat{\pi}^N}{\partial s}\left\{\frac{\partial\hat{s}^N}{\partial\lambda}\bigg|_{\text{stage 2 reaction}} dN + \frac{\partial\hat{s}^N}{\partial\lambda}\bigg|_{\text{stage 2 reaction}} d\lambda\right\}\\ &+\{P(\cdot)+P'(\cdot)NF[\cdot]\}(N-1)\hat{s}^N\,\mathrm{d}\lambda = 0. \end{aligned} \tag{5.17}$$

Here the middle expression acknowledges that stage 1 entrants recognize in Stackelberg fashion the stage 2 effects on stocking rate that take the number of entrants as given.

Expression (5.17) may be written as

$$\begin{aligned}&\{P(\cdot)F_2[\cdot]\lambda\hat{s}^N + P'(\cdot)NF_2[\cdot]\lambda\hat{s}^N F[\cdot] + P'(\cdot)(F[\cdot])^2\}\frac{\mathrm{d}N}{\mathrm{d}\lambda} + \frac{\partial\hat{\pi}^N}{\partial s}\frac{\partial\hat{s}^N}{\partial N}\bigg|_{\text{stage 2 reaction}}\frac{\mathrm{d}N}{\mathrm{d}\lambda}\\ &= -\frac{\partial\hat{\pi}^N}{\partial s}\frac{\partial\hat{s}^N}{\partial\lambda}\bigg|_{\text{stage 2 reaction}} - \{P(\cdot)+P'(\cdot)NF[\cdot]\}F_2[\cdot](N-1)\hat{s}^N. \end{aligned} \tag{5.18}$$

If demand is infinitely elastic then (5.18), with use of (5.16), simplifies to

$$\left(\hat{s}^N + (N-1)\frac{\partial\hat{s}^N}{\partial N}\bigg|_{\text{stage 2 reaction}}\right)\lambda\frac{\mathrm{d}N}{\mathrm{d}\lambda} = -(N-1)\left(\hat{s}^N + \lambda\frac{\partial\hat{s}^N}{\partial\lambda}\bigg|_{\text{stage 2 reaction}}\right). \tag{5.19}$$

When demand is infinitely elastic then, by (5.15),

$$\begin{aligned}\frac{\partial\hat{s}^N}{\partial N}\bigg|_{\text{stage 2 reaction}} &= -\frac{P(\cdot)F_{12}[\cdot]\lambda\hat{s}^N}{P(\cdot)\{F_{11}[\cdot]+\lambda(N-1)F_{12}[\cdot]\} - c''(\hat{s}^N)},\\ \frac{\partial\hat{s}^N}{\partial\lambda}\bigg|_{\text{stage 2 reaction}} &= -\frac{P(\cdot)F_{12}[\cdot](N-1)\hat{s}^N}{P(\cdot)\{F_{11}[\cdot]+\lambda(N-1)F_{12}[\cdot]\} - c''(\hat{s}^N)}\end{aligned} \tag{5.20}$$

so that

$$\begin{aligned}\hat{s}^N + (N-1)\frac{\partial\hat{s}^N}{\partial N}\bigg|_{\text{stage 2 reaction}} &= \hat{s}^N + \lambda\frac{\partial\hat{s}^N}{\partial\lambda}\bigg|_{\text{stage 2 reaction}}\\ &= \frac{\{P(\cdot)F_{11}[\cdot] - c''(\hat{s}^N)\}\hat{s}^N}{P(\cdot)\{F_{11}[\cdot]+\lambda(N-1)F_{12}[\cdot]\} - c''(\hat{s}^N)} > 0.\end{aligned} \tag{5.21}$$

It follows from (5.19) that $\mathrm{d}N/\mathrm{d}\lambda < 0$ and the number of entrants increases as the production system is closed off.

Proposition 3. *Given Assumption 1, the monotonicity assumptions, $L_1(s;\lambda) < 0$, and infinitely elastic demand then the number of entrants increases as the production system is closed off.*

Closing off farms to external disease increases profit directly, and so promotes entry. As laid out in Proposition 2, closure increases the stocking rate on each farm. By itself, this would indirectly decrease entry. The direct effect of closing out disease dominates. Thus, closing off the production system both increases stocking rate per farm and the number of farms in a region. In other words, the growth of animal production in response to clearing a region of an infectious disease occurs at both the extensive and intensive margins.

Model 2: Antibiotics, Capitalization, and Scale Come as a Package

Antibiotics facilitate control in that their use reduces product heterogeneity. Control is important because automation requires consistency (Chandler 1992). Machines cannot be readily adapted to the heterogeneities that nature allows, even among progeny. In what follows we adapt Hennessy's (2005) model of how animal heterogeneity affects processing efficiency to study how the antibiotics technology might impact the labor capital relation in animal production. There he studied the effect of animal heterogeneity on time allocated to food and worker safety on the packing line. Our present interest is with the effects that a uniformity-promoting technology, such as antibiotic treatments, have on capitalization in animal grow-out.[6]

We characterize labor as being the more flexible resource in that humans can intervene to accommodate disease induced irregularities when machines cannot. There are two animal types, A and B, in respective proportions θ and $1-\theta$. The labor requirement for each animal is ψ hours regardless of type under a flexible technology, which we refer to as FLEX. This technology uses no capital. The hourly wage rate is w so that the cost per animal is $C^{\mathrm{FLEX}} = \psi\, w$ under FLEX.

There are two components to the labor requirement per animal under an alternative technology, referred to as CAP. The baseline component arising from intensive capitalization is $\psi - \varsigma$ where $\psi - \varsigma > 0$ and $\varsigma > 0$. Here machinery settings must be adjusted whenever the animal type encountered changes, where each adjustment consumes time α. We may imagine the machine-assisted worker moving down the

[6] Capital-labor substitution is but one aspect of the profound effects that uniformity-promoting technologies can have in protein and other bulk commodity markets. Other aspects, dealt with elsewhere, are their effects on the efficient extent of value added processing and technological experimentation (Hennessy et al. 2004; Hennessy 2007).

barn attending to (drawing) animal types that are drawn independently where, for the sake of concreteness, the task at issue might involve using a Döppler machine to ascertain pregnancy.

If the animal being treated at any time is type A, which occurs with probability θ, then the probability that the machine has to be adjusted is $1-\theta$. If the animal being treated is type B, which occurs with probability $1-\theta$, then the probability that the machine has to be adjusted is θ. On average, the number of adjustments per animal treated is $\Pr(A)\Pr(B \text{ follows } A)+\Pr(B)\Pr(A \text{ follows } B)=\Theta$, $\Theta\equiv 2(1-\theta)\theta$. So the expected number of adjustments is zero whenever animals are uniformly of either type. The expected labor cost per animal under CAP is then $(\psi-\varsigma)w+\Theta\alpha w$. Of course, capitalization comes at a cost where we specify the per-animal cost for capital intensity at $\vartheta(s)$ and s is herd size as before. The total cost per animal under CAP is $C^{\text{CAP}}(\Theta)=(\psi-\varsigma)w+\Theta\alpha w+\vartheta(s)$. We defer consideration on how herd size is chosen until later.

Finally, a uniformity-promoting technology (antibiotics) exists that can replace animal heterogeneity parameter θ with either 0 or 1, where the choice among 0 and 1 is of no consequence. The technology costs τ per animal.[7] In this case, the technology is $C^{\text{ANT}}=C^{\text{CAP}}(\Theta)|_{\Theta=0}+\tau=(\psi-\varsigma)w+\vartheta(s)+\tau$. That is, the antibiotics technology complements capitalization as it mitigates the adjustment costs that can attend the industrial approach. In a sense, industrialization is a form of deskilling technology (Vandeman 1995) in that the animal husbandry skill of conditioning a response to the animal type is obviated. The FLEX technology does not benefit from the antibiotics technology because there are no adjustment costs to mitigate. The two parameters we will focus on are wage w and heterogeneity status Θ. In this light, the overall cost function may be written as

$$(s;w,\Theta)=\min\begin{cases}\psi w & \text{(labor flexible, or FLEX)}\\ \vartheta(s)+(\Theta\alpha+\psi-\varsigma)w & \text{(industrialization only or CAP)}\\ \vartheta(s)+\tau+(\psi-\varsigma)w & \text{(industrialization + antibiotics or ANT).}\end{cases} \tag{5.22}$$

The technology with the highest non-wage cost is ANT, at $\vartheta(s)+\tau$ per animal. FLEX has the lowest non-wage cost, at 0 per animal. ANT is the technology with the lowest wage cost, as represented by $(\psi-\varsigma)w$. If $\Theta\alpha<\varsigma$ then $\Theta\alpha+\psi-\varsigma<\psi$ and CAP has the second lowest wage cost.[8] If $\Theta\alpha>\varsigma$, or there is a large amount of heterogeneity, then the FLEX technology dominates CAP in that the per-animal

[7] Antibiotics also increase feed conversion efficiency and so reduce feed costs. We might reduce τ by saved feed costs to accommodate this effect, and the result could be negative in which case feed cost savings alone would justify technology adoption. This effect is not related to the industrialization phenomenon as we study it here. For the sake of focus, we ignore it.

[8] Knife-edge cases of indifference, such as when $\Theta\alpha=\varsigma$, are ignored as the implications warrant no additional comments.

fixed cost and the wage bill are both smaller. In that case, industrialization will only occur in the presence of the complementary antibiotics technology. The complementary technologies come in quantum packages. Thus, we consider two cases.

Case I: Under $\Theta\alpha > \varsigma$ we need to only compare FLEX with package technology ANT. The breakeven wage such that ANT is chosen is $w^I = [\vartheta(s) + \tau]/\varsigma$. Were $w > w^I$ then ANT would be preferred as the gains from industrialization in the presence of the uniformity-promoting technology more than compensate for the additional fixed costs associated with the technology package. Were $w < w^I$ then FLEX would be preferred as labor cost savings do not justify the capital investment made.

Case II: Under $\Theta\alpha < \varsigma$ it is possible, but not assured, that CAP is the chosen technology at some wage. Were $w < \tau/(\Theta\alpha)$ then CAP would dominate labor-saving ANT because the labor-saving benefits of making raw materials more uniform are small when compared with the per-animal cost. Then, we need to only compare CAP with FLEX. If $w > \tau/(\Theta\alpha)$ then ANT dominates CAP and we need to only compare ANT with FLEX.

Were $w < \min[\tau/(\Theta\alpha), \vartheta(s)/(\varsigma - \Theta\alpha)]$ then FLEX, being the least labor-saving technology, would be preferred overall as the reduction in the wage bill under CAP, $(\varsigma - \Theta\alpha)w$, does not justify the capital needed to effect this reduction, $\vartheta(s)$. Were $\vartheta(s)/(\varsigma - \Theta\alpha) < w < \tau/(\Theta\alpha)$ then CAP would be preferred. Here, wages are low enough to prefer CAP over labor-saving ANT but high enough to prefer CAP over FLEX, which is labor intensive relative to CAP.

Under wage interval $[\vartheta(s) + \tau]/\varsigma > w > \tau/(\Theta\alpha)$, then wages are sufficiently low that FLEX dominates labor-saving ANT yet sufficiently high that ANT dominates CAP, so that FLEX is preferred overall. If $w > \max[\tau/(\Theta\alpha), [\vartheta(s) + \tau]/\varsigma]$ then wages are sufficiently high that ANT, being the most labor-saving technology of all, dominates.

What can be stated about the industrialization process from the above? First, as w increases then the incentive to switch out of FLEX increases; witness the collapse of peasant farming in China over the past two decades. A more interesting story surrounds the role of biotechnology in industrialization. Here Θ is likely to decrease upon adoption of such genetic technologies as artificial insemination and sex sorting. It is useful to write

$$U(s; w, \Theta) = \psi w + \min[0, \vartheta(s) + (\Theta\alpha - \varsigma)w, \vartheta(s) + \tau - \varsigma w]. \tag{5.23}$$

Figure 5.1, with wage on the horizontal axis and $U(s; w, \Theta) - \psi w$ on the vertical axis, shows what happens as the value of Θ declines. Particular attention is paid to CAP cost $\vartheta(s) + (\Theta\alpha - \varsigma)w$. The breakeven wage when comparing FLEX with ANT is at $w = [\vartheta(s) + \tau]/\varsigma$, as identified by $\times$ in the figure. The value of Θ such that $\vartheta(s) + (\Theta\alpha - \varsigma)w = 0$ at $w = [\vartheta(s) + \tau]/\varsigma$ is $\hat{\Theta} = \varsigma\tau/([\vartheta(s) + \tau]\alpha)$. Were $\Theta > \hat{\Theta}$ then CAP would not be the optimal technology at any wage. But any wage above $w = \vartheta(s)/(\varsigma - \Theta\alpha)$ will support package technology ANT while any wage below that will support FLEX. Two lines are provided for the CAP

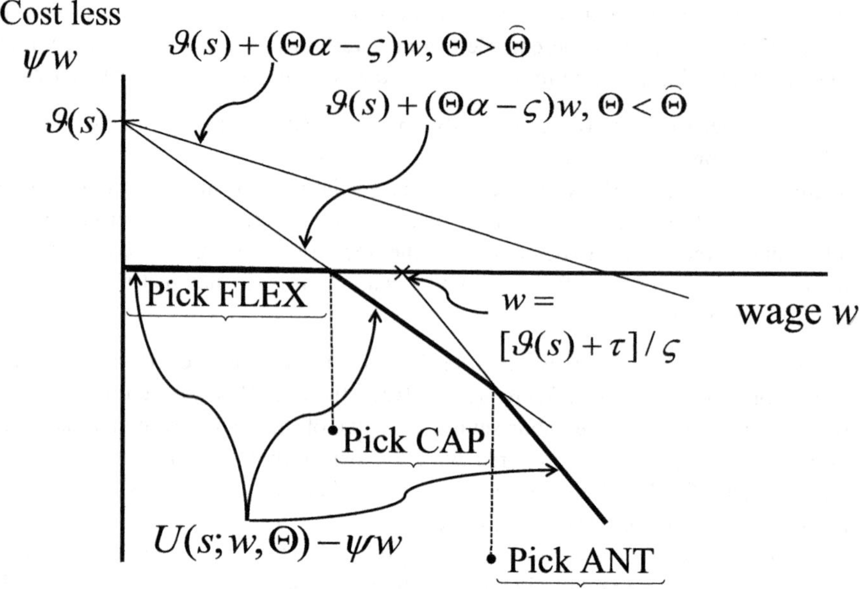

Fig. 5.1 Wage, technology choice and animal uniformality

cost, one where $\Theta > \widehat{\Theta}$ and one where $\Theta < \widehat{\Theta}$. Were $\Theta < \widehat{\Theta}$ then there would be a positive wage interval for which CAP is optimal. If wages are low enough, but not too low, and animal biotechnology is sufficiently well developed then industrialization can occur without the use of antibiotics. But in a high wage economy, the package technology will be chosen unless antibiotics are banned.

Now we consider optimal herd size s. Let unit capacity cost be $\kappa(s)$, which is assumed to be U shaped. These costs pertain to environmental compliance, agency costs due to management control and input acquisition costs. It is quite plausible that scale economies are associated with capitalization so we assume that $\vartheta(s)$ is decreasing such that total unit costs, $U(s; w, \Theta) + \kappa(s)$, remain U shaped. Unit costs are then

$$U(s; w, \Theta) + \kappa(s) = \psi w + \min[0, \vartheta(s) + (\Theta\alpha - \varsigma)w, \vartheta(s) + \tau - \varsigma w] + \kappa(s). \quad (5.24)$$

Under technology homogeneity among a large number of farms and free entry, equilibrium scale involves minimizing average cost. If Θ and τ are large enough then the production format is FLEX and the scale problem is that of choosing s to minimize $\kappa(s)$. If Θ and/or τ fall due to animal health or biotechnology innovations, or if wages rise, then the problem will change to that of choosing s to minimize $\vartheta(s) + \kappa(s)$with $\arg\min \vartheta(s) + \kappa(s) > \arg\min \kappa(s)$. A discrete increase in scale can attend the adoption of the uniformity-promoting technology, as suggested by Finlay (2004). So antibiotics, capitalization, and scale can come as a complementary package the adoption of which depends on external forces.

Model 3: Biosecurity and Scale

Biosecurity costs take many forms, including design and use of procedures, guarding against animal and other pathogen vectors, using quarantine and showering facilities, and vaccinating. Most are likely to display declining costs per unit production capacity. For example, truck wash facilities will be mostly idle for a small production unit, an annual employee training endeavor may as easily accommodate five or ten attendees, and it may be possible to obtain additional feed from the same source without scaling up attribute tests.

We develop a model of how scale economies in biosecurity costs affect size of operation and use of biosecurity inputs. Hennessy et al. (2005) study scale in a unit cost minimization model where each animal brings an independent and identically distributed disease risk while any diseased animal infects the entire feedlot. It is distinguished from the model to be presented in that (a) the biosecurity input's unit cost is scale neutral, (b) the independent, identical distribution feature creates a source of scale diseconomy rather than a source of scale economy, and (c) it is less articulated in characterizing the roles of biosecurity inputs.

Our model assumes that the animal output market is perfectly competitive and all producers are identical in the technology available to them. Let $G(z;v) > 0$ represent expected output from one animal. Here z is some biosecurity input that comes with scale economies and v is some pertinent beneficial external natural or socioeconomic factor. Possibilities for v include the state of biosecurity on other farms, the quality of animal public health infrastructure, or animal transportation regulations. Both arguments are beneficial so that $G_z(\cdot) > 0$ and $G_v(\cdot) > 0$.

With Q animals, expected output is $QG(\cdot)$. We require the input's marginal productivity to decline, or $G_{zz}(\cdot) < 0$. Since we set $G(z;v)|_{z=0} \equiv f(v) > 0$, the input is not essential for production. The effective unit price of z is $J(Q)r$ where r>0 is a cost parameter reflecting the state of knowledge on producing the biosecurity input and where $J(Q) > 0$. Conditions $0 < J_Q(Q) < \bar{J}(Q)$, $\bar{J}(Q) \equiv J(Q)/Q$, are imposed to reflect scale economies in the biosecurity input. This means that $\bar{J}(Q)$, which reflects how biosecurity unit costs change with scale, is decreasing. Costs other than the biosecurity input amount to $C(Q)$, which can be viewed as the standard minimized cost of farming Q animals and $\bar{C}(Q) \equiv C(Q)/Q$ is the unit cost for these nonbiosecurity inputs.

As in Model 2, equilibrium scale is that which minimizes unit costs. The objective function is therefore

$$T(Q,z;r,v) = \frac{\bar{B}(Q,z;r)}{G(z;v)}, \quad \bar{B}(Q,z;r) \equiv \bar{C}(Q) + \bar{J}(Q)rz. \tag{5.25}$$

For example, if $J(Q) = Q$, a case we have ruled out, then the biosecurity indicator is scale neutral such that scale and biosecurity choices will be determined independently. Expression $\bar{B}(Q,z;r)$ is held to be strictly convex in Q at an optimum, i.e., $C_{QQ}(\cdot) + J_{QQ}(\cdot)rz > 0$ where it is standard to assume $C_{QQ}(\cdot) > 0$. The optimality conditions are

$$T_Q(\cdot) = \frac{\bar{B}_Q(\cdot)}{G(\cdot)} = 0, \tag{5.26a}$$

$$T_z(\cdot) = \frac{G(\cdot)\bar{J}(\cdot)r - G_z(\cdot)\bar{B}(\cdot)}{[G(\cdot)]^2} = 0. \tag{5.26b}$$

The optimizing choices are written as z^* and Q^*. Second-order conditions are developed in the appendix and convexity is assumed. It is noteworthy that

$$T_{Qz}(\cdot) = \frac{\bar{J}_Q(\cdot)r}{G(\cdot)} < 0. \tag{5.27}$$

This means that an increase in the biosecurity input makes average cost less sensitive to scale. The decisions are technical complements in that more of z tends to make an increment of scale more beneficial in reducing average cost.[9]

Effect of Subsidy or Biosecurity Innovation

Intuition would suggest that the optimal choice of biosecurity input level should increase with a decrease in the input's price while the effect on scale of production is less clear. In the appendix we show that either of (a) a biosecurity input subsidy; or (b) an improvement in the biosecurity input production technology that leads to a reduction in its market price; have ambiguous effects on input use and production scale.

Consider the effect of an increase in biosecurity price r on biosecurity level z^* when production scale is fixed. Input choice z^* declines with an increase in own price, as one might expect. However, the indirect effect on choice z^* when mediated through scale is positive. This is because, from (5.27) above and due to scale economies, an increase in scale Q lowers the unit cost $J(Q)r$ of the biosecurity input z. Which effect dominates in practice is a matter that needs to be established by empirical analysis.

Regarding scale, the direct effect of an increase in r when holding the value of input z fixed is positive because an increase in scale reduces the input's effective unit cost. However, the indirect effect on scale when allowing biosecurity level z to adjust optimally is negative. This is because relation (5.27) shows that the input

[9] As we shall show, technical complementarity does not imply economic complementarity. An increase in biosecurity price r may increase the optimum level of scale.

complements production scale so that an increase in price r should decrease the incentive to scale up through a direct negative effect on use of the biosecurity input.

Effect of Improvement in External Biosecurity Environment

In the appendix, we show that improvements in the external biosecurity environment in which the firm operates are likely to increase both (1) operation scale and (2) use of the biosecurity input. The finding hinges on a technical assumption, the requirement that[10]

$$R(z;v) \equiv \frac{\partial^2 \ln[G(z;v)]}{\partial z \partial v} \geq 0, \tag{5.28}$$

where defined. If productivity function $G(z;v)$ satisfies (5.28) then it is said to be log-supermodular (Athey 2002), where $R(z;v) \geq 0$ may be viewed as a complementary technical relation between the levels of internal and external biosecurity.[11] The condition requires that an increase in argument v generate an increase in the input's productivity when expressed in fractional terms. In view of monotonicity assumptions $G_z(\cdot) > 0$ and $G_v(\cdot) > 0$, a necessary condition for (5.28) is $G_{zv}(\cdot) > 0$.[12]

There is reason to believe that $R(\cdot) \geq 0$. We have previously presented v as an external factor that increases productivity. It could do so stochastically, and so could represent the probability of avoiding some external source of productivity loss. Let the realization of an adverse event lead to constant productivity level $G(z;v)|_{v=0} = \widehat{G} \geq 0$. Absent the adverse event, productivity would be $G(z;v)|_{v=1} = \hat{G}(z)$, an increasing function. The productivity function is then $G(z;v) = (1-v)\widehat{G} + v\hat{G}(z)$ so that $R(\cdot) = \widehat{G}\hat{G}_z(z)/[G(z;v)]^2 \geq 0$.

Further support for property $R(\cdot) \geq 0$ arises from an economic perspective on species genetic manipulation. Animals bred to perform well in confinement can be viewed as allocating feed inputs largely toward market outputs (meat, eggs, milk). Free-range animals will need to allocate more energy toward other bodily functions, such as a robust skeleton and immune system. This is the resource allocation hypothesis.[13] Thus, breeding innovations may be Hicks biased in favor of high biosecurity use, tilting productivity function $G(\cdot)$ up when z is high but down when

[10] The derivatives are partial to acknowledge that an optimally chosen value of z will indeed depend on v. So a total derivative with respect to v would account for this indirect effect.

[11] Of course, (5.27) does not imply that $dz^*/dv>0$ as scale effects need to be considered.

[12] The condition is satisfied whenever $v_2>v_1$ and $G_z(z;v_2)$ is larger than $G_z(z;v_1)$ in the monotone likelihood ratio sense, an ordering widely used in information economics (p. 485 in Mas-Colell et al. 1995). Note that z can be a concretely defined quantity, as with veterinarian hours per 1,000 animals. So too can v, perhaps as a government animal public health infrastructure metric at the national level. A rich data set would allow for statistical tests on condition (5.28).

[13] See p. 254 in Greger (2007) for a brief review of evidence on the hypothesis.

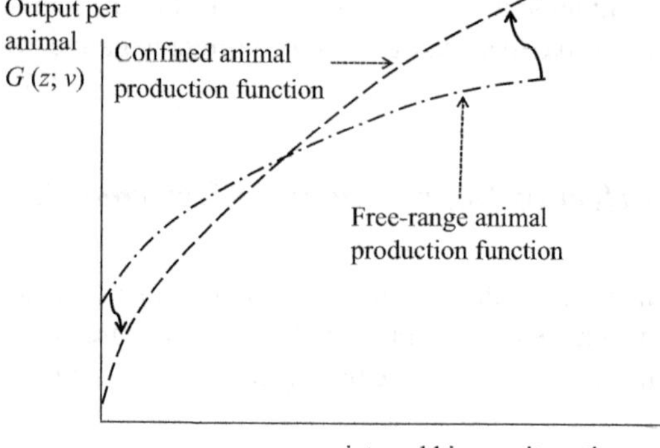

Fig. 5.2 Relation between biosecurity input and productivity under resource allocation hypothesis

z is low. Figure 5.2 illustrates requirement $R(z;v) \geq 0$. In it, output for confined animals will be higher only if on-farm biosecurity is high. Free-range animals are presented as performing well in comparison when biosecurity is low because they are hardier. But they perform poorly in comparison when biosecurity inputs reduce the need for allocating energy toward bodily protection.

Why should an increase in external biosecurity increase both internal biosecurity and scale? If external biosecurity is poor then a grower may reckon her herd will almost inevitably contract the disease and may see little point in seeking to protect her farm. If external biosecurity is good then the grower may feel more in control of the farm's destiny. The grower will increase use of the biosecurity input, and complementarity relation (5.27) then justifies an increase in scale. A similar point has been made in Ifft et al. (2011), who find validation in behavior on Viet Nam poultry farms.

We conclude that our model provides some support for the idea that a producer who is more confident an exogenous biosecurity risk to productivity will not materialize is more likely to increase (a) own-farm biosecurity and also (b) scale of operation. Point (a) suggests that public animal health inputs may not crowd out their privately provided counterparts while point (b) posits a role public policy may unwittingly play in the industrialization process.[14] Such firm-level complementarities have long been studied in more general contexts, see Milgrom and Roberts (1990a). What is most interesting in our model is that the responses can be mutually reenforcing across farms in a region. Parameter v may represent a summary statistic of biosecurity activities by other farms in the region. Then the $R(\cdot) \geq 0$ condition ensures that the best strategic response of a given farm to higher external protection is to increase on-farm protection and a virtuous circle plays out. The game is one of strategic complementarities (i.e., between firms), as studied in Milgrom and Roberts (1990b). In general, there is no guarantee that any equilibrium is unique (Echenique 2007).

[14] To obtain a better sense for the responses developed upon in discussions to this point, bear in mind that these are market equilibrium responses where output price is endogenous and equal to unit cost.

Conclusion

The path from pastoral agriculture toward industrial format animal farming has not been smooth. It has been faster for some farmed species, where avian produce acquiesced more quickly than hogs. Ruminants in general are still grown pastorally for at least some of their life. Nonetheless, from a distance the rough-stroke outline of the paths taken have been quite similar where prominence must be assigned to technologies that control how inputs perform so that capital can operate efficiently. Animal health inputs must rank high among technologies that control input attributes as mortality, morbidity, and cross infection can disrupt throughput and lead to complete shutdowns for extended periods. However, the quintessential control input is genetic composition for it allows the nature of primary input to be rendered homogeneous, so that only environmental considerations remain.

This chapter has sought to characterize aspects of how animal health inputs fit with production format. The interactions involved are many and varied so, rather than try dine with a Swiss Army knife, we tailor three separate models. Model 1 shows that the reference view of infectious disease as a form of commons problem involving excess stocking is incomplete. The disease also induces socially suboptimal entry. Consequently, disease management innovations such as fencing commonage and regulating livestock marts are likely to have intricate consequences for productivity and welfare. Although admittedly very simplistic, this model warrants further scrutiny.

Model 2 looks inside the farm by focusing on an innovation that mitigates disease externalities. In particular, we emphasize the role that antibiotics play as a means of reducing the irregularities that inhibit capital substitution for labor. We show how antibiotics enable capitalization to substitute for labor. We also show how antibiotics can come as part of a complementary package involving capital inputs, to be adopted entirely or not at all, when the extent of nonuniformities that antibiotics can mitigate is large. In this sense, the removal of antibiotics could reduce the incentive to capitalize in high-wage countries. Innovations in genetic control, however, make this possibility less likely, at least under our model assumptions. One final comment on the model is that, in light of scale economies in physical asset investments, an external shock in technology or wage levels can see a fundamental transformation in production paradigm to the industrial format.

The third model is motivated by the observation that biosecurity investments are likely to exhibit scale economies so that scale and biosecurity investments are likely technical complements. We show that a decline in the unit cost of a biosecurity investment can reduce optimal scale. This is because one motive for increasing scale in the presence of biosecurity investments is to take best advantage of scale economies and this incentive weakens when the biosecurity investment becomes less costly. So while intuition might suggest that any subsidy or innovation in technologies associated with capital intensive production will ensure more of it, we beg to differ. The model is also used to argue that public investments in securing a region from infectious disease can complement both scale and internal biosecurity investments. If this is the case, then an external shock could generate mutually reinforcing beneficial effects rippling through a decentralized production system. The sorts of shocks at issue could include efforts to improve technical proficiency or to reduce government corruption in the provision of public sector veterinary health services.

Stepping back to overview the recent evolution of animal production, we readily acknowledge a variety of shortcomings in our general framework. The approach is entirely production-driven, without reference to environmental bads or demands for goods other than cheap animal produce. It ignores the role of feedstuffs. Cheap in situ forage complements low-input pastoral systems that are open to disease, and health inputs in such systems might be best embodied in hardy genetics rather than drugs or thick perimeter walls.

Finally, a tale of structural dynamics may have been missed when pointing to technology spillovers as the source of technological change. Once an industrial format takes root, much subsequent innovation is endogenous. After all, innovations translated from manufacturing processes have typically dealt with nonbiological subjects while those from human medicine pertain to a very different market environment. Looking forward, the sector's horizon is overcast with environmental, animal welfare, zoonotic disease, and other concerns that affect the sector in unique ways. Reservoirs of capacity for endogenous innovation will be required if animal production is to succeed in adapting.

Appendix

Second-order conditions: Compute

$$T_{QQ}(\cdot) = \frac{\bar{B}_{QQ}(\cdot)}{G(\cdot)} > 0, \quad T_{zz}(\cdot) = -\frac{G_{zz}(\cdot)\bar{B}(\cdot)}{[G(\cdot)]^2} > 0, \tag{5.29}$$

where optimality conditions $T_Q(\cdot) = 0$ and $T_z(\cdot) = 0$ have been applied. The cross-derivative for the average cost function is $T_{Qz}(\cdot) = \bar{J}_Q(\cdot)r/G(\cdot) < 0$, where optimality condition (5.26a) and computation $\bar{J}_Q(\cdot) = [QJ_Q(\cdot) - J(\cdot)]/Q^2 \overset{\text{sign}}{=} J_Q(\cdot) - \bar{J}(\cdot) < 0$ have been applied. Finally,

$$\Phi \equiv T_{QQ}(\cdot)T_{zz}(\cdot) - [T_{Qz}(\cdot)]^2 = -\frac{\overbrace{G_{zz}(z)\bar{B}(\cdot)\bar{B}_{QQ}(\cdot)}^{<0}}{[G(\cdot)]^3} - \frac{\overbrace{[\bar{J}_Q(\cdot)]^2 r^2}^{>0}}{[G(\cdot)]^2}. \tag{5.30}$$

Convexity on $T(\cdot)$ requires $\Phi > 0$. If cost parameter $r > 0$ is small (i.e., the biosecurity innovation has been well developed) or economies of scale are small (i.e., $|\bar{J}_Q(Q)|$ is small) then $\Phi > 0$ is likely. Otherwise (5.30) may be violated, so that discontinuous scale and input responses result from a small parameter change. Henceforth we assume $\Phi > 0$.

Subsidy and innovation: Define $H(\cdot) = G(\cdot) - G_z(\cdot)z > 0$ where $H(\cdot) > 0$ is due to $G_z(\cdot) < G(\cdot)/z$ on $z \geq 0$ whenever $G(\cdot)$ is increasing and concave in z with $G(0; v) > 0$. Using (5.26a, b):

$$T_{Qr}(\cdot) = \frac{\bar{J}_Q(\cdot)z}{G(\cdot)} < 0, \quad T_{zr}(\cdot) = \frac{\bar{J}(\cdot)H(\cdot)}{[G(\cdot)]^2} > 0. \tag{5.31}$$

Differentiate system (5.26a, b) completely with respect to (Q, z, r) and then invert to obtain:

$$\frac{dz^*}{dr} = \frac{T_{Qz}T_{Qr} - T_{QQ}T_{zr}}{\Phi} = \overbrace{\frac{[\bar{J}_Q(\cdot)]^2 rz}{[G(\cdot)]^2\Phi}}^{<0} + \overbrace{\frac{\bar{B}_{QQ}(\cdot)\bar{J}(\cdot)H(\cdot)}{[G(\cdot)]^3\Phi}}^{>0};$$

$$\frac{dQ^*}{dr} = \frac{T_{Qz}T_{zr} - T_{zz}T_{Qr}}{\Phi} = \overbrace{\frac{\bar{J}_Q(\cdot)r\bar{J}(\cdot)H(\cdot)}{[G(\cdot)]^3\Phi}}^{<0} + \overbrace{\frac{\bar{J}_Q(\cdot)\bar{B}(\cdot)G_{zz}(\cdot)z}{[G(\cdot)]^3\Phi}}^{>0}. \tag{5.32}$$

Both expressions are indeterminate in sign without further information. Plausible functional forms are readily identified such that both derivatives are positive.

For the own-price response, $-T_{QQ}T_{zr}/\Phi < 0$ characterizes the direct effect. The indirect effect, through the effect on Q, is $T_{Qz}T_{Qr}/\Phi > 0$. This is because an increase in scale lowers the unit cost of the biosecurity input due to scale economies emphasizing that input's cost. For the scale response, the direct effect is represented by $-T_{zz}T_{Qr}/\Phi > 0$. The indirect effect on scale when mediated through the biosecurity input is $T_{Qz}T_{zr}/\Phi < 0$.

External effects. Differentiate system (5.26a, b) completely with respect to (Q, z, v) and then use the optimality conditions:

$$\begin{pmatrix} \frac{\bar{B}_{QQ}(\cdot)}{G(\cdot)} & \frac{\bar{J}_Q(\cdot)r}{G(\cdot)} \\ \frac{\bar{J}_Q(\cdot)r}{G(\cdot)} & -\frac{G_{zz}(\cdot)\bar{B}(\cdot)}{[G(\cdot)]^2} \end{pmatrix} \begin{pmatrix} \frac{dQ^*}{dv} \\ \frac{dz^*}{dv} \end{pmatrix} = \begin{pmatrix} 0 \\ \frac{R(\cdot)\bar{B}(\cdot)}{G(\cdot)} \end{pmatrix}. \tag{5.33}$$

So

$$\frac{dQ^*}{dv} = -\frac{\bar{J}_Q(\cdot)r\bar{B}(\cdot)}{\Phi[G(\cdot)]^2}R(\cdot) \overset{\text{sign}}{=} R(\cdot), \quad \frac{dz^*}{dv} = \frac{\bar{B}_{QQ}(\cdot)\bar{B}(\cdot)}{\Phi[G(\cdot)]^2}R(\cdot) \overset{\text{sign}}{=} R(\cdot). \tag{5.34}$$

Explicit solution

Set $C(Q) = Q^{\beta_1+1}, \beta_1 > 0,$ and $J(Q) = Q^{1-\beta_2}, \beta_2 \in (0,1).$[15] Writing $\lambda = \beta_2/\beta_1$, (5.26a) provides $Q^{\beta_1-1} = \lambda Q^{-(1+\beta_2)}rz$ and (5.26a) and (5.26b) solve as:

$$\frac{G(z^*;v)}{G_z(z^*;v)} = (1+\lambda)z^*, \tag{5.35a}$$

$$Q^*(z^*) = (\lambda r z^*)^{1/(\beta_1+\beta_2)}. \tag{5.35b}$$

So for these functional forms, and regardless of the choice of some $G(z)$ function that supports an interior solution, the z that minimizes unit cost is independent of unit cost parameter r.

Finally, and as a specification distinct from $G(\cdot) = (1-v)\widehat{G} + v\hat{G}(\cdot)$ previously discussed, let $G(z;v) = z/[\mu + \sigma(v)z]$. This is the logistic function form. With constant $\mu > 0$ and $\sigma(v) > 0$, the external factor influences this distribution through $\sigma_v(v) < 0$ so that $G_v(\cdot) > 0$ and an increase in the factor reduces the unit cost of output. Then $G_z(\cdot) = \mu/[\mu + \sigma(\cdot)z]^2$, $G_z(\cdot)/G(\cdot) = \mu/\{z[\mu + \sigma(\cdot)z]\}$, and $R(\cdot) = -\mu\sigma_v(\cdot)/[\mu + \sigma(\cdot)z]^2 > 0$. An improvement in external biosecurity increases productivity, and also the marginal productivity of the biosecurity input when calculated in percent terms.

System (5.35a, b) then solves as

$$(Q^*, z^*) \in \left\{(0,0), \left(\left(\frac{\lambda^2 \mu r}{\sigma(\cdot)}\right)^{1/(\beta_1+\beta_2)}, \frac{\mu\lambda}{\sigma(\cdot)}\right)\right\}. \tag{5.36}$$

It will be shown later that $(Q^*, z^*) = (0,0)$ is not optimal, leaving only the interior solution. Notice that the equilibrium value for productivity is $G(\cdot) = \lambda/[(1+\lambda)\sigma(v)]$, which is increasing in the beneficial external natural or socioeconomic factor, v.

To understand why $\mathrm{d}Q^*/\mathrm{d}r > 0$ in (5.36), consider the direct and indirect effects of an increase in the price of biosecurity. An increase in r increases the incentive to reduce unit costs by increasing scale. This is the direct effect on scale. The indirect effect, via complementarity relation (5.27), is that a higher value of r reduces the incentive to use the biosecurity input and this reduces the incentive to increase scale. Relation (5.36) shows that the direct effect wins out.

Expression z^* in (5.36) also indicates that the direct and indirect effects of an increase in the value of r exactly offset. Although (5.36) does not show $\mathrm{d}z^*/\mathrm{d}r > 0$, this is possible, i.e., there could conceivably be a positive own-price effect in long-run equilibrium. The possibility arises because the endogenous scale choice alters

[15] Were $r = 0$, which we rule out as both unrealistic and uninteresting, then $\bar{B}(\cdot) = Q^{\beta_1}$. The scale and biosecurity choice variables would be separable and $Q^* = 0$. This is the classical setting whereby an infinity of firms each produce an infinitesimal amount, and no long-run equilibrium actually exists (p. 337 in Mas-Colell et al. 1995).

the effective unit cost of the input in a manner broadly similar to how the income effect mediates price response in demand theory. As with Giffen goods, the indirect effect can overwhelm the direct effect.

It follows from $\sigma_v(\cdot) < 0$ that $\mathrm{d}z^*/\mathrm{d}v > 0$ and $\mathrm{d}Q^*/\mathrm{d}v > 0$, so that more external biosecurity coaxes out more internal biosecurity and increases scale. The example supports the hypothesis that more public health inputs do not crowd out (i.e., do complement) private activities to safeguard animal health but do promote large-scale production. In the example, strengthening the external biosecurity environment complements incentives for internal biosecurity and also encourages a scaling up of production activities. With better external biosecurity, the animal production format is more likely to become industrial than back-yard.

Finding the optimum in the explicit solution: Pose the problem as having two stages:

$$\min_{z} \min_{Q|z} T(\cdot). \tag{5.37}$$

Upon inserting (5.35b) into the given $C(\cdot)$ and $J(\cdot)$ functions, cost function (5.25) becomes:

$$\begin{aligned}\widehat{T}(\cdot) &= \frac{(Q^*)^{\beta_1} + (Q^*)^{-\beta_2} r z^*}{z^*/[\mu + \sigma(\cdot)z^*]} = \mu\frac{(Q^*)^{\beta_1}}{z^*} + \sigma(\cdot)(Q^*)^{\beta_1} + (Q^*)^{-\beta_2} r[\mu + \sigma(\cdot)z^*] \\ &= \Gamma \times [\mu(z^*)^{-\beta_2/(\beta_1+\beta_2)} + \sigma(\cdot)(z^*)^{\beta_1/(\beta_1+\beta_2)}], \\ \Gamma &\equiv [\lambda^{\beta_1/(\beta_1+\beta_2)} + \lambda^{-\beta_2/(\beta_1+\beta_2)}] r^{\beta_1/(\beta_1+\beta_2)} > 0,\end{aligned} \tag{5.38}$$

which is the solution to the inner conditional optimization problem in (5.37). When $z^* = 0$ then $(z^*)^{\beta_1/(\beta_1+\beta_2)} = 0$ but $(z^*)^{-\beta_2/(\beta_1+\beta_2)} \to \infty$ so that $(Q^*, z^*) = (0,0)$ maximizes unit cost and cannot be optimal. To establish the problem's overall convexity, differentiate the final expression for $\widehat{T}(\cdot)$ in (5.38) twice with respect to the remaining endogenous variable, z, and insert $z^* = \mu\lambda/\sigma(v)$ to obtain

$$\begin{aligned}&\frac{\Gamma\beta_2[\mu(\beta_1 + 2\beta_2) - \sigma(\cdot)\beta_1 z^*](z^*)^{-(2\beta_1+3\beta_2)/(\beta_1+\beta_2)}}{(\beta_1 + \beta_2)^2} \\ &= \frac{\Gamma\beta_2\mu(z^*)^{-(2\beta_1+3\beta_2)/(\beta_1+\beta_2)}}{\beta_1 + \beta_2} > 0.\end{aligned} \tag{5.39}$$

Thus the problem is indeed convex and the interior solution is optimal.

References

Athey, S. 'Monotone Comparative Statics Under Uncertainty', *Quarterly Journal of Economics*, Vol. 117 (1, February, 2002) pp. 187–223.

Beach, R.H., Poulos, C. and Pattanayak, S.K. 'Farm Economics of Bird Flu', *Canadian Journal of Agricultural Economics*, Vol. 55 (4, December, 2007) pp. 471–483.

Besanko, D., Dranove, D., Shanley, M. and Schaefer, S. *Economics of Strategy, 3rdedn.* (Hoboken NJ: John Wiley & Sons, 2004).

Chandler, A.D. 'Organizational Capabilities and the Economic History of the Industrial Enterprise', *Journal of Economic Perspectives*, Vol. 6 (3, Summer, 1992) pp. 79–100.

Ciftci, I. and Ercan, A. 'Effects of Diets of Different Mixing Homogeneity on Performance and Carcass Traits of Broilers', *Journal of Animal and Feed Sciences*, Vol. 12 (1, 2003) pp. 163–171.

Cottingham, J. "A Brute to the Brutes?': Descartes' Treatment of Animals', *Philosophy*, Vol. 53 (206,October, 1978) pp. 551–559.

Dolberg, F., GuerneBleich, E. and McLeod, A. *Summary of Project Results and Outcomes.*Food and Agriculture Organization of the United Nations.Final report on the project TCP/RAS/3010 (E) 'Emergency Regional Support for Post Avian Influenza Rehabilitation', Rome, February 2005.

Echenique, F. 'Finding all Equilibria in Games of Strategic Complements', *Journal of Economic Theory*, Vol. 135 (1, July, 2007) pp. 514–532.

Finlay, M.R. 'Hogs, Antibiotics, and the Industrial Environments of Postwar Agriculture', in S. R. Schrepfer and P. Scranton (eds), *Industrializing Organisms: Introducing Evolutionary History*, pp. 237–260 (London: Routledge, 2004).

Galor, O. and Moav, O. 'Natural Selection and the Origin of Economic Growth', *Quarterly Journal of Economics*,Vol. 117 (4, November, 2002) pp. 1122–1191.

Gibbons, R. *Game Theory for Applied Economists* (Princeton NJ: Princeton University Press, 1992).

Graham, J.P., Boland, J.J. and Silbergeld, E. 'Growth Promoting Antibiotics in Food Animal Production: An Economic Analysis', *Public Health Reports*, Vol. 122 (January/February, 2007) pp. 79–87.

Greger, M. 'The Human/Animal Interface: Emergence and Resurgence of Zoonotic Infectious Diseases', *Critical Reviews in Microbiology*,Vol. 33 (4, 2007) pp. 243–299.

Hardin, G. 'The Tragedy of the Commons', *Science*, Vol. 162 (13 December, 1968) pp. 1243–1248.

Hayes, D.J. and Jensen, H.H. *Lessons from the Danish Ban on Feed-Grade Antibiotics, Food Control* (Ames, Iowa: Center for Agricultural and Rural Development, Briefing Paper 03-BP 41, June 2003).

Hennessy, D.A. 'Slaughterhouse Rules: Animal Uniformity and Regulating for Food Safety in Meat Packing', *American Journal of Agricultural Economics*,Vol. 87 (3, August, 2005) pp. 600–609.

Hennessy, D.A., Roosen, J. and Jensen, H.H. 'Infectious Disease, Productivity, and Scale in Open and Closed Animal Production Systems', *American Journal of Agricultural Economics*,Vol. 87 (4, November, 2005) pp. 900–917.

Hennessy, D.A. 'Informed Control over Inputs and Extent of Industrial Processing', *Economics Letters*,Vol. 94 (3, March, 2007) pp. 372–377.

Hennessy, D.A., Miranowski, J.A. and Babcock, B.A. 'Genetic Information in Agricultural Productivity and Product Development', *American Journal of Agricultural Economics*,Vol. 86 (1, February, 2004) pp. 73–87.

Ifft, J., Roland-Holst and Zilberman, D. 'Production and Risk Prevention Response of Free Range Chicken Producers in Viet Nam to Highly Pathogenic Avian Influenza Outbreaks', *American Journal of Agricultural Economics*,Vol. 93 (2, January, 2011) pp. 490–497.

Key, N., McBride, M. and Mosheim, R. 'Decomposition of Total Factor Productivity Change in the U.S. Hog Industry', *Journal of Agricultural and Applied Economics*, Vol. 40 (1, April, 2008) pp. 137–149.

Li, P.J. 'Exponential Growth, Animal Welfare, Environmental and Food Safety Impact: The Case of China's Livestock Production', *Journal of Agricultural and Environmental Ethics*, Vol. 22 (3, June, 2009) pp. 217–240.

MacDonald, J.M. and McBride, W.D. *The Transformation of U.S. Livestock Agriculture: Scale, Efficiency, and Risks* (Washington DC: U.S. Department of Agriculture, Electronic Information Bulletin 43, January 2009).

Madsen, T.G. and Pedersen, J.R. 'Broiler Flock Uniformity', *Feedstuffs*, Vol. 82 (31, August 2, 2010) pp. 12–13.

Mankiw, N.G. and Whinston, M.D. 'Free Entry and Social Inefficiency', *Rand Journal of Economics*, Vol. 17 (Spring 1986) pp. 48–58.

Mas-Colell, A., Whinston, M.D. and Green J.R. *Microeconomic Theory* (Oxford: Oxford University Press, 1995).

Milgrom, P. and Roberts, J. (1990a) 'The Economics of Modern Manufacturing – Technology, Strategy, and Organization', *American Economic Review*,Vol. 80(3, June, 1990) pp. 511–528.

Milgrom, P. and Roberts, J. (1990b) 'Rationalizability, Learning, and Equilibrium in Games with Strategic Complementarities', *Econometrica*, Vol. 58 (6, November, 1990) pp. 1255–1277.

Miller, G.Y., McNamara, P.E. and Singer, R.S. 'Stakeholder Position Paper: Economist's Perspectives on Antibiotic Use in Animals', *Preventive Veterinary Medicine*, Vol. 73 (February 24, 2006) pp. 163–168.

Mortensen, S., Stryhn, H., Sgaard, R., Boklund, A., Stärk, K. D. C., Christensen, J. and Willeberg, P. 'Risk Factors for Infection of Sow Herds with Porcine Reproductive and Respiratory Syndrome (PRRS) Virus', *Preventive Veterinary Medicine*, Vol. 53 (February 14, 2002) pp. 83–101.

Mosheim, R. and Knox Lovell, C.A. 'Scale Economies and Inefficiency of U.S. Dairy Farms', *American Journal of Agricultural Economics*, Vol. 91 (August 2009) pp. 777–794.

Proctor, M. 'Antibiotic Use Risk Unlikely', *Feedstuffs*, Vol. 82 (19, May 10, 2010) p. 9.

Rose, N. and Madec, F. 'Occurrence of Respiratory Disease Outbreaks in Fattening Pigs: Relation with the Features of a Densely and a Sparsely Populated Pig Area in France', *Veterinary Research*,Vol. 33 (2, March-April, 2002) pp. 179–190.

Smil, V. *Feeding the World: A Challenge for the Twenty-First Century* (Cambridge, MA: MIT Press, 2000).

Steinfeld, H., Mooney, H.A., Schneider, F. and Neville, L.E., eds. *Livestock in a Changing Landscape: Drivers, Consequences, and Responses,* Vol. 1 (Washington, D.C.: Island Press, 2010).

Trostle, R. *Global Agricultural Supply and Demand: Factors Contributing to the Recent Increase in Food Commodity Prices* (Washington DC: U.S. Department of Agriculture, Economic Research Service report WRS-0801, July 2008).

Vandekerchove, D., De Herdt, P., Laevens, H. and Pasmans, F. 'Risk Factors Associated with Colibacillosis Outbreaks in Caged Layer Flocks', *Avian Pathology*, Vol. 33 (3, June, 2004) pp. 337–342.

Vandeman, A.M. 'Management in a Bottle: Pesticides and the Deskilling of Agriculture', *Review of Radical Political Economics*, Vol. 27 (3, September, 1995), pp. 49B59.

Vives, X. *Oligopoly Pricing: Old Ideas and New Tools* (Cambridge, MA: MIT Press, 1999).

World Bank. *Enhancing Control of Highly Pathogenic Avian Influenza in Developing Countries through Compensation* (Washington DC: The International Bank for Reconstruction and Development, World Bank, 2006).

Chapter 6
Dynamic Perspectives on the Control of Animal Disease: Merging Epidemiology and Economics

Richard D. Horan, Christopher A. Wolf, and Eli P. Fenichel

Introduction

The literature on managing animal diseases has its roots in mathematical epidemiology, which focuses on understanding the dynamics of infectious populations (Kermack and McKendrick 1927; Anderson and May 1979). Mathematical epidemiology models can be used to predict the conditions under which disease prevalence will diminish over time and eventually be eradicated from the animal system. Management in this context generally is viewed as a sequence of exogenous perturbations designed to produce the required conditions for prevalence decline and, when possible, eradication (Heesterbeek and Roberts 1995).

Operationally, the conditions for prevalence decline or eradication are expressed in terms of a disease ecology metric, known as R_t, that can guide management (e.g., Roberts and Heesterbeek 2003; Hethcote 2000; Chowell and Brauer 2009). Specifically, R_t is the number of secondary infections caused by an infectious individual, and this number must be less than one for disease prevalence to be declining (this is a tautology; Chowell and Brauer 2009). A special case of R_t is R_0, which is the number of secondary infections caused by the first infected individual (Anderson and May 1979) – a measure of invasibility of the pathogen. Most of the epidemiology and disease ecology literature focuses on R_0 as an approximation for R_t, with the management objective being to reduce R_0 below one. Indeed, R_0 is considered "the most pervasive and useful concept in the mathematical epidemiology of infectious diseases" due to its perceived role in guiding disease management (Roberts and Heesterbeek 2003).

R.D. Horan (✉) • C.A. Wolf
Department of Agricultural, Food, and Resource Economics, Michigan State University, East Lansing, MI, USA
e-mail: horan@msu.edu

E.P. Fenichel
School of Life Science and ecoServices Group, Arizona State University, Tempe, AZ, USA

D. Zilberman et al. (eds.), *Health and Animal Agriculture in Developing Countries*, Natural Resource Management and Policy 36, DOI 10.1007/978-1-4419-7077-0_6

While R_0, or more generally, R_t, is an important metric, economic concerns arise from making R_t the only – or even the principal – metric to guide management choices. Studies that focus on R_t provide no guidance on how much less than one this metric should be at any point in time, nor do they explore whether eradication is economically desirable. Indeed, by itself, R_t does not provide any indication as to the economic benefits or costs of disease control. Moreover, in the absence of an economic behavioral model to accompany the epidemiological model, R_t does not fully account for feedbacks between the epidemiological and economic systems (Fenichel et al. 2010).

This chapter provides an introduction to constructing bioeconomic models of animal disease management. We begin by outlining an epidemiological model incorporating human management choices, and briefly discuss the relation between human choices and traditional management metrics. Next, we introduce a bioeconomic framework for management. We do not focus on solving any particular disease management problem. Rather, we discuss how epidemiological features of the problem, as well as limitations on management choices, complicate the problem and may affect optimal management outcomes. We conclude with a brief discussion of future research needs in this area.

A Simple Epidemiological Model

Classical epidemiological models divide potential disease host populations into subpopulations called compartments (Kermack and McKendrick 1927; Anderson and May 1979). The SIR model is the most well-known compartmental model. It has three epidemiological compartments for each host: healthy but susceptible, S, infected and infectious, I, and recovered (or resistant) and immune, R (Anderson and May 1979).[1] Additional compartments, such as demographic factors associated with the host population (e.g., spatial location, age, sex), can also be incorporated. In this chapter, we keep things simple and focus on the basic susceptible infected, or SI, model (i.e., there is no immunity or recovery from disease). Nevertheless, we allow the aggregate population to change over time, a feature common in animal epidemiology but rare in analogous models of human epidemiology. The assumption of no immunity means vaccination will not be modeled as an explicit control here, though that is not a huge omission since our focus here is on presenting some basic insights that will be relevant for more complex specifications as opposed to

[1] Note $R(t)$ differs from R_t or R_0, which were mentioned in the introduction and are described in greater detail below. Though these measures differ, the use of this similar notation is conventional. Furthermore, in some cases, the R in SIR stands for removed.

developing the most general of models.[2] We describe our model as being for a wildlife disease problem, though many of the results and insights also apply to livestock problems.

Consider an *SI* model consisting of a single wildlife host population within a particular location. The total host population at time t is $N(t) = S(t) + I(t)$. System dynamics are described by the following differential equations (with t henceforth suppressed)[3]:

$$\dot{S} = F(N, \mathbf{x}) - \delta(\mathbf{x})S - \tau(S, I, \mathbf{x}) - hS, \tag{6.1}$$

$$\dot{I} = \tau(S, I, \mathbf{x}) - (\delta(\mathbf{x}) + \alpha(\mathbf{x}))I - hI. \tag{6.2}$$

The susceptible population grows due to reproduction, $F(\cdot)$, and declines due to natural mortality, at the rate $\delta(\cdot)$, disease transmission, $\tau(\cdot)$, and population controls or harvests, applied at the rate h. The infected population is increased by disease transmission, and declines due to natural mortality, disease mortality, at the rate $\alpha(\cdot)$, and also harvests, which occurs at the rate h. This formulation assumes that all animals are born susceptible, a common assumption but not a necessary one. The vector $\mathbf{x}$ represents human choices influencing reproduction and also disease transmission and mortality parameters (described below).

---

[2] Vaccines are more common for human diseases than animal diseases. When vaccines are available for animals, they are not always used. Vaccines may not be given to livestock because healthy, vaccinated animals may test positive for disease, thereby resulting in trade restrictions that lower the value of the animals in trade (USDA-APHIS 2002). Wildlife populations are seldom vaccinated because vaccines are unavailable or ineffective for many diseases (Smith and Cheeseman 2002), or vaccination of wildlife may be controversial – particularly for threatened and endangered species. This concern has arisen, at least in part, because a study population of Serengeti wild dogs became extinct following interventions that used vaccination (Burrows 1992; Burrows et al. 1994).

[3] As indicated in the introduction, the basic population-based epidemiology model (6.1) and (6.2) could be expanded to include ecologically defined compartments like resistant and exposed or carrier subpopulations (Hethcote 2000), or human-defined compartments such as subpopulations under quarantine or in reserve areas. The model can also be expanded to incorporate populations in additional locations, with animal movement between locations (Fulford et al. 2002; Rowthorn et al. 2009). Wild animal movement could depend on ecological and human factors. For instance, humans can manage population densities and construct habitat corridors that encourage movement to areas of greater resource abundance, and of reduced competition and predation (e.g., Gichohi 2003; Kaiser 2001; Ewing 2005). Often, though not always (e.g., see Sanchirico and Wilen 1999; Horan et al. 2005), spatial metapopulation models define the state variables as the number of "patches" of habitat, or the number of farms, in each disease state (cf. Levins 1969) (as opposed to animal densities). "Patch-based" models are simply based on a coarser unit of analysis, achieved by rescaling the system (6.1) and (6.2), to focus on changing disease patterns across the landscape. These models are well suited for farm sector analysis, where authorities are primarily concerned about the number of infected farms (Barlow et al. 1998). However, as population-based and patch-based models are related, the insights for both types of models are similar.

The specification for net growth assumes all animals contribute to reproduction. Suppose F takes a standard density-dependent form, such that $F_{NN} < 0$ and $F(0, \mathbf{x}) = F(K(\mathbf{x}), \mathbf{x}) - \delta K(\mathbf{x}) = 0$.[4] Here $K(\mathbf{x})$ is the carrying capacity of the population, absent disease and harvests. The carrying capacity potentially depends on the vector $\mathbf{x}$.

Disease transmission reduces the susceptible population but increases the infected population. The disease transmission function, $\tau(S,I,\mathbf{x})$, is assumed to be increasing and quasi-concave in both S and I, and that transmission cannot occur when there are no animals to transmit the infection or to become infected (i.e., $\tau(0, I, \mathbf{x}) = \tau(S, 0, \mathbf{x}) = 0$). Transmission may also depend on human actions, such as supplemental feeding or habitat modification, affecting the frequency of animal contacts. Momentarily ignoring the role of human choices, a general form for transmission is $\tau(S,I) = \beta C(N) I f(S/N)$, where β is the likelihood of infection conditional on a contact, C is a contact function defining the number of contacts made by each infected individual (so that CI is total contacts), and f is a function describing the density of susceptible individuals that may be contacted (Barlow 1995). We make the common assumption that $f = S/N$ and adopt the specification $\tau(S,I,\mathbf{x}) = \beta(\mathbf{x}) C(N, \mathbf{x}) SI/N$. Later, we discuss how C may affect management.

Harvests in system (6.1) and (6.2) occur at the rate h, which is nonselective with respect to health status. Nonselectivity means harvest controls simultaneously impact healthy and infected animals. This situation occurs because many infected animals do not exhibit easily observable signs of infection (at least not until very late stages of infection, if even then), and so postmortem testing is often the only practical option to identify wildlife health status (Lanfranchi et al. 2003).[5] Denote $h(S, I, \mathbf{p})$ as the harvest rate made in a decentralized economy, when human agents act in their own best interests and ignore any associated positive or negative externalities not internalized via current economic or policy variables, $\mathbf{p}$. Specifically, this harvest rate is a feedback response that depends on current population levels in addition to $\mathbf{p}$. The dependence on S and I reflects the fact that these values can affect the benefits and costs (including any disease risks) of harvesting (e.g., Clark 2005). Denote $\mathbf{p}_0$ to be the policy vector when $I = 0$ and there are no foreseeable disease risks.

The vector $\mathbf{x}$ represents human choices, apart from population controls, that affect population and disease dynamics. These may include activities such as supplemental feeding or land-use decisions affecting habitat. Some of these

[4] Subscripts denote partial derivatives.

[5] By the LeChatelier principle, selective harvests would reduce disease management costs, resulting in greater social welfare, but selective harvests are not always feasible. Harvests may be selective with respect to disease status for diseases exhibiting easily observable outward signs, or after diagnostic testing to elicit an animal's health status. Diagnostic testing has been proposed for some disease problems where field tests are quick and accurate. For instance, diagnostic testing has been proposed for Devil Facial Tumor Disease (DFTD) in Tasmanian devils (STTD 2009; Platt 2009), brucellosis in bison (Bienen and Tabor 2006), and chronic wasting disease in mule deer (Watry et al. 2004; Wolfe et al. 2004).

behaviors may be selective with respect to health status, while others may be nonselective. For instance, supplemental feeding of wildlife is nonselective with respect to an individual animal's health status (i.e., one cannot set out food that will only be eaten by healthy animals), so that supplemental feeding can influence both healthy and infected populations. Generally, supplemental feeding can increase disease transmission by bringing animals into closer contact, and it can decrease disease mortality by supporting sick animals (Hartup et al. 2000; Schmitt et al. 2002; Daszak et al. 2000; Wobeser 2002). Denote $\mathbf{x}(S, I, \mathbf{p})$ as the vector of human choices when human agents act in their own best interests, as described above for the case of harvests.

Disease Ecology Metrics and Management Recommendations

Bioeconomic approaches to management, which we discuss in the following section, are quite different than conventional disease management strategies that are rooted solely in epidemiology. Still, there is value to understanding how conventional strategies are derived, both to fully understand and appreciate the epidemiological literature and to communicate with epidemiologists and policy makers who may only be familiar with the conventional approach.

Conventional strategies are generally founded on the assumptions that disease eradication is the appropriate objective, and that this outcome is best attained using guidance based on the basic reproduction ratio of a pathogen, R_0 (Roberts and Heesterbeek 2003). R_0 can be thought of as the growth rate of infection at the point when the pathogen is introduced (Diekmann et al. 1990). Accordingly, R_0 is a measure of invasibility at the disease-free equilibrium (Diekmann et al. 1990; Gubbins et al. 2008).

We can derive R_0 for our model by first noting that $\dot{I}<0$ when the following condition holds:

$$\frac{\beta(\mathbf{x})C(N,\mathbf{x})S/N}{\delta(\mathbf{x})+\alpha(\mathbf{x})+h}<1. \tag{6.3}$$

R_0 is generally defined as a special case of the left hand side (LHS) of (6.3), under three assumptions: (1) the system is at the disease-free equilibrium, so that $I = 0$ and $S = N_0$, where N_0 is the equilibrium value of N when there is no disease; (2) no disease controls are implemented (i.e., $\mathbf{x} = \mathbf{x}(N_0, 0, \mathbf{p}_0)$; $h = h(N_0, 0, \mathbf{p}_0)$), since there are generally considered to be no disease controls in a disease-free equilibrium; and (3) behaviors are fixed: there are no behavioral responses to S, I, or $\mathbf{p}$.[6] Assumption (4) stems from the fact that behaviors of individuals not directly

[6] More generally, R_0 is derived as the dominant eigenvalue of the linearized epidemiological system at the disease-free equilibrium.

involved in the management process are ignored. Accordingly, the parameters that depend on these variables are generally assumed fixed. To emphasize this in what follows, we define $\beta_0 = \beta(\mathbf{x}(N_0, 0, \mathbf{p}_0))$, $\alpha_0 = \alpha(\mathbf{x}(N_0, 0, \mathbf{p}_0))$, $\delta_0 = \delta(\mathbf{x}(N_0, 0, \mathbf{p}_0))$, and $h_0 = h(\mathbf{x}(N_0, 0, \mathbf{p}_0))$, and treat these as fixed parameters – both prior to and after infection occurs. We also define $C_0(N) = C(N, \mathbf{x}(N_0, 0, \mathbf{p}_0))$, so as to suppress the reliance of C on $\mathbf{x}$. Given these assumptions, we have the standard form

$$R_0 = \frac{\beta_0 C_0(N_0)}{\delta_0 + \alpha_0 + h_0}. \tag{6.4}$$

Using (6.3) and (6.4), we see that the pathogen cannot invade the system when $R_0 < 1$. The pathogen could invade the system, under assumptions (2) and (3), when $R_0 > 1$. The conventional management objective is therefore to reduce R_0 below one, so as to prevent pathogens from invading naïve (i.e., predisease) systems, or to eliminate pathogens from a previously infected system.

Given that the parameters in (6.4) are treated as fixed, how is it possible to satisfy the $R_0 < 1$ criterion? Conventional management strategies suggest undertaking actions to alter the equilibrium value N_0.[7] When N_0 is viewed as variable, the $R_0 < 1$ criterion often implies a threshold effect (Dobson and Foufopoulos 2001; Roberts and Heesterbeek 2003; Dobson 2004; Heffernan et al. 2005). Specifically, we can often derive a threshold value of N_0, denoted N_0^{T} and referred to as the host-density threshold, to satisfy $R_0 = 1$.

The most common example stems from the assumption of density-dependent transmission, $C_0(N) = N$. The $R_0 < 1$ criterion is satisfied in this case whenever (Anderson and May 1986)

$$N < N_0^{\mathrm{T}} = \frac{\delta_0 + \alpha_0 + h_0}{\beta_0}. \tag{6.5}$$

N_0^{T} is generally defined as the value of N that prevents pathogen invasion into an uninfected population. Moreover, it is generally understood that maintaining $N < N_0^{\mathrm{T}}$ in already-infected populations will ultimately eradicate the pathogen (McCallum et al. 2001; Holt et al. 2003). The intuition is straightforward: the disease is eradicated by maintaining the population at levels sufficiently low that infectious contacts do not occur frequently enough to sustain the disease. The rule is

[7] For problems where vaccination and immunity is possible, S will not equal N in the predisease equilibrium. Rather, S will equal $(1 - \upsilon)N$, where υ is the proportion of the population that is immune due to vaccination. (Note that υ is not a control, but a state variable. We use υ to proxy for the immune population, which we previously defined as R, since $R = \upsilon N$. Here, we have used υ to avoid confusion between R and R_0.) In that case, the $R_0 < 1$ criterion could be satisfied by manipulating either N or υ. The basic insights are not affected by this complication: the approach is still to manage R_0 by affecting the pre-infection state of the world, rather than through current-period control choices.

also simple to apply: managers only need to focus on N_0^{T}, which is fixed given the assumptions that behavior is fixed and external to the disease system.

Several problems arise with the interpretation that N_0^{T} represents an invasion threshold.

First, the only way to reduce N in this model is to increase harvest rates beyond h_0, which is treated as a fixed parameter. These additional harvests, which could be encouraged through the use of the policy vector $\mathbf{p}$, are part of (6.2) and hence should be recognized as part of (6.5) as well:

$$N < N_0^{\mathrm{T}} = \frac{\delta_0 + \alpha_0 + h_0(\mathbf{p})}{\beta_0}. \tag{6.6}$$

Now the interpretation of (6.6) becomes muddled: we need to keep N below N_0^{T}, but this is accomplished by altering $\mathbf{p}$ to increase harvests, h_0, which in turn increases the threshold, so that N could remain at a larger value. In other words, the larger the harvest we use to reduce N below the threshold, the fewer harvests are needed to reduce N. This seems contradictory. Second, the threshold N_0^{T} is not fixed because $\mathbf{x}_0$ and h_0 are not fixed once we take behavioral feedbacks into account, which likely also depend on N. Indeed, incorporating these feedbacks would affect the derivation of the threshold. Moreover, the regulatory authority could also change the policy vector $\mathbf{p}$ to encourage behavioral changes in $\mathbf{x}$ to affect the threshold. For instance, perhaps the host-density threshold could be maintained at higher levels if supplemental feeding of wildlife, which tends to congregate animals, was reduced.

Problems also arise with the interpretation of N_0^{T} as a disease control threshold for an already-infected system. In this case with density-dependent transmission, $\dot{I} = 0$ implies $\beta_0 S/[\delta_0 + \alpha_0 + h_0] < 1$ must hold to reduce disease prevalence. If we substitute the relation $S = N - I$ for S, then we can solve this new threshold criterion for a threshold value of N that depends on I:

$$N^{\mathrm{T}}(I) = \frac{[\delta_0 + \alpha_0 + h_0]}{\beta_0} + I > N_0^{\mathrm{T}}. \tag{6.7}$$

Hence, the threshold is now state dependent, and larger than N_0^{T}. This means the rule defined by (6.5) is more than sufficient to reduce disease prevalence – at least when behaviors are fixed (Fenichel et al. 2010).[8] Moreover, the rule defined by (6.7) is inadequate just as the rule defined by (6.5) was shown to be inadequate for the predisease case. Indeed, the same issues described above arise here, with the additional complication that behaviors may now depend on S and I as opposed to

[8] Fenichel et al. (2010) find the opposite result for the case of multiple hosts: host-density thresholds predicted by R_0 may be insufficient to eradicate a pathogen from an already-infected system.

just N. Even if $N^{\mathrm{T}}(I)$ were the appropriate eradication threshold, the analysis still does not provide guidance as to how far N should be held below the threshold. A smaller N would presumably eradicate the disease faster, but at a greater cost.

At the other end of the transmission spectrum lies frequency-dependent transmission, which arises when the contact function is defined as $C_0(N) = 1$. In this case, R_0 becomes

$$R_0 = \frac{\beta_0}{\delta_0 + \alpha_0 + h_0} \tag{6.8}$$

which is independent of N when behavior is not considered. Assuming all parameters are perceived as constants, (6.8) suggests there is no role for management: the value of R_0 is entirely parameter dependent, and will therefore always remain more or less than one. However, if we recognize that the parameters do depend on human choices, $\mathbf{x}$, and in some cases these may be state dependent, then we may find an important role for managers after all. Horan and Wolf (2005) examine such a case, whereby supplemental feeding affects both β and α.

The discussion above indicates that the host-density threshold is endogenous and of the form $\hat{N}^{\mathrm{T}}(S, I, \mathbf{p})$, particularly in settings where the disease has already established. The implications of this result are best illustrated using an example. Assuming transmission takes the density-dependent form βSI, that β and δ are constant parameters, and that h is not a feedback expression but rather is chosen directly by a management authority, the endogenous threshold takes the form $\hat{N}^{\mathrm{T}}(I, h)$ and (6.2) can be rewritten as

$$\dot{I} = \beta[N - \hat{N}^{\mathrm{T}}(I, h)]I. \tag{6.9}$$

Equation (6.9) indicates disease prevalence falls when N is less than the host-density threshold, but also recognizes that the threshold is endogenous. More complex relations arise for more complex transmission specifications and when β and δ are endogenous, but the implication is the same: disease management occurs not by managing N relative to a fixed threshold, but by jointly managing the population *and* the threshold. Equation (6.9) (or more complex variations of it) provides no guidance as to how choices should be made to accomplish this, or even as to how much disease control should be pursued. We now turn to an economic criterion to help guide these choices.

A Bioeconomic Model of Disease Management

We now present a bioeconomic framework for optimal disease management, in which an economic criterion is used in conjunction with the epidemiological model (6.1) and (6.2) to determine the optimal allocation of disease control efforts.

Dynamic bioeconomic models relax the assumptions (1)–(3) that were introduced to construct R_0. Rather than focusing on a single ecological metric, such as R_0 or a related ecological threshold that does not account for behavioral feedbacks, the bioeconomic framework is based on the full suite of ecological dynamics, (6.1) and (6.2), including economic behaviors (Clark 2005). Moreover, the bioeconomic framework does not necessarily impose disease eradication as an explicit objective, though eradication can emerge as an optimal outcome.

We will focus on an equivalent, but modified, form of the system (6.1) and (6.2) for our bioeconomic analysis. Specifically, we use the relation $S = N - I$ to focus on the state variables N and I as opposed to S and I. The reason for focusing on N instead of S is that this approach will eliminate the transmission term in one of the equations of motion, thereby simplifying the analysis. Specifically, the equation of motion for N is derived from adding (6.1) and (6.2):

$$\dot{N} = F(N, \mathbf{x}) - \delta(\mathbf{x})N - \alpha(\mathbf{x})I - hN. \tag{6.10}$$

We focus on the social planner's problem of maximizing discounted social net benefits, as this represents an efficiency benchmark. Taking h and $\mathbf{x}$ to be the feedback response functions described earlier, we define the social net benefits in each period by $\mathrm{SNB}(N, I, \mathbf{p}) = \mathrm{SNB}(N, I, h(N, I, \mathbf{p}), \mathbf{x}(N, I, \mathbf{p}), \mathbf{p})$. The function SNB represents economic net benefits that people derive from the species and pathogens in question. These net benefits could include the benefits and costs of harvesting the resource (e.g., in the case of deer), existence or ecotourism values for the resource (e.g., in the case of bison or lions), and damage costs the animals – particularly infected animals – impose on other sectors of the economy (e.g., when infected wildlife put livestock or people at risk of infection). We assume SNB is concave in the state and control variables.

Given this general specification and assuming a discount rate of ρ, the management problem is written as

$$\begin{aligned} \underset{\mathbf{p}}{\text{Max}} \quad & \mathrm{PVSNB} = \int_0^\infty \mathrm{SNB}(N, I, \mathbf{p}) \mathrm{e}^{-\rho t}\, \mathrm{d}t, \\ \text{s.t.} \quad & (6.10), (6.2), \quad N(0), I(0). \end{aligned} \tag{6.11}$$

A first-best optimum would effectively entail choosing h and $\mathbf{x}$ directly (i.e., $\mathbf{p} = [h\ \boldsymbol{x}]$), whereas the choice of $\mathbf{p}$ would only imperfectly control the human choices for h and $\boldsymbol{x}$ in a second-best setting. We focus our attention on the first-best case, for simplicity, but we note some potential implications of second-best management in our discussion below.

The current-value Hamiltonian associated with problem (6.11), in a first-best setting with h and $\mathbf{x}$ as the control variables, is

$$\begin{aligned} H = {} & \mathrm{SNB}(N, I, h, \mathbf{x}) + \lambda[F(N, \mathbf{x}) - \delta(\mathbf{x})N - \alpha(\mathbf{x})I - hN] \\ & + \mu[\beta(\mathbf{x})C(N, \mathbf{x})[N - I]I/N - (\delta(\mathbf{x}) + \alpha(\mathbf{x}))I - hI], \end{aligned} \tag{6.12}$$

where λ is the co-state variable associated with N, representing the marginal value of a small increase in population N, and μ is the co-state variable associated with I, representing the marginal cost of a small increase in population I. Assuming population N is generally valuable for harvests, ecotourism, or existence, then $\mathrm{SNB}_N > 0$ and we expect $\lambda > 0$. But assuming infected animals are costly to society, $\mathrm{SNB}_I < 0$, then $\mu < 0$.

The necessary conditions for optimality (discussed below) are both necessary and sufficient when the Hamiltonian is concave in the states and controls (Leonard and van Long 1993, Theorem 4.6.2, p. 163). If the Hamiltonian is not concave, then the standard optimality conditions are not sufficient for an optimum and multiple optimality candidates may arise, with some solutions possibly being only locally optimal (or even minima). A negative co-state often raises concerns about nonconvexities, though it is not just the co-state that matters. If $\lambda > 0$, then concavity requires the equation of motion for N to be concave. This condition is satisfied assuming $F - \delta N$ is concave in N (e.g., logistic growth). Since $\mu < 0$, concavity of the Hamiltonian is only ensured if the equation of motion for I is convex. This depends on the transmission function, which is nonlinear in both states. Transmission is generally assumed to occur somewhere within the spectrum of frequency and density dependence (McCallum et al. 2001), and it is readily verified that either of these specifications is concave in the states. Hence, the Hamiltonian is not guaranteed to be concave and the possibility of multiple solutions arises. Note that the Hamiltonian may also be nonconcave in the choice variables $\mathbf{x}$, depending on how these choices affect growth and transmission relations. Nonconvexity issues may be exacerbated in a second-best setting with h and $\mathbf{x}$ being nonlinear feedback responses of N, I,and $\mathbf{p}$.

The following necessary conditions for problem (6.11) are used to determine the efficient allocation of control efforts over time and across the various control activities:

$$\frac{\partial H}{\partial h} = 0 \Rightarrow \mathrm{SNB}_h(N, I, h, \mathbf{x}) = \lambda N + \mu I, \tag{6.13}$$

$$\begin{aligned}\frac{\partial H}{\partial x_i} = 0 \Rightarrow \mathrm{SNB}_{x_i}(N, I, h, \mathbf{x}) = & -\lambda[F_{x_i}(N, \mathbf{x}) - \delta_{x_i}(\mathbf{x})N] + \mu[\alpha_{x_i}(\mathbf{x})I] \\ & - \mu[\beta_{x_i}(\mathbf{x})C(N, \mathbf{x}) + \beta(\mathbf{x})C_{x_i}(N, \mathbf{x})][N - I]I/N \\ \forall x_i \in \mathbf{x}, &\end{aligned} \tag{6.14}$$

$$\begin{aligned}\dot{\lambda} = & \rho\lambda - \mathrm{SNB}_N(N, I, h, \mathbf{x}) - \lambda[F_N(N, \mathbf{x}) - \delta(\mathbf{x}) - h] \\ & - \mu[C_N(N, \mathbf{x})[N - I] + C(N, \mathbf{x})(I/N)]\beta(\mathbf{x})I/N,\end{aligned} \tag{6.15}$$

$$\begin{aligned}\dot{\mu} = & \rho\mu - \mathrm{SNB}_I(N, I, h, \mathbf{x}) - \lambda\alpha(\mathbf{x}) + \mu[\delta(\mathbf{x}) + \alpha(\mathbf{x}) + h] \\ & - \mu[C_N(N, \mathbf{x})I + C(N, \mathbf{x})([N - I]/N)]\beta(\mathbf{x})[N - I]/N,\end{aligned} \tag{6.16}$$

$$\lim_{t\to\infty} \mathrm{e}^{-\rho t}\lambda(t)S(t) = 0, \quad \lim_{t\to\infty} \mathrm{e}^{-\rho t}\mu(t)I(t) = 0. \tag{6.17}$$

Conditions (6.13) and (6.14) are first-order conditions that equate the marginal current-period net benefits (costs) of an action (the left hand side, or LHS) with the marginal intertemporal costs (benefits) of that action (the right hand side, or RHS). Conditions (6.15) and (6.16) are arbitrage conditions governing how the co-states evolve over time, and condition (6.17) represents the transversality conditions for problem (6.11).

Long-Run Stability

It is difficult to go into significant detail about the optimality conditions without greater specification of the model. However, some insight is gained by looking at conditions (6.13) and (6.14). The RHS of each condition contains both co-states and both states. Accordingly, conditions (6.13) and (6.14) must be solved simultaneously to yield expressions for the optimal control functions, which depend on both co-state variables and, in particular, *both* state variables: $h(N, I, \lambda, \mu)$ and $\mathbf{x}(N, I, \lambda, \mu)$. The dynamic system can be rewritten using these relations:

$$\begin{aligned}\dot{N} &= F(N,\mathbf{x}(N,I,\lambda,\mu)) - \delta(\mathbf{x}(N,I,\lambda,\mu))N - \alpha(\mathbf{x}(N,I,\lambda,\mu))I - h(N,I,\lambda,\mu)N \\ &= \Phi(N,I,\lambda,\mu),\end{aligned} \tag{6.18}$$

$$\begin{aligned}\dot{I} &= \beta(\mathbf{x}(N,I,\lambda,\mu))C(N,\mathbf{x}(N,I,\lambda,\mu))[N-I]I/N \\ &\quad -(\delta(\mathbf{x}(N,I,\lambda,\mu)) + \alpha(\mathbf{x}(N,I,\lambda,\mu)))I - h(N,I,\lambda,\mu)I] \\ &= \Gamma(N,I,\lambda,\mu).\end{aligned} \tag{6.19}$$

Though economists often look for steady-state solutions to be optimal, Wirl (1992, 1995) indicates that we should not necessarily expect steady states when managing multiple, interacting species. In particular, and in the context of our model, the following condition is necessary for the optimality of a cyclical strategy (Wirl 1992):

$$\frac{\partial\Phi}{\partial I}\frac{\partial\Gamma}{\partial N} < 0. \tag{6.20}$$

Condition (6.20) indicates one state variable generates positive spillovers on the other, while the reverse is true for the second state variable. Wirl (1992) uses a predator-prey system as an example of systems likely to satisfy (6.20), and our disease ecology model has many similarities to predator-prey models. Using (6.18) and (6.19), we can derive

$$\frac{\partial \Phi}{\partial I} = -\alpha - \frac{\partial h}{\partial I}N + \sum_i [F_{x_i} - \delta_{x_i}N - \alpha_{x_i}I]\frac{\partial x_i}{\partial I}, \tag{6.21}$$

$$\frac{\partial \Gamma}{\partial N} = \left[C_N[N-I] + C\left(\frac{I}{N}\right)\right]\frac{\beta I}{N} - \frac{\partial h}{\partial N}I + \sum_i \left[(\beta_{x_i}C + \beta C_{x_i})\frac{[N-I]I}{N} - (\delta_{x_i} + \alpha_{x_i})\right]\frac{\partial x_i}{\partial N}. \tag{6.22}$$

Absent the $\partial h/\partial j$ and $\partial x_i/\partial j$ terms ($j = I,N$), the relations in (6.21) and (6.22) are of opposite signs ($\partial\Phi/\partial I<0$ and $\partial\Gamma/\partial N>0$) and therefore satisfy (6.20). This means the potential for a cyclical optimum also exists when the $\partial h/\partial j$ and $\partial x_i/\partial j$ terms are included. Consider the $\partial h/\partial j$ terms, each of which we might expect to have a positive sign (i.e., there are greater incentives to harvest when animals are more abundant, as harvest costs are generally reduced in this case, and there may be greater incentives to harvest when there are more infected animals causing damages).[9] Assuming this is the case, then $\partial\Phi/\partial I$ remains negative, whereas the sign of $\partial\Gamma/\partial N$ becomes ambiguous. $\partial\Gamma/\partial N$ will remain positive if an increase in N has a bigger impact on disease transmission than on harvest responses, and it will switch signs otherwise. Harvest controls could therefore have a stabilizing effect. However, any increase in stability due to harvests could be offset by the other choices, **x**, which have ambiguous impacts on the sign of (6.22).

Several studies have found cyclical strategies, in which cycling occurs around an interior equilibrium, to be optimal when wildlife harvests and supplemental feeding are nonselective (Horan and Wolf 2005; Fenichel and Horan 2007b; Horan et al. 2008). A cycle is the result of the system overshooting and undershooting the equilibrium as a result of having nonselective controls. When I is large, marginal damages from the disease are large and there are incentives to reduce N below the host-density threshold. When I is small, marginal damages from the disease are small and there are incentives to increase N, even if that results in increases in I. Interestingly, Fenichel and Horan (2007b) found that controlling harvests alone (so that h is a choice variable while **x** is not) did optimally have a stabilizing effect, so that the system optimally moved to a steady state. However, cycles were found to be optimal when supplemental feeding (**x**) was also included as a control. In this case, the presence of the additional control had a de-stabilizing effect – even though the use of this control improved welfare. These results highlight the need to explore the dynamics of the optimal solution and to not simply focus on steady-state outcomes.

[9] For instance, if we assume harvest benefits take the form $B(hN) - ch$, where B is willingness to pay and costs correspond to Schaefer-type costs (Clark 2005) since h is a rate, then $\partial h/\partial N > 0$ and $\partial h/\partial I > 0$.

We should emphasize that instability is not necessarily due to the use of nonselective controls. For instance, we argued above that instability could be overcome if the term $\partial h/\partial N$ in (6.22) was positive and sufficiently large. Now suppose harvests were selective and that harvests of infected animals yielded only costs, $c(h_I I, I)$, where h_I is the harvest rate for infected animals. The first-order condition in this case would yield $-c_h(h_I I, I) = \lambda + \mu$, as harvests of infected animals impact the growth of both N and I.[10] This results in $h_I = h_I(I, \lambda, \mu)$, which means the term $\partial h/\partial N$ in (6.22) now vanishes. Absent other controls, **x**, the result is that (6.22) is unambiguously positive and instability remains a concern.

Targeting Controls to Improve Efficiency

Though perhaps stabilizing in some instances, the nonselective application of a particular control is generally efficiency reducing compared to selective application of the same control. Accordingly, there are incentives to seek out alternative controls that are better targeted at managing pathogen risks. Diagnostic tests that reveal an animal's disease status are sometimes available to improve targeting, particularly for livestock. This is not a perfect option, as the tests may be costly, subject to error, and the results may not be immediately available. But it may be an economically desirable option in some instances. Bicknell et al. (1999) examine the problem of bovine tuberculosis (bTB) in a wildlife-livestock system involving cattle and possums, and find adopting a test-and-slaughter program for cattle represents a selective and hence a low-cost control measure relative to nonselectively controlling possum populations. Indeed, diagnostic field tests are only available for a few wildlife disease problems (e.g., brucellosis in bison: Bienen and Tabor 2006; and devil facial tumor disease (DFTD) in Tasmanian devils: STTD 2009; Platt 2009). The result is that livestock harvests of anything but valuable breeding stock are semiselective, and wildlife harvests are generally nonselective with respect to disease status. Therefore, population controls will likely impact healthy and infected animals, culling disproportionately more healthy animals if prevalence is low.

Biosecurity efforts that reduce transmission risks across farms, or from wildlife to farms, directly target transmission risks and therefore may be an efficient alternative to nonselective population controls. Horan et al. (2008) analyze a wildlife-livestock disease system where transmission within livestock herds can be reduced via a selective test-and-slaughter program for livestock, by reducing the susceptible livestock population to reduce the number of animals at risk of infection, by using nonselective wildlife controls to reduce the force of infection from wildlife, and by implementing biosecurity such as fencing that limits wildlife-livestock contacts.

[10] Suppose harvests are selective and that SNB takes the form $B(S, I, h_S) - c(h_I I, I)$. Then the current-value Hamiltonian is $H = B(S, I, h_S) - c(h_I I, I) + \lambda[F(N) - \delta N - \alpha I - h_S[N - I] - h_I I] + \mu[\tau(N, I) - (\delta + \alpha)I - h_I I]$.

They find biosecurity is preferred to both wildlife controls and reductions in healthy livestock. Biosecurity reduces the expected damages from infected wildlife, which reduces the incentives to implement wildlife controls and to mitigate on-farm disease risks in other ways (e.g., by reducing healthy livestock populations). Farm size and location may also affect the efficiency of biosecurity (Hennessy 2007a, b), as may market access (Hennessy et al. 2005; Horan et al. 2008), as these features have a bearing on producer vulnerabilities to infection.

Improved signaling that reduces information limitations about animal health status can also improve risk management. Hennessy et al. (2005) find improved signaling about animal health can reduce infection risks and improve efficiency in the context of animal trade. Fenichel and Horan (2007a) show wildlife disease management can be improved using signals observed in nature. Specifically, deer behavior affecting bTB risks varies based on an animal's sex, with male deer being significantly more likely to be infected than female deer. As a deer's sex is observable, sex-based population management can improve the likelihood that infected animals will be culled, improving efficiency. Fenichel and Horan (2007a) find there are greater incentives to target male deer because male deer are more likely to become infected and spread the disease. However, as the host-density threshold for male deer is initially quite low, significant and costly reductions in the male population would be required for infection levels to decline. Fenichel and Horan find population controls that reduce the disease incidence among females also reduces the infection pressures on males. This has the effect of increasing the male host-density threshold, which means male infection levels will decline at larger male population levels. Even though females are a smaller source of infection, targeting some controls toward females can help reduce disease control costs for males.

Vaccination, when available, is another means of targeting infection risks. However, vaccination can result in trade restrictions since vaccinated animals may produce antibodies that can make them appear to be infected or exposed to infection (USDA-APHIS 2002). When there is a threat of trade restrictions, Mahul and Gohin (1999) indicate culling infected herds may be a more efficient option, at least initially. They find managers should wait to see how bad the outbreak might become before switching strategies from culling to vaccination. In related work, Mahul and Durand (2000) illustrate emergency vaccination in response to a foot and mouth disease (FMD) outbreak in Brittany is only optimal if the additional cost of delayed slaughter of vaccinated animals (required to re-open trade) is offset by avoided losses from a shorter duration epidemic. These analyses emphasize the need to focus on welfare effects and not the number of infected individuals.

Eradication May Not Be Optimal

Though disease eradication is often the focus of disease ecology research (e.g., Roberts and Heesterbeek 2003) and of disease management authorities (Bicknell et al. 1999), the optimality of eradication is an empirical matter. To be optimal, the marginal benefits of eradication should exceed the marginal costs of eradication (McInerney et al. 1992). But the marginal damages of a disease generally decline while marginal control costs often increase as infection levels diminish.

The optimality of eradication depends on many factors, including whether infection generates large fixed damage costs (e.g., a ban on livestock exports), the available control options and their effectiveness, epidemiological characteristics, productivity impacts, and the current state of infection (Barrett and Hoel 2007). For instance, Bicknell et al. (1999) find eradicating bTB from cattle and possums in New Zealand is suboptimal. Eradication would only be optimal if possums were not part of the disease system: in that case, the test-and-slaughter program represents a selective, low-cost control measure.

Horan and Wolf (2005) and Fenichel and Horan (2007b) examine optimal management of bTB in Michigan deer that infect cattle. Horan and Wolf (2005) find eradication may be optimal if there are large fixed costs associated with infection, such as if livestock producers in affected areas lose market access. Otherwise, bTB would optimally be managed as an endemic disease. Horan et al. (2008) explicitly incorporate the cattle sector and find the incentives for eradication decline. The reason is that biosecurity and on-farm mitigation efforts can reduce harm to the cattle sector, which in turn reduces the incentives for controlling bTB among deer.

Fenichel and Horan (2007a) find eradication becomes optimal, even without fixed costs, when sex-specific controls are better able to target infected deer. This allows for more efficient management of deer populations and host-density thresholds, as described above, thereby reducing the marginal cost of control at the point of eradication. Though eradication is optimal, this outcome is only achieved after considerable time.

Conclusion

Bioeconomic modeling is an important tool for examining the management of animal disease problems. In this chapter, we have provided a general introduction to the issues associated with explicitly including human economic decisions into animal disease systems. Whereas traditional epidemiological models take human behaviors as fixed, we have shown these matter. We have focused on social planner problems, which represent a particular class of bioeconomic models associated with animal disease, but it is important to recognize that human behavioral feedbacks usually matter even without socially optimal decisions. In the case of wildlife, recreational hunters generally respond to changes in population levels. Also, land-use decisions are sometimes made with impacts to wildlife populations in mind. Even if these population-based feedback responses do not directly reflect changes in disease risks, they will affect disease dynamics. For most livestock systems, feedback responses will directly reflect some (though perhaps not all) changes in disease risks. Indeed, producers are forward looking and often will make biosecurity investments to protect their own animals from disease risks, thereby affecting disease dynamics.

The bioeconomic literature on animal diseases has only scratched the surface of problems needing researched. Few studies incorporate multiple host species and spatial interactions, though these are important features of many animal disease problems. The transmission of animal diseases to humans has also not been addressed, though zoonotics remain one of the biggest risks to humans. These problems are complex and will require the development of new theories and methods. But to gain acceptance, predictive models that can show the value of human behavioral feedbacks are likely needed in addition to management models. Moreover, estimates that illustrate differences in traditional ecological metrics from jointly determined epidemiological–economic metrics would be useful to better illustrate the role of behavior in epidemiological processes.

References

Anderson, RM, and RM May. 1979. Population Biology of Infectious Diseases: Part I. *Nature* 280: 361–67.

Anderson, RM, and RM May. 1986. The invasion, persistence, and spread of infectious disease within animal and plant communities. *Philosophical Transactions of the Royal Society B: Biological Science* 314: 533–70.

Barlow, N.D. "Critical Evaluation of Wildlife Disease Models", in *Ecology of Infectious Diseases in Natural Populations* (eds. B.T. Grenfell & A.P. Dobson), Cambridge University Press, New York, 1995.

Barlow, ND, JM Kean, NP Caldwell, and TJ Ryan. 1998. Modelling the regional dynamics and management of bovine tuberculosis in New Zealand cattle herds. *Preventive Veterinary Medicine* 36(1): 25–38.

Barrett, S, and M Hoel. 2007. Optimal disease eradication. *Environmental and Development Economics* 12: 627–52.

Bicknell, KB, JE Wilen, and RE Howitt. 1999. Public policy and private incentives for livestock disease control. *Australian Journal of Agricultural and Resource Economics* 43: 501–21.

Bienen, L, and G Tabor. 2006. Applying an ecosystem approach to brucellosis control: can an old conflict between wildlife and agriculture be successfully managed? *Frontiers in Ecology and the Environment* 4: 319–27.

Burrows, R. 1992. Rabies in Wild Dogs. *Nature* 359: 277–277.

Burrows, R., H. Hofer, and M.L. East. 1994. Demography, Extinction and Intervention in a Small Population: the Case of the Serengeti Wild Dogs. *Proceedings: Biological Sciences* 256: 281–292.

Chowell, G. and F. Brauer 2009. The basic reproduction number of infectious diseases: computation and estimation using compartmental epidemic models, in *Mathematical and Statistical Estimation Approaches in Epidemiology* (G. Chowell et al., eds), Springer.

Clark, CW. 2005. *Mathematical Bioeconomics Optimal Management of Renewable Resources Second Edition*. Hoboken: Jon Wiley & Sons.

Daszak, P, AA Cunningham, and AD Hyatt. 2000. Emerging infectious diseases of wildlife-threats to biodiversity and human health. *Science* 287: 443–49.

Diekmann, O, JAP Heesterbeek, and JAJ Metz. 1990. "On the definition and computation of the basic reproduction ratio R_0 in models of infectious disease in heterogeneous populations." *Journal of Mathematical Biology* 28:365–82.

Dobson, A. 2004. Population dynamics of pathogens with multiple hosts species. *The American Naturalist* 164: s64–78.

Dobson, A, and J. Foufopoulos. 2001. Emerging infectious pathogens of wildlife. *Philosophical Transactions of the Royal Society B: Biological Science* 356: 1001–12.

Ewing, J. 2005. The Mesoamerican Biological Corridor: a bridge across the Americas. *EcoWorld Magazine*. Dec. 19. available at http://www.ecoworld.com/home/articles2.cfm?tid=377 (downloaded November 3, 2006).

Fenichel, EP, and RD Horan. 2007a. "Gender-based harvesting in wildlife disease management", *American Journal of Agricultural Economics* 89: 904–20.

Fenichel, EP, and RD Horan. 2007b. Jointly-determined ecological thresholds and economic trade-offs in wildlife disease management. *Natural Resource Modeling* 20: 511–47.

Fenichel, E.P., R.D. Horan, and G. Hickling. 2010. "Management of Infectious Wildlife Diseases: Bridging Conventional and Bioeconomic Approaches", *Ecological Applications*, 20: 903–914.

Fulford GR, MG Roberts, and JAP Heesterbeek. 2002. The metapopulation dynamics of an infectious disease: tuberculosis in possums. *Theoretical Population Biology* 61: 15–29.

Gichohi, HW. 2003. *Direct payments as a mechanism for conserving important wildlife corridor links between Nairobi National Park and its wider ecosystem: the Wildlife Conservation Lease program*, Presented at Vth World Parks Congress, Sustainable Finance Stream, Durban South Africa.

Gubbins, S., S. Carpenter, M. Baylis, J.L.N. Wood, and P.S. Mellor. 2008. Assessing the risk of bluetongue to UK livestock: uncertainty and sensitivity analyses of a temperature-dependent model for the basic reproduction number. *Journal of the Royal Society Interface* 5:363–371.

Hartup, BK, GV Kollias, and DH Ley. 2000. Mycoplasmal conjunctivitis in songbirds from New York. *Journal of Wildlife Diseases* 32: 257–64.

Heesterbeek, JAP, and MG Roberts. 1995. Mathematical models for microparasites of wildlife. In *Ecology of Infectious Diseases in Natural Populations*. eds. BT Grenfell, AP Dobson. New York: Cambridge University Press.

Heffernan, JM, RJ Smith, and LM Wahl. 2005. Perspectives on the basic reproductive ratio. *Journal of the Royal Society Interface* 2: 281–93.

Hennessy, D.A. "Behavioral incentives, equilibrium endemic disease, and health management policy for farmed animals", *American Journal of Agricultural Economics* 89(2007a): 698–711.

Hennessy, D.A. "Biosecurity and Spread of an Infectious Animal Disease", *American Journal of Agricultural Economics* 89(2007b): 1226–1231.

Hennessy, D.A., J. Roosen, and H.H. Jensen. "Infectious Disease, Productivity, and Scale in Open and Closed Animal Production Systems", *American Journal of Agricultural Economics* 87 (2005): 900–917.

Hethcote, HW. 2000. The mathematics of infectious diseases. *SIAM Review* 42: 599–653.

Holt, RD, AP Dobson, M Begon, RG Bowers, and EM Schauber. 2003. Parasite establishment in host communities. *Ecology Letters* 6: 837–42.

Horan, RD, and CA Wolf. 2005. The economics of managing infectious wildlife disease. *American Journal of Agricultural Economics* 87: 537–51.

Horan, RD, CA Wolf, EP Fenichel, and KH Mathews Jr. 2005. Spatial management of wildlife disease. *Review of Agricultural Economics* 27: 483–90.

Horan, RD, CA Wolf, EP Fenichel, and KH Mathews Jr. 2008. Joint management of wildlife and livestock disease. *Environmental and Resource Economics* 41: 47–70.

Kaiser, J. 2001. Conservation biology: bold corridor project confronts reality. *Science* 21: 2196–99.

Kermack, WO, and AG McKendrick. 1927. Contributions to the mathematical theory of epidemics, part I. *Proceedings of the Royal Society of Edinburgh A* 115: 700–21.

Lanfranchi, P, E Ferroglio, G Poglayen, and V Guberti. 2003. Wildlife vaccination, conservation and public health. *Veterinary Research Communications* 27:567–74.

Leonard, D. and N. van Long. 1993. *Optimal Control Theory and Static Optimization in Economics*, Cambridge: Cambridge University Press.

Levins, R. 1969. Some demographic and genetic consequences of environmental heterogeneity for biological control. *Bulletin of the Entomological Society of America* 15: 237–40.

Mahul, O., and B. Durand. "Simulated economic consequences of foot-and-mouth disease epidemics and their public control in France." *Preventive Veterinary Medicine* 47, no. 1-2 (2000): 23–38.

Mahul, O., and A. Gohin. "Irreversible decision making in contagious animal disease control under uncertainty: an illustration using FMD in Brittany." *European Review of Agriculture Economics* 26 (1999): 39–58.

McCallum, H, N Barlow, and J Hone. 2001. How Should Pathogen Transmission be Modelled? *Trends in Ecology & Evolution* 16:295–300.

McInerney, J.P., K.S. Howe, and J.A. Schepers. "A Framework for the Economic Analysis of Disease in Farm Livestock", *Preventive Veterinary Medicine* 13(1992): 137–154.

Platt, J. 2009. New test creates hope for cancer-plagued Tasmanian devils. *Scientific American* April 3. Accessed October 27, 2009, available online at http://www.scientificamerican.com/blog/post.cfm?id=new-test-creates-hope-for-cancer-pl-2009-04-03.

Roberts, MG, and JAP Heesterbeek. 2003. A new method for estimating the effort required to control an infectious disease. *Proceedings of the Royal Society of London B* 270:1359–64.

Rowthorn, RE, R Laxminarayan, and CA Gilligan. 2009. Optimal control of epidemics in metapopulations. *Journal of the Royal Society Interface* 6: 1135–44.

Sanchirico, JN, and JE Wilen. 1999. Bioeconomics of spatial exploitation in a patchy environment. *Journal of Environmental Economics and Management* 37: 129–50.

Save the Tasmanian Devil (STTD). 2009. *Tasmanian Devil Facial Tumor Disease*. Online at: http://tassiedevil.com.au/disease.html. Accessed June 17, 2009.

Schmitt, SM, DJ O'Brien, CS Bruning-Fann, and SD Fitzgerald. 2002. Bovine tuberculosis in Michigan wildlife and livestock. *Annals of the New York Academy of Sciences* 969: 262–68.

Smith, G.C., and C.L. Cheeseman. 2002. A Mathematical Model for the Control of Disease in Wildlife Populations: Culling, Vaccination and Fertility Control, *Ecological Modeling* 150: 45–53.

United States Department of Agriculture, Animal and Plant Health Inspection Service (USDA-APHIS). 2002. *Foot-and-Mouth Disease Vaccine Factsheet*, January.

Watry, M.K., LL Wolfe, JG Powers, and M.A. Wild. 2004. Comparing Implementation of a Live Test-and-Cull Program for Chronic Wasting Disease in Wildland and Urban Settings. *Proceedings: Health and Conservation of Free-Ranging Wildlife, Joint Conference of the American Association of Zoo Veterinarians, American Association of Wildlife Veterinarians, and Wildlife Disease Association*, San Diego, CA.

Wirl, F. 1992. Cyclical strategies in two-dimensional optimal control models: necessary conditions and existence. *Annals of Operations Research* 37: 345–56.

Wirl, F. 1995. The cyclical exploitation of renewable resource stocks may be optimal. *Journal of Environmental Economics and Management* 29:252–261.

Wobeser, G. 2002. Disease management strategies for wildlife. *Revue scientifique et technique Office International des Epizooties* 21: 159–78.

Wolfe,L.L., M.W. Miller, and E.S. Williams. 2004. Feasibility of "Test and Cull" for Managing Chronic Wasting Disease in Urban Mule Deer. *Wildlife Society Bulletin* 32: 500–505.

Chapter 7
Border Inspection and Trade Diversion: Risk Reduction vs. Risk Substitution

Qiong Wang, Eli P. Fenichel, and Charles A. Perrings

Introduction

International trade increasingly brings previously separated geographical regions into contact with one another and increases the frequency of those contacts. These trends bring many benefits to the trading partners involved, but increasing international trade also facilitates the spread of pathogens and increases disease risks. The rapid growth of trade, transport, and travel across national borders has increased the frequency of introduction, establishment, and spread of invasive infectious pathogens (Jones et al. 2008). The development of new trade pathways and the growth in the number and volume of commodities traded increase the likelihood that novel infectious pathogens are introduced to importing or stop-over countries. The growth in trade volumes has increased the risk that introduced pathogens establish and spread, because it has increased the frequency with which infectious pathogens are introduced (Cassey et al. 2004; Dalmazzone 2000; Semmens et al. 2004). Other factors such as the bioclimatic similarities between trading partners, the vulnerability of ecosystems in the importing countries, and risk management policies adopted by both importing and exporting countries also influence the risks of invasive infectious pathogens (Wiens and Graham 2005).

The impact of infectious diseases on human wellbeing may be direct or indirect. Infectious pathogens cause indirect effects by successfully establishing in novel environments, which can lead to irreversible functional and structural change in local ecosystems including extinction of species (Collins and Storfer 2003; Gurevitch and Padilla 2004; Lips et al. 2006; Williamson 1996). For example, avian pox and

Q. Wang
School of Sustainability, Arizona State University, Tempe, AZ, USA

E.P. Fenichel (✉) • C.A. Perrings
School of Life Sciences and ecoServices Group, Arizona State University, Tempe, AZ, USA
e-mail: eli.fenichel@asu.edu

D. Zilberman et al. (eds.), *Health and Animal Agriculture in Developing Countries*, Natural Resource Management and Policy 36, DOI 10.1007/978-1-4419-7077-0_7,

avian malaria significantly contributed to the extinction of endemic Hawaiian birds following the pathogens introduction from Asia (Simberloff 1996). Chytrid fungus, an emerging infectious disease of amphibians, is playing a major role in the global decline of amphibian populations (Daszak et al. 1999).

Many emerging diseases are zoonoses, they can be transmitted from non-human to human hosts, and cause direct impacts on people (Cleaveland et al. 2001). Examples include H5N1 (Kilpatrick et al. 2006), West Nile virus (Lanciotti et al. 1999), and SARS (Guan et al. 2003; Li et al. 2005). In addition to the potential for zoonoses, animal pathogens impose substantial economic costs on agricultural producers. Invasive pathogens have caused substantial economic losses in agriculture, forestry, and other segments of the U.S. economy (Pimentel et al. 2000, 2002, 2005). Losses include both the direct costs of control, and the forgone benefits of trade interruptions. The outbreak of foot-and-mouth disease (FMD) in the U.K. in 2001, for example, caused the European Union to ban Great Britain from exporting livestock, meat, and animal products to non-British member countries, resulting in substantial trade losses. By the time the outbreak had ended, FMD had cost the U.K. £8 (approximately $16) billion. Thirty-nine percent of the costs, £355 (approximately $710) million, were borne by agricultural producers (DEFRA 2002). Losses due to trade restrictions and business interruption can motivate disease eradication programs even in cases when the direct productivity losses are not great (Horan and Wolf 2005). The potential economic damage associated with animal pathogens motivates significant investment in mitigation strategies that prevent the introduction of pathogens or control their spread. For example, the U.S. spends approximately $106 million each year on the inspection and control (detection, eradication, etc.) of bovine tuberculosis costs (Wolf et al. 2008).

The primary source of novel animal disease risk is international trade in animal products (Otte et al. 2007; Pavlin et al. 2009). Other things being equal, higher trade volumes increase the cumulative probability that infectious animal pathogens appear at the border, while higher levels of screening effort at the border reduce the probability that pathogens that appear at the border are dispersed within the country (Perrings et al. 2010). To address these risks, importing countries typically adopt a range of border control measures extending from outright bans on trade in risky goods, or with risky trading partners, to inspection and interception regimes designed to reduce the likelihood of the introduction of infectious pathogens. Such unilateral defensive measures are the only measures allowed under the General Agreement on Tariffs and Trade (GATT) and the 1994 Sanitary and Phytosanitary (SPS) Agreement (Perrings et al. 2010). In all cases, the cost of these measures is borne by the trading parties, and hence directly affects the cost of imported goods. If we define the landed postinspection cost of imports to be the c.i.f.i. price – cost, insurance, freight, inspection – then any positive level of inspection (and interception) implies that the c.i.f.i. price of imports is strictly greater than the c.i.f. price.

There is a considerable body of work addressing the appropriate level of inspection and interception effort that relates the direct costs of inspection to the expected damage costs of introductions (Leung et al. 2002; Caley et al. 2006; Batabyal 2007; Finnoff et al. 2007; Keller et al. 2007; Springborn et al. 2010). Others have focused on the design of institutional measures to reduce invasion risk. Perrings et al. (2010) show that the risks of invasive infectious pathogens depend both on factors that influence trade volumes and the SPS measures undertaken by trading partners. Horan and Lupi (2005) develop a model of tradable risk permits for invasive species. They show that tradable risk permit systems potentially have a greater risk reducing affect than uniform quotas or taxes do to the heterogeneous risks that different sources provide. McAusland and Costello (2004) analyze the effectiveness of potential policies such as tariffs and inspection and interception to combat the potential risk of invasive infectious pathogens introduction. They conclude that tariffs do not always safeguard against the introduction of invasive infectious pathogens. Kohn and Capen (2002) show that institutional policies discouraging international trade may increase environmental damage by invasive infectious pathogens if change in the composition and volume of international trade unexpectedly creates new opportunities for invasive infectious pathogens.

There is relatively little work on the effect of direct risk-reducing measures such as inspection and interception on the structure of trade. Inspection and interception changes the cost of importing good, and any change in the cost of imports will induce a response on the part of importers. Ameden et al. (2007) present a model of strategic firm behavior where importers respond to inspection regulations. Their model suggests that importing firms may decrease imports and engage in avoidance behavior (e.g., smuggling or port-shopping – the process of importing through ports with less stringent regulation). Many third-party logistics providers maintain databases to help identify least cost and least time ports, attributes importers might uses as proxies for regulations. These data are intended to identify the most efficient shipping routes from the firm's perspective, which likely means avoiding more stringent regulations. Ameden et al. (2007) also suggest that firms may engage in lower protection investments with more stringent inspection because the expected profit from importing goods is reduced by the expect inspection costs. Empirical work also confirms that firms respond to different border inspection effort by choosing alternative ports with less border inspection (Ameden et al. 2009). Alternatively, importers with nimble supply chains may respond by sourcing substitute commodities from elsewhere, which will affect the risk of importing an animal disease. Trade diversion affects the structure of risks and since the risks from invasive infectious pathogens depend on the origin of the traded commodities, such trade diversions affect the efficiency of inspection and interception measures adopted by importing countries.

We fill a gap in the literature by considering how inspection and interception regimes induce both supply and demand responses for the commodities affected (Fig. 7.1). We consider the impact of these responses on disease risks, and outline the importance of incorporating market responses into the choice of inspection and interception regimes. We build on the work of Ameden et al. (2007) and specifically show the need to consider the importer's response to border inspection regimes.

Infectious Disease Risks Are an Externality of Trade

Trade creates private benefits for consumers and importing firms. However, at the same time trade may introduce infectious pathogens that cause damage both to the local economy and the natural environment – a public risk. The private nature of

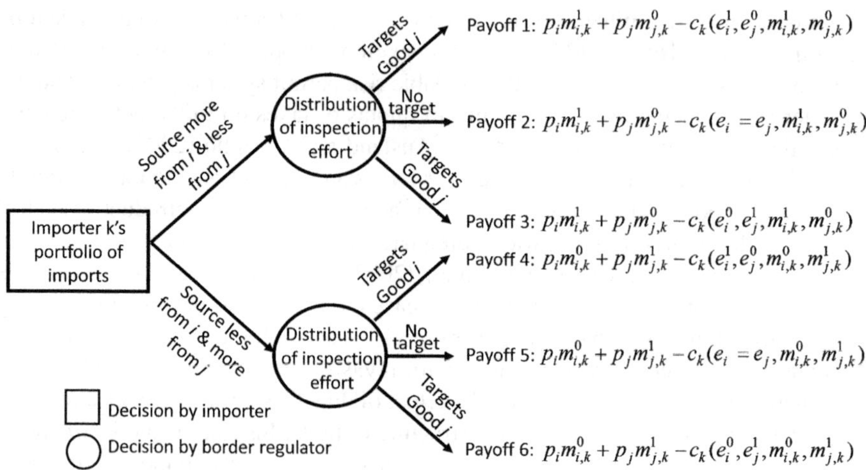

Fig. 7.1 A decision tree illustrating importing firm k's decision: p_i is the market price of good i, $m_{i,k}$ is the import volume of good i by firm k, superscripts indicate more (1) or less (0). The term c is firm k's private costs of importing, where e_i and e_j are costs associated with the inspection. The importer's decision is illustrate with the *box*, and the importer generally takes the border regulator's decision (*circle*) as a stochastic event with a given distribution (though in the long-run the importer may attempt to influence the regulator) and forms expectations over e_i and e_j. The firm chooses the import level to maximize the payoff conditional on its own cost structure and its expectation of inspection. If the regulator recognizes this structure, and if the goal is to minimize total social costs, then the regulator takes the firm behavior as endogenous when solving for its inspection effort program. If the regulator never targets, then the firm only considers payoffs 2 and 5, and the import portfolio is determined by market conditions. When the regulator differentially targets goods (or sources) i and j, then trade diversion will occur and the firm compares the expectation of payoff 1 and 3 with the expectation of payoff 4 and 6 and chooses the allocation that leads to a greater expected payoff

the benefits and public nature of the costs implies that the infectious disease risks of trade will be excessive from a social perspective – the risk is an externality. Private consumers and importers will not generally consider the public risks that their private activities produce.

An importer's choice of supply sources depends on the expected costs of importing (supply costs) – including the costs associated with inspection and interception. Demand for substitute commodities depends on the relative prices of those commodities and the elasticity of substitution between pairs of commodities, which is dependent on consumer preferences. Market interactions drive what ultimately gets imported – including the pathogens that accompany traded goods.

Risk is the product of the probability and consequences of a stochastic event. Risk can be reduced by reducing either the probability of the event or the costs it is expected to impose. Public border protection agencies engage in inspection and interception activities to reduce the probability of pathogen introduction – a risk mitigation strategy (Perrings 2005). Post invasion control or adaptation strategies are often left to other governmental agencies. However, if a public border inspection agency is to balance the social benefits (e.g., consumer and producer surplus) from trade with the public risk of a novel infectious animal disease, then the agency must account both for the actions of other agencies and decentralized importer and consumer behavior. The border inspection agency does not directly control the quantity of goods imported, the price of those goods, or the management response to a pathogen introduction. The border agency can, however, consider the importing firms' incentives. Firms' choices with respect to trade volume, source, and commodity affect the risk of pathogen introduction directly, but the border inspection agency can influence this decision by adjusting inspection effort and hence the cost of various goods or goods from various trading partners.

Springborn et al. (2010) discuss two potential inspection strategies for managing the risks posed by potentially invasive infectious pathogens. The first involves focusing inspection effort on known high-risk imports. There are two potential problems with this strategy. First, it ignores the risk posed by new commodities or new trade routes – the problem of fundamental uncertainty. Second, it is not clear a priori whether inspection is best allocated to low probability but high potential cost imports or to imports for which the cost of inspection is low. Springborn et al.'s second strategy is to allocate a fraction of prevention effort on unknown risky imports, and use a Bayesian updating process to information future allocation of inspection effort. In practice, this second strategy implies a broader inspection effort where learning is explicitly part of the management objective. The second strategy is essentially an adaptive management strategy (Walters 1986). Neither of Springborn et al.'s strategies explicitly account for substitution and trade diversion effects, though both could be adapted to do so. A reasonable hypothesis is that targeted inspection regimes would have greater affect on the structure of trade creating a moving target.

Following Springborn et al. (2010) and Ameden et al. (2007), we explore the case where the border inspection agency's objective is to minimize the social cost of an introduced disease that results from imports of either particular commodities or from particular countries. The social costs comprise the expected damage costs of an introduced pathogen including expected control costs, the direct cost of border inspections, and the expected indirect costs to consumers and importers of the inspected goods in terms of lost surplus. Inspection has two effects on risk. The intended effect is that inspection directly reduces the probability that a novel pathogen will be imported. This constitutes a direct benefit. A secondary effect follows from the fact that the direct cost of border inspection functions like a tax on imports. Importing firms incur additional costs due to inspection (e.g., inspection fees, time delays, and the risk that some goods will test positive and will be rejected). This "tax" affects the market clearing price of the inspected good. This in turn affects the quantity demanded of the inspected good and the quantity demanded of substitute goods. The substitution effect influences trade composition and volume and has implications for the expected damages of associated with substitute commodities or substitute suppliers in the resulting supply and demand shifts generate indirect benefits and costs.

A Model of Risk Substitution

We develop a simple model to illustrate the potential for risk substitution to show how substitution between goods can affect both the cost of inspection regimes, and the risks faced by the importing country. We consider a country that imports goods from two sources (i.e., there are two exporting countries, country $i \in \{1, 2\}$, $j \neq i$ each supplying a single good) to simplify analysis. We assume that importers operate in a competitive market that clears, but neglect the social costs of disease risk. Furthermore, we assume that the goods imported from countries 1 and 2 are substitutes, that countries 1 and 2 have different characteristics θ_i, and that these characteristics affect the probability that the country is the source of an infectious pathogen. The risk to the importing country depends on trade volumes, m_i, inspection effort, e_i, and country-specific factors such as bio-climatic conditions. If the border inspection agency in the importing country had perfect information about the probability that a specific exporter would introduce a pathogen, then the inspection service might be expected to target inspection by exporter – i.e., adopt the first strategy identified by Springborn et al. (2010). However, for the inspection program to be efficient, the inspection regime should account for substitution effects on the volume of imports from other countries. This should affect the distribution of inspection effect across risk sources.

A border inspection agency chooses a vector of inspection efforts, $\mathbf{e}$ for the vector of imported good $\mathbf{m}$ (where $\mathbf{e} = \{e_i, e_j\}, \mathbf{m} = \{m_i, m_j\}$) to minimize the sum of three cost components, C: the direct costs of border inspection and interception; $W(\mathbf{e})$: the forgone producer and consumer surplus from the tax effect of the inspections; $-S(\mathbf{x})$, where $\mathbf{x}$ is a vector of goods consumed, $\mathbf{x} = \{x_i, x_j\}$; and the expected cost or the risk of disease introduction $\gamma(\mathbf{e}, \mathbf{m}, \theta)Z$, where γ is the probability that pathogen is introduced conditional on inspection effect, $\mathbf{e}$ is the volume of imports, $\mathbf{m}$ is the exporter specific characteristics, θ and Z is the cost associated with any control program and disease related damages. Formally, the border inspection agencies problem is

$$\underset{\mathbf{e}}{\text{Min}} C = W(\mathbf{e}) - S(\mathbf{x}) + \gamma(\mathbf{e}, \mathbf{m}, \theta)Z. \tag{7.1}$$

We make two general observations before discussing each term in detail. First, the prices paid by consumers for good x_i must adjust so that the market clears and $x_i = m_i$. The quantity x_i denotes goods demanded, and m_i denotes goods supplied. Second, in practice, optimal detection, prevention, and control of an invasive animal pathogen are a dynamic problem. Prior work (reviewed in Horan et al. 2010 and Horan et al. 2011) and chap 6 in this book (Horan et al. 2012) discuss the optimal control of an animal pathogen once it has been introduced and detected. To show the effect of inspection on the substitution between imported goods, however, we assume that damage costs are fixed at Z, and that these reflect the optimal level of control after introduction of infectious pathogens occurs.[1] This enables us to treat the border agency's problem in static framework. Our approach can, however, easily be extended to the dynamic context using the approach of Springborn et al. (2010).

The direct cost of inspection, $W(\mathbf{e})$, depends on the level of inspection effort invested at the border. We assume that more effort leads to higher costs, $W'(\mathbf{e}) > 0$. The border inspection agency must also consider the losses from any reduction of trade or price effects caused by the increased transactions costs associated with boarder inspection. This is the loss of surplus, $-S(x_i, x_j)$, associated with the "tax" effect of the inspections. To simplify the problem, we assume that there is no local production of substitute goods (though an extension to include a local market would be straightforward).

First, consider the effect of border inspection if surplus were only a function of one good, x_i. Assume that good x_i is imported by a competitive group of price-taking importing firms, such that the marginal firm earns zero economic profit. Inspection regimes may directly charge importers for goods inspected or may impose additional storage costs and longer transit times. This shifts the supply curve for the good upwards and reduces the quantity of good x_i demanded while increasing the price of the good (Fig. 7.2).

In a two good system, the demand for good x_j depends on the prices of the goods i and j. Consumers make decisions about how much of each good to consume as a result of the new prices (Fig. 7.3), the resulting allocation reflecting both income and substitution effects. The income effect results in a loss of surplus; however, the ability of consumers to substitute an alternative good offsets some of the potential utility loss. It follows that, the loss of consumer surplus depends on the elasticity of substitution between the goods. If the goods are perfect substitutes, then only the lower priced good would be consumed, and when the price of good x_i increases society reduces consumption of good x_i up to the point where good x_j becomes the lower priced good. After which good x_i is not consumed and $m_i = 0$. When the elasticity of substitution is finite, the consumption of both goods will change.[2] This change in the demand for good x_j has a feedback effect on the demand for good x_i so there is also a demand shift for good x_i (Fig. 7.4).

The third term in the border inspection agency's problem is the expected damage from introduced pathogens or the risk of disease. The probability of introducing a pathogen from country i is a function of the volume of trade from country i, the characteristics of country i, and inspection effort directed at goods from country i. Let γ_i represent the probability that an infectious pathogen is introduced from country i into the country i, so that $\gamma_i = \gamma_i(m_i, e_i, \theta_i)$, with $\partial\gamma_i/\partial m_i > 0$ and

[1] Even if the likely response were to be suboptimal, the expected damage of the future response that should be determined prior to the inspection effort needs to be determined.

[2] While we focus on substitute goods, if goods and x_i and x_j were complements, then the reduced quantity demanded for x_i would also reduce the demand for x_j potentially creating greater welfare loses, but also potential additional risk reductions.

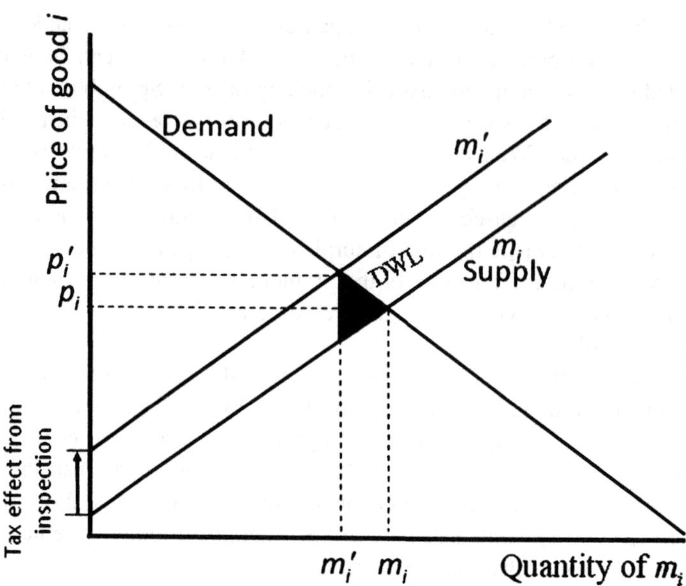

Fig. 7.2 The market for imported good m_i. The line m_i indicates the import quantities and prices of good m_i at an initial level of inspections. The line m_i' indicates the import quantities and prices of good m_i in with a more stringent level of border. With inspections the domestic market price of m_i increases from p to p'. The loss of produced and consumer surplus is illustrated by the *black triangle*

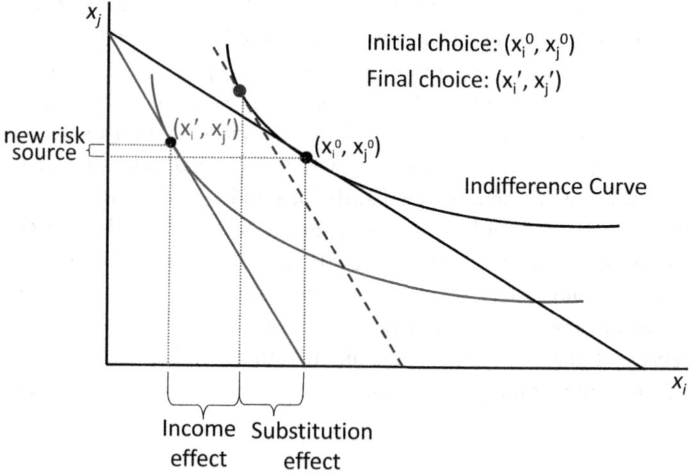

Fig. 7.3 How a representative consumer substitutes between goods. Summing over consumers can lead to increases in import volume of other goods. (x_i^0, x_j^0) are a representative customer's initial preferences of trade goods from country i and j. (x_i', x_j') are the final preference choices of good i and j after a price increase due to higher inspection cost. This preference change includes substitution effect and income effect. Trade reduction from country i, x_i will lower the risk of pathogen introduction; while at the same time, trade increase from country j, x_j may cause new risk of pathogen introduction

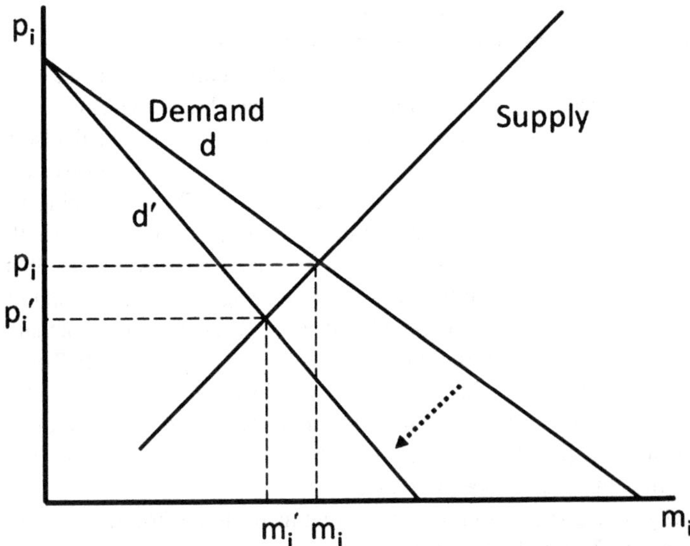

Fig. 7.4 The market for demand good m_i. When inspection on good from country i increase, market price of good i increase too. This leads to customer to look for substitutable goods. And hence demand for x_i shifts from d to d'; correspondingly, trade volume decrease from m_i to m_i'

$\partial\gamma_i/\partial e_i < 0$. We note, however, that $m_i = m_i(\mathbf{e}, \mathbf{m}_{-i}, d(x_i))$ where $\mathbf{m}_{-i}$ is the vector of all goods except i.

If the probability of pathogen introduction by countries i and j is statistically independent, then

$$\gamma(\mathbf{e}, \mathbf{m}, \theta) = \gamma_i + \gamma_j - \gamma_i\gamma_j. \tag{7.2}$$

Since the cost of infection is not dependent of the source, (7.2) indicates that the risk of infection is jointly determined by the volume of trade in and inspection of goods from all sources.

In much of the animal health and invasive species literature, risk is assumed to be independent of human prevention activities or exogenous (e.g., Leighton 2002; Caley et al. 2006; Gubbins et al. 2008). Therefore, if the surplus effects are ignored (i.e., $S(\mathbf{x}) = 0$), then one might reason that the marginal cost of increased inspection, $W'(e_i)$ should be set equal to the marginal benefits from reducing risk, $\partial\gamma/\partial e_i$ for all countries independently. However, the same forces that change demand and supply resulting from the costs imposed by border inspections, also drive changes in $\mathbf{m}$. Assuming market clearing conditions, $\mathbf{m} = \mathbf{x}$, importing firms and consumers cause the market to reach a new equilibrium in terms of quantities, $\mathbf{x}$, and prices, given a level of inspection effort. Therefore, the inspection agency's choice of $\mathbf{e}$ affects both the cost of inspection and consumer surplus, S, and has both a direct and an indirect effect on the probability of pathogen introduction through γ.

It follows that the choice of **e** has three effects on risk. First, an increase in inspection effort directly reduces that probability that an infectious agent makes it into the country. This reduces the risk of infection. Second, an increase in e_i increases the price of x_i decreasing the quantity of x_i demanded, and therefore the volume of good *i* imported (m_i). This decrease in m_i is expected to reduce risk, all else equal. This implies that a given level of risk reduction may require less inspection effort when the price effects of the inspection program are considered. Third, assuming that goods *i* and *j* are substitutes, then an increase in the price of good *i* will increase the demand for good *j* from country *j*. If country *j*'s specific characteristics are such that country *j* has a significantly higher probability of introducing a pathogen conditional trade volume than country *i*, then it is possible that increases in e_i could actually increase γ through γ_j, consistent with the finding of Kohn and Capen (2002). The opposite is also true if country *j* is considerably less risky, since ignoring the substitution effect would result in over-investment in inspection. Therefore, in addition to country specific characteristics, the elasticity of substitution among goods is important for devising efficient border inspection programs.

Formally, it is possible to find the optimal inspection effort directed at good *i* by differentiating (7.1) with respect to e_i.

$$\frac{\partial C}{\partial e_i} = \frac{\partial W}{\partial e_i} - \frac{\partial S}{\partial \mathbf{x}}\frac{\partial \mathbf{x}}{\partial e_i} + Z\frac{\partial \gamma}{\partial e_i} + Z\frac{\partial \gamma}{\partial \mathbf{m}}\frac{\partial \mathbf{m}}{\partial e_i} = 0. \tag{7.3}$$

The first right-hand-side (RHS) term is the marginal direct cost of increasing inspection effort. The second RHS term is the indirect cost associated with a loss of producer and consumer surplus as a result of the costs imposed on the market by the inspection program. The third RHS term is the direct marginal risk reduction associated with an increased probability of intercepting a pathogen with increased effort – a benefit. The final RHS term is the indirect effect on risk of the shifts in market supply and demand as a result of the inspection effort. This term includes own price effects that reduce risk, and cross-price effects that have an ambiguous effect on risk.

Now consider what happens as the number of source countries and goods gets large. If a country has a large number of trading partners and imports a large number of goods, then it is likely that there will be a close substitute for inspected goods. If the highest probability goods are targeted, then we may expect inspections to induce substitution, i.e., inspections reduce the probability that a pathogen crosses the border and the substitution effect reduces the volume of trade in the risky good. If this were the only effect, then we would expect little welfare loss because a large number of substitutes increases the probability of finding a close substitute. However, as the number of trading partners increases, the differences among the θ_i's likely decrease. This implies that the substitute good becomes more likely to also introduce a pathogen (though perhaps a different pathogen). Spillover effects of inspection efforts increase, and targeting risk becomes more complicated with a large number of trading partners and commodities.

Case Studies

To provide a preliminary test of the strength of substitution effects on disease risks we consider three cases. The first of these is bovine tuberculosis in the U.S. Wolf et al. (2008) analyze effect of different levels of bovine tuberculosis inspection on cattle imports from Mexico to the U.S. Bovine tuberculosis is endemic in Mexico, but only occurs in isolated areas in the U.S. that are close to the Mexican border or associated with a wildlife reservoir (e.g., in Michigan where bovine tuberculosis is endemic in wild deer). Wolf et al. showed that more prevention effort leads to an increase of the price of cattle in U.S. markets and changes the quantity demanded of cattle and other goods. They considered multiple levels of border inspection intensity, including a ban on all Mexican cattle. Their analysis showed that a ban on importing Mexico cattle would increase U.S. prices for beef and live steers. A ban lasting 5 years would increase the retail price of live steers by \$6.02 per cwt and the retail price for beef by \$8.78 per cwt. Supply and consumption of beef were projected to decline by 3,135 million pounds. Wolf et al. considered an alternative strategy that allowed imported Mexican cattle subject to increased testing for bovine tuberculosis at the border. This increased the cost of each animal imported, the cost of additional inspection being estimated to be 2.4% of the value of the animal (Federal Register, Vol. 69, No. 138).

A second case is bovine spongiform encephalopathy (BSE). Most U.S. cattle imports come from Mexico and Canada. It is a reasonable hypothesis that a ban on cattle from Mexico would increase demand for Canadian cattle. In December 2003, a cow in the U.S. tested positive for BSE. This cow had been imported from Canada (CDC 2010), which – as of February 2010 – had reported 18 additional cases of BSE (CDC 2010). The detection of a BSE positive cow in the U.S. induced trade restrictions on U.S. cattle that resulted in significant trade losses as well as domestic adaptation costs; these losses were not "disastrous" due to a combination of unique market conditions (Mathews et al. 2006). Nevertheless, it seems feasible that in addition to the costs of additional inspection of Mexican cattle reported by Wolf et al., there is also the potential for increased disease risk from cattle from other sources.

The detection of a single infected cow in the U.S. led almost 15 countries to ban U.S. beef and cattle products. In 2004, U.S. beef export declined dramatically (Fig. 7.5). However, the world beef export volume continued to increase (Fig. 7.5), which suggests that countries substituted beef and cattle products from other sources. Three of the four largest importers of U.S. beef (Japan, South Korea, and Mexico) increased imports from Australia, New Zealand, and South America to substitute for U.S. beef and beef products (Mathews et al. 2006) and total exports from these countries increased (Fig. 7.6). Interestingly, New Zealand has suffered from ongoing bovine tuberculosis infections and South America has suffered from ongoing foot and mouth infection (OIE Handistatus database II, http://www.oie.int/hs2/report.asp?lang=en). Therefore, these changes in sourcing beef and cattle products may have increased the overall exposure of these countries to reinfection from these pathogens. This is particularly true for Mexico, which had experienced

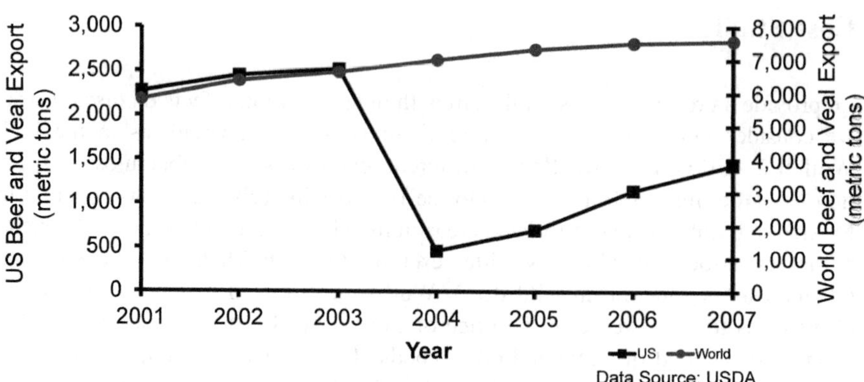

Fig. 7.5 U.S. and World beef and veal exports from 2001 to 2007. Though U.S. beef and veal exports declines dramatically in 2004, world exports increased instead

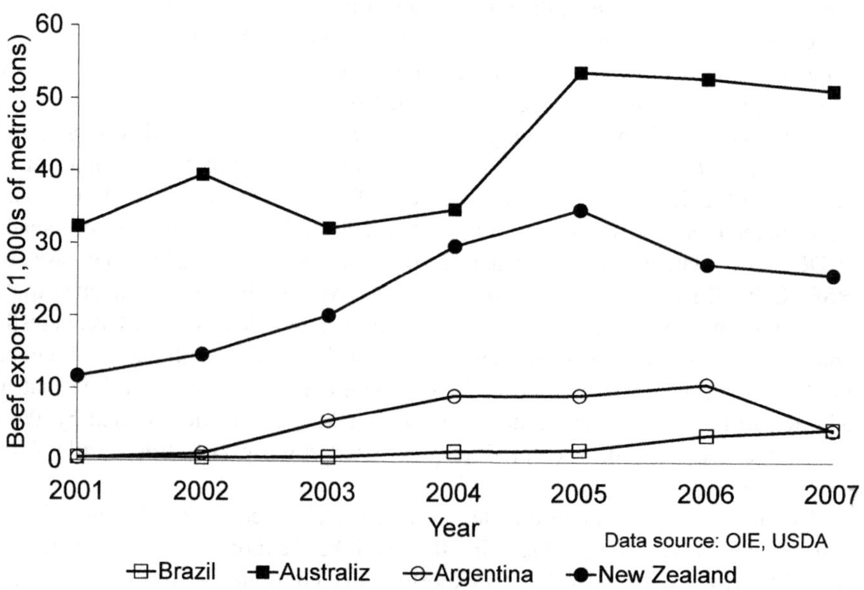

Fig. 7.6 Cattle exports from Brazil, Australia, Argentina, and New Zealand from 2001 to 2007. These are BSE free countries during these years. All export volumes increased after 2004, after the U.S. detected BSE

its last FMD outbreak in 1954 (OIE Handistatus database II, http://www.oie.int/hs2/report.asp?lang=en).

BSE affected trade within Europe (Mathews et al. 2003). Outbreaks of BSE in the U.K. caused importing countries to ban the imports of live cattle form the U.K.

and increased expenditures for BSE inspection and interception. Data show that after outbreaks of BSE in U.K. in 1996, the export of beef and live cattle from U.K. dramatically declined. However, data also show that world beef imports remained stable in that period. This implies that beef imports were resourced. Following the ban on U.K. cattle, the range of the livestock disease bluetongue expanded into northwestern Europe. This range expansion is generally attributed to climate change (Purse et al. 2008). However, Jones et al. (2008) found little evidence for climate related shifts in vector-borne pathogens such as blue tongue, and shifts in trade patterns cannot be ruled out.

Conclusion

To establish inspection policies that aim to minimize the sum of expected costs and damage, it is important to consider the market response to any inspection program. Prior work has largely focused on the direct costs and direct expected benefits associated with increases in expected inspection efforts and disease risk mitigation. Inspection programs have the potential to cause indirect effects through changes in importer behavior (see Ameden et al. 2007 for other types of firm response then those discussed here). The incentives that inspection efforts provide for importers to re-source goods or change import volumes have received less attention. To date, there appears to be two distinct literatures. The first literature largely focuses on how to balance the marginal cost of inspection effort with the marginal risk reductions from increased probability of mitigating against pathogen introduction. The second literature focuses on the use of market-based policy instruments that provide incentives for individual agents to internalize and reduce risk. Yet, inspection programs likely do both. In this chapter, we have laid out how inspection programs may also provide incentives for importing agents to alter their importing behavior. We also show that if these incentives are not considered, there is the possibility that additional inspection effort could increase risk.

Costello et al. (2007) provide evidence that biogeographic similarity and trade volume is important for identifying the probability that a trading partner will introduce an invasive species. Perrings et al. (2010) find similar results for invasive animal pathogens. The ability to identify such probabilities through easily observed proxies has lead to many recommendations for invasive species inspection programs (e.g., Caley et al. 2006; Keller et al. 2007). Although extremely useful, this information is insufficient to design an efficient inspection and interception program. It is important to account for the alternative goods that will arrive as a result of trade diversions from the increased cost of importing goods that are perceived to be more risky. Not considering these effects likely results in too much effort being spent on some goods or some sources and not enough on others. For this reason, we believe that it is important to build substitution and trade diversion effects into the design of inspection programs.

References

Ameden, H. A., S. B. Cash, and D. Zilberman (2007) Border enforcement and firm response in the management of invasive species. Journal of Agricultural and Applied Economics 39:35–46.

Ameden, H. A., P. C. Boxall, S. B. Cash, and D. A. Vickers (2009) An agent-based model of border enforcement for invasive species management. Canadian Journal of Agricultural Economics 57:481–496.

Batabyal A.A. (2007) International aspects of invasive species management: a research agenda. Stochastic Environmental Research and Risk Assessment 21:717–727.

Caley P, W.M. Lonsdale, and P.C. Pheloung (2006) Quantifying uncertainty in predictions of invasiveness. Biological Invasions 8:277–286.

Cassey P. et al. (2004) Influences on the transport and establishment of exotic bird species: an analysis of the parrots (Psittaciformes) of the world. Glob. Change Biol. 10, 417–426.

CDC (Centers for Disease Control). BSE (Bovien Spongiform Encephalopathy, or Mad Cow Disease). http://www.cdc.gov/ncidod/dvrd/bse/. Downloaded 6/8/2010.

Cleaveland, S., M. K. Laurenson, and L. H. Taylor (2001) Diseases of humans and their domestic mammals: pathogen characteristics, host range and the risk of emergence. Philosophical Transactions of the Royal Society of London B 356:991–999.

Collins, J. P. and A. Storfer (2003) Global amphibian declines: sorting the hypotheses. Diversity and Distributions. Vol 9(2): 89–98.

Costello C, M. Springborn, C. McAusland, and A. Solow (2007) Unintended biological invasions: Does risk vary by trading partner? Journal of Environmental Economics and Management 52: 262–276.

Dalmazzone, S. (2000) Economic Factors affecting vulnerability to biological invasions, in C. Perrings, M. Williamson and S. Dalmazzone (eds) The Economics of Biological Invasions, Cheltenham, Edward Elgar: 17–30.

Daszak, P., L. Berger, A. A. Cunningham, A. D. Hyatt, D. E. Green, and R. Speare (1999) Emerging infectious diseases and amphibian population declines. Emerging Infectious Diseases. Vol 5: 735–748.

Economic cost of foot and mouth disease in the U.K. (2002) A report to Department for Environment, Food and Rural Affairs.

Finnoff, D., J.F. Shorgren, B. Leung, et al. (2007) Take a risk: Preferring prevention over control of biological invaders. Ecological Economics. Vol 62(2): 216–222.

Guan, Y., B. J. Zheng, Y. Q. He, et al. (2003) Isolation and characterization of viruses related to the SARS coronavirus from animals in Southern China. Science. 302:276–278.

Gubbins S, S. Carpenter, M. Baylis, J.L.N. Wood, and P.S. Mellor (2008) Assessing the risk of bluetongue to U.K. livestock: uncertainty and sensitivity analyses of a temperature-dependent model for the basic reproduction number. Journal of the Royal Society Interface 5: 363–371.

Gurevitch, J. and D. K. Padilla (2004) Are invasive species a major cause of extinctions? Trends in Ecology and Evolution. Vol 19(9): 470–474.

Horan, R. D., E.P. Fenichel, C.A. Wolf, and B.M. Gramig (2010) Managing infectious animal disease systems. Annual Review of Resource Economics. Vol 2: 101–124, doi:10.1146/annurev.resource.012809.103859.

Horan, R. D., E. P. Fenichel, and R. T. Melstrom, (2011) Wildlife disease bioeconomics. International Review of Environmental and Resource Economics. Vol 5: 23–61, doi.org/10.1561/101.00000038.

Horan, R. D, and F. Lupi (2005) Tradeable risk permits to prevent future introductions of invasive alien species into the Great Lakes. Ecological Economics 52:289–304.

Horan, R. D., and C. A. Wolf (2005) The economics of managing infectious wildlife disease. American Journal of Agricultural Economics 87:537–551.

Horan, R. D., C. A. Wolf, and E. P. Fenichel, (2012) Dynamic Perspectives on the Control of Animal Disease: Merging Epidemiology and Economics. In Zilberman, O. Joachim,

D. Roland-Host, D. Pfeiffer (eds.) Health and Animal Agriculture in Developing Countries, Springer, 135–143.

Jones, K. E., N. G. Patel, M. A. Levy, A. Storeygard, D. Balk, J. L. Gittleman, and P. Daszak. (2008) Global trends in emerging infectious diseases. Nature 451:990–994.

Lanciotti, R. S., J. T. Roehrig, V. Deubel, et al. (1999) Origin of the West Nile virus responsible for an outbreak of Encephalitis in the Northeastern United. Science. Vol 286: 2333–2337.

Lips, K.R., F. Brem, R. Brenes, et al. (2006) Emerging infectious disease and the loss of emerging infectious disease and the loss of biodiversity in a Neotropical amphibian community. Proceedings of the national academy of sciences of the United States of America. Vol 103 (9): 3165–3170.

Keller, R. P., D. M. Lodge, and D. Finnoff (2007) Risk assessment for invasive species produces net bioeconomic benefits. Proceedings of the National Academy of Sciences 104(1):203–207.

Kilpatrick, A.M., A.A. Chmura, D.W. Gibbons, R.C. Fleischer, P. P. Marra, P. Daszak, (2006) Predicting the global spread of H5N1 avian influenza. Proceedings of the National Academy of Sciences of the United States of America 103, 19368–19373.

Kohn, R. and P.D. Capen (2002) Optimal volume of environmentally damaging trade. Scottish Journal of Political Economy. 49(1): 22–38. http://papers.ssrn.com/sol3/cf_dev/AbsByAuth.cfm?per_id=27435; http://papers.ssrn.com/sol3/cf_dev/AbsByAuth.cfm?per_id=334786.

Leighton, F. A. (2002) Health risk assessment of the translocation of wild animals. Scientific and Technical Review International Office of Epizootics 21: 187–195.

Leung, B., et al. (2002) An ounce of prevention or a pound of cure: bioeconomic risk analysis of invasive species. Proceedings of the Royal Society, London B. 269:2407–2413.

Li, W. D., Z. L. Shi, M. Yu, et al. (2005) Bats are natural reservoirs of SARS-like coronaviruses. Science. 310:676–679.

Mathews, K., J. Bernstein, and J. Buzby (2003) International Trade of Meat/Poultry Products and Food Safety Issues In Buzby, J. (eds). International Trade and Food Safety: Economic Theory and Case Studies. Agricultural Economic Report No. 828. U.S. Department of Agriculture, Economic Research Service.

Mathews, Jr. K.H., M. Vandeveer, and R.A. Gustafson (2006) An economic chronology of bovine spongiform encephalopathy in the North America. Electronic outlook report from the Economic Research Service, USDA, LDP-M-143-01.

McAusland, C., and C. Costello (2004) Avoiding invasives: Trade-related policies for controlling unintentional exotic species introductions. Journal of Environmental Economics and Management 48: 954–977.

Otte, J., D. Roland-Holst, and D. Pfeiffer, et al. (2007) Industrial Livestock Production and Global Health Risks. A Living from Livestock. Research Report. 1. RR Nr 07–09; June 2007. Available at http://www.fao.org/AG/AGAINFO/programmes/en/pplpi/docarc/rep-hpai_industrialisationrisks.pdf.

Pavlin, B. I., L.M. Schloegel, and P. Daszak (2009) Risk of importing zoonotic diseases through wildlife trade. United States Emerg Infect Dis. 15(11):1721–1726.

Perrings, C. (2005) Mitigation and adaptation strategies for the control of biological invasions. Ecological Economics 52:315–325.

Perrings, C., E.P. Fenichel, and A.P. Kinzig (2010) Globalization and Invasive Alien Species: Trade, Pests, and Pathogens. In: Perrings C, Mooney H and Williamson M (eds) Bioinvasions and Globalization, pp 42–55, Oxford University Press, Oxford.

Pimentel, D., L. Lach, R. Zuniga, and D. Morrison. (2000) Environmental and economic costs of nonindigenous species in the United States. Bioscience, 50(1): 53–56.

Pimentel, D. (2002) Biological invasions: economic and environmental costs of alien plant. Boca Raton, Fla.; London: CRC Press.

Pimentel D., R. Zuniga, and D. Morrison. (2005) Update on the environmental and economic costs associated with alien-invasive species in the United States. Ecological Economics 52: 273–288.

Purse, B.V., H.E. Brown, L. Harrup, P.P.C. Mertens, and D. J. Rogers (2008) Invasion of bluetongue and other orbivirus infections into Europe: the role of biological and climatic processes. Scientific and Technical Review International Office of Epizootics 27: 427–442.

Simberloff, P. (1996) Impacts of introduced species in the United States. Consequences Vol. 2(2): Achieved from http://www.gcrio.org/CONSEQUENCES/vol2no2/article2.html.

Semmens, B.X. et al. (2004) A hotspot of non-native marine fishes: evidence for the aquarium trade as an invasion pathway. Mar. Ecol. Prog. Ser. 266, 239–244.

Springborn, M., C. Costello, and P. Ferrier (2010) Optimal Random exploration for trade-related nonindigenous species risk. In C. Perrings, H. Mooney and M. Williamson, eds. Bioinvasions and Globalization: Ecology, Economics, Management, and Policy, Oxford University Press, Oxford.

Walters, C. (1986) Adaptive management of renewable resources. MacMillan Publishing, New York.

Wiens, J.J. and C.H. Graham (2005) Niche Conservatism: Integrating evolution, ecology, and conservation biology. Annual review of ecology evolution and systematics. Vol 36: 519–539.

Williamson, M.H. (1996) Biological invasions. Chapman &Hall, London.

Wolf, C., J. Hadrich, R. Horan, P. Paarlberg, and J. Kaneene (2008) Economic analysis of US TB eradication program options. A report to the U.S. Department of Agriculture, Animal and Plant Health Inspection Service.

Chapter 8
The Cost of Saving a Statistical Life: A Case for Influenza Prevention and Control

Thomas W. Sproul, David Zilberman, David Roland-Holst, and Joachim Otte

Since the beginning of the new millennium, governments and international organizations have spent billions of dollars on containing and controlling the emergence and spread of influenza viruses originating in domesticated animals, such as recent outbreaks of Avian Flu and Swine Flu. Much of this funding has been allocated to educational activities, subsidization and support of disease-prevention efforts, and compensation to farmers for culled animals. With globalization, contagious diseases can spread very rapidly and extensively, so controlling them benefits the world as a whole. Thus, investments in disease-prevention suppression and control activities have properties of a global public good, which have to be shared by the global community. Policy makers need relevant metrics to assess benefits and costs of flu-prevention investments, particularly in such a multilateral context.

The purpose of this chapter is to provide economic guidance for determining how much should be spent and how expenses could be allocated among nations. We start by evaluating the expected benefits of flu-prevention expenditures and then

---

T.W. Sproul (✉)
Department of Environmental & Natural Resource Economics,
University of Rhode Island, Kingston, RI, USA
e-mail: sproul@mail.uri.edu

D. Zilberman • D. Roland-Holst
Department of Agricultural and Resource Economics, University of California, Berkeley, CA, USA

J. Otte
Food and Agriculture Organization of the United Nations, Rome, Italy

D. Zilberman et al. (eds.), *Health and Animal Agriculture in Developing Countries*, Natural Resource Management and Policy 36, DOI 10.1007/978-1-4419-7077-0_8,

develop a sharing rule based on these calculations. Our analysis is indicative in nature because of data limitations regarding the likelihood, extent (in terms of how many people are affected) and severity (in terms of how seriously they are affected) of outbreaks, effectiveness of prevention and control measures, and the overall expenditures.

Conceptual Foundations

There are two key concepts behind our analysis—global public good and value of life. Control of animal diseases, such as influenza, has properties of a global public good. An activity is a global public good if it generates benefits simultaneously to people throughout the world and you cannot exclude people from getting these benefits. Diseases, such as Avian Flu, originate in certain source countries and are then spread globally. Activities that suppress their origination and spread are global public goods. Based on the work in Lichtenberg and Zilberman (1988), health risk is measured by the probability of death from a disease (Avian Flu) in country i, denoted by R_i, and is a function of the probability of disease origination as well as the spread of the disease. Both these processes are affected by prevention and control activities. So $P_i(I_A + I_B) = f_A(I_A)g_{Bi}(I_B)$, where $f_A(I_A)$ is the probability of origination of a disease at a source location affected by investment in prevention and treatment at the source, defined by I_A, and $g_{Bi}(I_B)$ is the probability of spread of the disease to country i, which depends on efforts to control the spread of the disease, denoted by I_B. If the population of country i is denoted by N_i, then the expected number of people who will die from the disease is P_iN_i. The expenditures in both prevention and transmission reduction are public goods because they affect many countries.

The management of global disease control requires raising the appropriate funds and allocating them optimally between prevention and control of the spread of disease. If a disease is affecting K nations, then the expected number of deaths from the disease is $\sum_{i=1}^{K} P_iN_i = f_A(I_A)\sum_{i=1}^{K} N_ig_{Bi}(I_B)$. To solve this optimization problem, we rely on a concept called the statistical value of a life saved (SVLS) (Viscusi and Aldy 2003). This metric imputes the expenditure needed to reduce mortality probabilities to save a statistical life. In our application, we will calibrate the metric nationally. We use the U.S. Environmental Protection Agency (EPA) for actuarial life value as a benchmark and adjust it to variations in per-capita gross domestic product (GDP) around the world.

Let V_i be the value of saving a statistical life in country i. Thus, the global optimization problem is

$$\max_{I_A, I_B} f_A(I_A)\sum_{i=1}^{K} V_iN_ig_{Bi}(I_B) - (I_A + I_B).$$

The first-order conditions are

$$\frac{\partial f_A(I_A)}{\partial I_A}\sum_{i=1}^{K} V_i N_i g_{Bi}(I_B) - 1 = 0$$

and

$$f_A(I_A)\sum_{i=1}^{K} V_i N_i \frac{\partial g_{Bi}(I_B)}{\partial I_B}(I_B) - 1 = 0.$$

The first optimality condition suggests that investment in prevention in the source country should be carried until the marginal gain (which is equal to the expected value of life lost because of increased investment in disease prevention) is equal to the marginal cost of the investment. The second condition suggests that the investment in control of spread should be carried until the marginal gain (which is equal to expected value of life lost because of the spread of the disease) is equal to the marginal cost of the investment. Thus, the macrolevel challenge requires assessing the marginal benefit from investment in different activities that will result in efficient allocation between activities. That will result in efficient overall fundraising to support disease control at the margin. The marginal gain that is equal to the expected value of reduced life lost is equal to the marginal cost of investment. Because we increase safety throughout the world where different countries elect to allocate different amounts of money based on their economic situation, we need to develop a calculus that weighs life saving differently among different countries. These differences do not reflect a moral difference but a practical economic difference in the ability of countries to pay to save the lives of their citizens.

Application to Avian Influenza

Influenza viruses of various types are endemic throughout the world. Every year, a new or existing strain circles the globe, killing about 37,000 people in the United States alone—mostly newborns, elderly, and infirm. Influenza pandemics, however, have occurred only three times in the last century, but they have resulted in substantially higher fatality rates, which include healthy people in their prime (Simonsen et al. 1998). For example, the Spanish Flu epidemic of 1918–1920 killed an estimated 40–50 million people according to the World Health Organization (WHO) (2009a). As a percentage of the world population, that figure scales to 140–175 million people today. However, advances in medical technology, such as antibiotics, to control secondary infection (Taubenberger and Morens 2006), combined with early warning systems and advance preparedness in developed countries, will likely reduce that number. In fact, WHO (2009b) estimates that a severe influenza pandemic occurring today will result in 2–7.4 million fatalities though the universe of mortality estimates for a contemporary pandemic range as high as 50–80 million (Murray et al. 2006).

Of course, this reduction in global vulnerability to a pandemic is the result of investments in disease prevention and preparedness (a global public good), which have already been made. Here, we will develop a simple model to assess the expected benefits from additional investment and, especially, the implied SVLS. The model will be followed by a numerical exercise using various estimates for expenditures and disease-prevention effectiveness to assess the scale of investments and to develop a sharing rule.

Pandemics are random events that can be characterized by a yearly distribution of fatalities. Since pandemics are infrequent, in most years, the number of fatalities is effectively zero. However, pandemics are severe; in a small number of years, numbers of fatalities can be quite large. Our simple model considers the effects of investments in terms of their impact on the average, or expected, number of yearly fatalities, N. For example, if an annual investment of I dollars will reduce the expected number of yearly fatalities by a fraction, F, then the cost of a statistical life saved is $C = I/FN$. Obviously, one can develop a more elaborate analysis that assumes a social-utility function where differing weights are given to higher loss of life.[1]

Using the estimates from above, with a fixed frequency of three times per century, the WHO's (2009a) yearly average number of fatalities from the influenza pandemic, N, is roughly between 60,000 and 222,000 (based on 2–7.4 million fatalities occurring every 33.3 years). These are relatively conservative numbers, compared to Murray et al.'s (2006), who suggest that N is in the range of 1.5–2.4 million.

Estimating the systemic relationship between F (the fraction of expected annual fatalities averted) and I (the annual investment in prevention and preparedness) is a challenge for future research. For simplicity, let F be a proportional function of I (in dollars), where $F = If/10^9$ and f is the fractional reduction of expected annual fatalities per billion dollars invested. For example, if $f = 0.01$ and $I = 10$, then an annual investment of \$10 billion will reduce the magnitude of a pandemic by 10% ($F = If = 0.1$). With this notation, the cost per statistical life saved is $C = 10^9/fN$, which suggests that the cost of saving a statistical life is inversely related to the expected number of annual fatalities and to the effectiveness of prevention efforts.

Under these assumptions, the cost of a statistical life saved is $C = 10^{11}/N$, which is roughly \$450,000–\$1.67 million for the WHO (2009a) pandemic estimates (\$$10^{11}$ divided by 222,000 and 60,000, respectively) and \$41,700–\$66,700 for the Murray estimates.[2]

[1] This might take account, for example, of the costs of attendant social disruption and collateral risk to human health and safety.

[2] However, it is possible that WHO (2009a) and Murray et al. (2006) would differ on the estimation of f as well.

To provide context for these numbers, the U.S. EPA uses $6.9 million as the SVLS. To obtain a global estimate for this figure, we take the U.S. EPA value as a benchmark and assume that the SVLS rises in proportion to per-capita GDP. The U.S. per-capita GDP is roughly $47,000 while the worldwide per-capita GDP is $10,400 [Central Intelligence Agency (CIA) 2008] implying a worldwide SVLS (based on the U.S. standard) of roughly $1.53 million. This figure suggests that a billion-dollar annual investment in safety is justified if it saves, on average, 654 people per year. If our assumptions are reasonable, and a $10 billion annual influenza safety investment reduces the expected fatalities of a pandemic by 10%, then the cost of a statistical life saved is in the ballpark of the worldwide value, according to the more conservative WHO (2009a) estimates. Thus, if a $10 billion annual investment instead reduces fatalities by 20%, or if the Murray et al. (2006) estimates are correct, then it is a real bargain for humanity.

While it is clear from this simple analysis that investment in flu control may be worthwhile from a global perspective, the value to individual players depends on the costs that they incur and the benefits that they gain. In this chapter, we have assumed that the benefits, in terms of risk reduction, are shared equally (of course, further research will need to account for varying benefits across countries). We continue to assume that the SVLS is proportional to GDP per capita. Thus, the expected benefit (EB_i) of investment in influenza control to a country is the expected risk reduction ($E\Delta R$ assumed constant across countries) times its SVLS times its population:

$$\mathrm{EB}_i = E\Delta R \times \mathrm{SVLS}_i \times \mathrm{pop}_i.$$

This formula suggests a rule for sharing expenditures, which is simply that each country pays its share of the total expected benefits. This share, by our assumptions, is exactly equal to each country's share of global GDP.[3] Table 8.1 illustrates contribution shares of major economies and selected countries of interest.

Conclusion

The design of a strategy to address Avian Influenza requires continuous investment in activities to monitor the state of the disease, suppress its probability of occurrence, and reduce the likelihood of spread. While the emergence of the disease is certain and random, the activities that are associated with suppression of its origination and spread are ongoing. The global community needs to invest in building the capacity to deal with this and other diseases. In this chapter, we found that the value of life on the one hand and likelihood of death on the other can be good guidance to determine how

[3] This is because $\mathrm{SVLS}_i = \mathrm{SVLS}_{\mathrm{US}} \times \mathrm{GDP}_i/\mathrm{pop}_i/\mathrm{GDPPC}_{\mathrm{US}}$, SCL where GDPPC is GDP per capita.

Table 8.1 Contribution shares of major economies and selected countries

Country	Population share (%)	GDP per capita (in dollars)	Contribution share (%)
United States	4.54	47,000	20.56
China	19.99	6,000	11.22
Japan	1.90	34,200	6.26
India	17.40	2,800	4.70
Germany	1.23	34,800	4.12
United Kingdom	0.91	36,600	3.21
Russia	2.09	15,800	3.20
France	0.96	32,700	3.02
Brazil	2.97	10,100	2.86
Mexico	1.66	14,200	2.37
Thailand	0.98	8,500	0.80
Pakistan	2.63	2,600	0.65
Egypt	1.24	5,400	0.64
Vietnam	1.30	2,800	0.35
Israel	0.11	28,200	0.29
Kenya	0.58	1,600	0.09

Source: CIA (2008) using GDP (purchasing power parity) and population.

much to spent on controlling the disease and its spread. These are the amounts of money that are spent for global public-good activities. Countries at the disease center that are more likely to suffer from the disease will probably be willing to spend extra to protect their citizens from the disease. So one can expect that, in addition to the global investment in disease control, there will be extra investment in countries, such as Vietnam, China, and Indonesia. Our analysis, here, is only instructive. It is the first phase of a project that should assess the cost of various disease-control activities and implications on risk in different countries and then use this information to assess overall investment and distribution of burden among nations.

References

CIA *World Factbook,* (2008). Cited 12 May, 2009. Available at https://www.cia.gov/library/publications/the-world-factbook/.

Lichtenberg, E., and D. Zilberman. "Efficient Regulation of Environmental Health Risks," "Quarterly J Econ," **CIII** (February, 1988): 167–178.

Murray,C.J., A.D. Lopez, B. Chin, D. Feehan, and K.H. Hill. "Estimation of Potential Global Pandemic Influenza Mortality On the Basis of Vital Registry Data from the 1918–20 Pandemic: A Quantitative Analysis." *Lancet* **368** (2006): 2211–2218.

Simonsen, L., M. Clarke, L. Schonberger, N. Arden, N. Cox, and K. Fukuda. "Pandemic versus Epidemic Influenza Mortality: A Pattern of Changing Age Distribution." *J Infect Dis* **178** (1998): 53–60.

Taubenberger, J., and D. Morens. "1918 Influenza: The Mother of All Pandemics." *Emerg Infect Dis* **12** (2006): 15–22.

Viscusi, V.K., and J.E. Aldy. "The Value of a Statistical Life: A Critical Review of Market Estimates Throughout the World." *J Risk and Uncertainty* **27** (2003): 5–76.

WHO. "Assessing the Severity of an Influenza Pandemic" (May, 2009a). Available at http://www.who.int/csr/disease/swineflu/assess/disease_swineflu_assess_20090511/en/index.html.

WHO. "Pandemic Preparedness" (May, 2009b). Available at http://www.who.int/csr/disease/influenza/pandemic/en/index.html.

Chapter 9
The Effect of Social Norms and Economic Considerations on Purchases of Chicken

Amir Heiman, Bruce McWilliams, and David Zilberman

Introduction

An important characteristic of modern life is the increasing diversity of products and services that save time. While, in the past, time-saving services were primarily limited to those who could afford to employ maids or servants, the modern economy has created a whole range of time-saving products and services that are accessible to people of any economic status.

In this study, we examine the effect of religion and time constraints on the consumption of convenience food products. This research is particularly relevant to the literature on the consumption of time-saving products and services when the two adults work outside of the household. Within that literature, the impact of religion has not been studied. Religion may be characterized in two ways: first, qualitatively in terms of affiliation to a general grouping and, second, in terms of intensity of belief or devoutness. Religion shapes entire cultures, family values, and personal lifestyles, which, in turn, affect the consumption choices that households make.

In order to study the effects of religion in depth, this study was carried out in Israel and the sample population includes Jews, Muslims, and Christians. Moreover, to identify the impact of religious intensity on the consumption decision, we account for the devoutness of individuals, with devoutness ranging from secular to ultra-Orthodox.

A. Heiman (✉)
Department of Agricultural Economics and Management,
The Hebrew University of Jerusalem, Rehovot, Israel
e-mail: heiman@agri.huji.ac.il

B. McWilliams
Autonomous Institute of Technology in Mexico (ITAM), Mexico City, Mexico

D. Zilberman
Department of Agricultural and Resource Economics, University of California, Berkeley, CA, USA

D. Zilberman et al. (eds.), *Health and Animal Agriculture in Developing Countries*, Natural Resource Management and Policy 36, DOI 10.1007/978-1-4419-7077-0_9,

We focus, here, on the consumption of unprocessed chicken. Chicken was chosen for analysis for two reasons: first, because chicken is the most consumed meat in Israel and, second, because it can be purchased in various ways that reflect convenience features—frozen and uncut, precut (whole), or in parts. Chicken that is frozen reduces the need for future shopping trips, and precut chicken and chicken parts reduce cooking time. Examining a single differentiated product in detail distinguishes this study from other research in this area and allows us to investigate how religion and time constraints interact in the choice of convenience food products.

This study is arranged in the following manner. We begin by reviewing the relevant literature in this area. We then present hypotheses to be tested. This is followed by the statistical model and a discussion of the data. We then present the estimation results and conclude with a summary of the findings and future areas of research.

Literature Review

The empirical analysis in this study will examine how religion and time constraints affect the demand for food items differentiated in the time-saving convenience features that they offer. The two areas of literature most related to this topic are the literature that examines the effect of religion on consumption decisions and the literature that looks at the consumption behavior of wives employed outside of the household. In this section, we discuss the primary findings of these two areas of the literature.

Religion is an important contributor to culture, personal identity, and values, all of which have consumption implications. Clearly, consumer behavior is affected by the enforcement of religious laws that restrict the consumption of particular goods or services. Bell's (1968) study of the fishing industry in the northeastern United States provides one empirical analysis of the impact of a religious law on consumption. He found that Pope Paul VI's 1966 apostolic decree that relaxed the Catholic Church's rules demanding abstinence from meat consumption on Fridays led to a significant decline in the demand for fish in New England and threatened the viability of the fishing industry in that area.

Less obvious is the impact of religion on the consumption of goods and services that are not directly restricted by religious laws. Most of the empirical research on the effect of religion on consumption focuses on the impact of religion on factors that affect consumer preferences and decision making. A general finding is that more devout believers have more traditional roles and value systems. Delener (1994) found that religious households had greater traditional sex-role orientation in the decision-making process when purchasing automobiles. Less religious households made relatively more joint decisions while in religious households males tended to make decisions, such as where the car would be purchased, and females decided the color. In addition, he found differences between Jewish and Catholic households with the differences depending on the degree of devoutness.

Schiffman et al. (1981) surveyed Nigerian students in the United States to examine how religion affected their acculturation to the new culture. They found that more religiously dogmatic respondents, whether Muslim or Christian, acculturated less to U.S. consumption patterns than did nondogmatic respondents. Differences in acculturation between the Muslim and Christian populations arose primarily because the proportion of religious dogmatists was much larger in the Muslim group.

The impact of religion on consumption can differ from one culture or country to another. In a crosscultural comparison of the effect of religiosity on general purchasing behavior, Sood and Nasu (1995) found that devoutness had no effect on buying behavior in Japan but had a significant effect in the United States.

Religion also appears to affect people's willingness to be exposed to risk and uncertainty. Wilkes et al. (1986) and Delener (1990) found that more religious consumers tended to be more concerned about the possible negative consequences of their purchasing decisions and buy more nationally known brands. Delener (1990) and Hirschman (1981) found that religious affiliation also affects willingness to try new products, with households of Jewish origin being more willing to seek out and try new products in the United States.

The primary findings in the above literature are that religion affects the consumption decision by affecting who will make the decisions about what to buy and by maintaining traditional consumption patterns. These studies do not specify the impact of religion or religiousness on the demand for a particular product. Our study contributes to developing this literature by examining the impact of religion on the demand for the convenience features (frozen, precut, parts) in a particular product category (chicken).

The demand for time-saving features has been extensively studied in the literature that looks at the consumption behavior in families where wives are employed outside of the household. Families with working wives are expected to purchase time-saving durable goods, such as dishwashers and microwave ovens; hire outside labor for household services, such as childcare; and purchase time-saving consumables, such as ready-to-eat foods to a greater extent than in households where wives do not work. Despite this logic, empirical studies show mixed and conflicting results. For example, while Oropesa (1993), Weinberg and Winer (1983), and Strober and Weinberg (1980) find no difference in purchases of time-saving durables by working wives, Kim (1989) does find a difference in these purchases. Nickols and Fox (1983) and Bellante and Foster (1984) find that families with working wives have higher purchases of childcare services, and Kim (1989), Nickols and Fox (1983), Bellante and Foster (1984), and Jacobs, Shipp, and Brown (1989) find that they consume more meals away from home.

Ethnicity and culture has also been shown to affect family purchases of these products. Bellante and Foster (1984) find that African Americans tend to spend less of their additional income on time-saving products, such as food away from home; childcare; and domestic services than do other Americans. Kim (1989) finds that French and English Canadians differ in the types of household appliances that they purchase while Douglas (1976) finds differences in behavior between U.S. and French consumers.

However, most relevant to the current analysis, past research has generally found little or no statistical differences in the purchases of convenience foods between households with or without working wives (Kim 1989; Reilly 1982; Douglas 1976). In a study of five consumption categories, Strober and Weinberg (1980) also found no difference in consumption except for frozen TV dinners and only among low-income families where frozen TV dinners were purchased more by working wives.

Schaninger et al. (1993) and Schaninger and Allen (1981) categorized families by occupational intentions of the wife: "career wife," "just-a-job wife," "plan-to-work wife," and "stay-at-home wife." Using these classifications, they found differences in purchasing behavior that cannot be represented in a dichotomous working/non-working classification. For example, Schaninger et al. (1993) found that career-wife families consume healthy staples more frequently than just-a-job or stay-at-home wife families, consume less junk food than just-a-job or plan-to-work wives, and consume the least amount of convenience foods of all of the wife groupings. Their underlying assumption is that the occupational status of wives is a reflection of their value systems and lifestyle—the factors that ultimately determine consumption choices and response to time constraints.

This study extends the analysis from the working wives literature by introducing religion as factor in consumption choice. Religion is a fundamental variable because it determines lifestyle and values whereas causality may be reversed in the case of wives' occupational status discussed above.

A key issue in the working wives literature is whether time constraints affect family consumption of time-saving features. We do not have information on the employment status of the wives but do have respondents' subjective evaluations of the availability of leisure time. Previous research suggests that wives' employment outside the home is related to both spouses' feelings of time pressure (Lavin 1993). Therefore, although our findings are based on time constraints rather than employment status, there are sufficient grounds for comparing our results with those that arise from the literature on working wives.

Hypotheses Construction

The focus of the demand analysis is on the purchases of convenience features in chicken, in particular the demand for precut, parts, and frozen chicken. Frozen chicken may be considered a convenience feature in that it allows consumers to purchase large quantities of chicken at one time, keep the meat in a freezer, and reduce the need for future shopping excursions. If a consumer is buying for immediate consumption, then the consumer is more likely to purchase fresh chicken since it does not require defrosting before use. Precut chicken and chicken parts are also convenience features since they reduce cooking time. Obviously, convenience is not the only aspect of these features. Fresh chicken may be considered by some to be of better quality, and is typically more expensive, than frozen chicken. The taste of fresh meat, including poultry, is perceived to be better than that of frozen chicken (http://www.fao.org/docrep/013/al700e/al700e00.pdf), and fresh is considered to be healthier (http://milk.mednet.co.il/SiteFiles/1/5076/68368.asp) although freezing is considered by some experts as a technology that lowers the risk of contamination (http://www.professorshouse.com/Food-Beverage/Topics/Meat/Articles/Chicken-Breasts/ and http://www.cdc.gov/ncidod/EId/vol12no02/pdfs/05-0936.pdf). In addition, chicken parts differ from precut chicken in that the household consumes a specific part of the chicken rather than the whole chicken. If the household has specific likes and dislikes in terms of chicken parts, they may prefer parts over whole chicken.

We now build on the discussion of the model from the previous section to develop specific hypotheses to test. The implicit understanding in the following hypotheses is that time-saving convenience features represent "modern" consumption while products without these features are "traditional." In the specific case of chicken, fresh uncut whole chicken is considered traditional while chicken that is frozen and/or that is purchased precut or in specific parts is considered nontraditional chicken with time-saving innovations. If we were to identify the extent to which a product is nontraditional, chicken that is both frozen and in parts may be considered less traditional than any of the other categories.

We now develop specific hypotheses. As identified in the literature review, previous research has found that more devout consumers tend to maintain traditional consumption patterns (Schiffman et al. 1981). This leads us to predict more traditional consumption patterns (i.e., more fresh uncut whole chicken) among the more devout households.

Hypothesis 1: More devout consumers will purchase fewer products with time-saving features.

Studies in the past have found mixed results when testing for the effects of religious affiliation of consumption decisions. Schiffman et al. (1981) found no difference between Christian and Muslim Nigerians once devoutness was accounted for while Hirschman (1981) and Delener (1990) found differences between American Jews and other Americans. No previous studies have looked at differences in consumption patterns between Muslims and Jews, particularly within Israel. We, therefore, do not make any a priori assumptions about how differences in religious affiliation will affect consumption.

Hypothesis 2: There is no difference between Jews, Muslims, and Christians in their purchase of time-saving features.

Previous studies on the purchase of convenience foods have not looked at the age of the respondents. Therefore, we do not have any a priori expectations about the effect of age on the purchasing behavior.

Hypothesis 3: The consumption of time-saving features will be independent of age.

Young children can have two effects on family consumption. The first is that families with young children will be busier and, therefore, have less time. This means that the families will be more time constrained and want to purchase time-saving features. However, in H_6, we will be controlling for time constraints, so families that are time constrained from children should already be accounted for when we test the effect of time constraints. The second effect of children on family consumption is that families with small children tend to be more conscious of what is being eaten. Nickols and Fox (1983) found that families with small children purchased less fast foods and ready-to-eat dinners. Hence, we expect that families with young children will be more concerned about the quality of food purchased and purchase less time-saving features that are perceived as lower quality. In the case of chicken, frozen chicken might be considered by some to be less healthy.

Hypothesis 4: Families with young children will consume fewer frozen chickens.

Income should be negatively correlated with the purchase of the cheaper goods, which, in the case of chicken, is uncut whole and frozen chicken.

Hypothesis 5: Families with low income will purchase more of the inexpensive goods, which are the uncut whole and frozen chickens.

Customers with very little leisure time will have a high value of time. So theory predicts that time-constrained households will purchase more time-saving convenience foods although, as discussed in the literature review, most of the findings contradict these expectations. We begin with the assumption provided by theory. Furthermore, we will examine how households of different religious affiliation and devoutness respond to these time constraints.

Hypothesis 6: Families with time constraints will purchase more time-saving products.

Empirical Model

We now describe the empirical analysis. The questionnaire asked shoppers the frequency with which they purchased different types of chicken, where the choices were "never," "sometimes," "often," or "always." Since these responses provide a qualitative measure of purchase, we cannot directly infer the relative consumption of the different types of chicken. Even when consumers responded that they always purchased a specific type of chicken, they also tended to give positive responses to consuming other types of chicken. In order to estimate how religion and other factors affect the household demand for different chicken types, we must first develop quantitative measures of consumption. We do this by decomposing total household demand for chicken into the demand for different chicken types, depending on the qualitative responses to the question about frequency of purchase. Therefore, the empirical analysis is done in a two-stage process: The first stage decomposes total demand into the demand for chicken types, and the second stage uses the estimated demand for chicken types in the dependent variable for estimating the effect of religion and time constraints on consumption of convenience features. In this section, we describe in greater detail this two-stage estimation process.

The total demand for chicken is known and is given by shoppers' response to the question: "How many times a week do you eat chicken?" This total demand for chicken can be decomposed into the demand for each type of chicken by taking into account the stated frequency of purchase of each chicken type by consumers. The types of chicken that are the focus of our analysis are: (1) fresh uncut whole chicken, (2) fresh precut whole chicken, (3) frozen uncut whole chicken, (4) fresh chicken parts, and (5) frozen chicken parts. We regress the total demand for chicken on all of the chicken types, accounting for whether customers purchased those types never, sometimes, often, or always. We use the following nonlinear demand equation to estimate the consumption of each chicken type:

$$\text{Total chicken consumption} = \sum_{i=1}^{5} \delta_i \cdot (\chi D_{iS} + \gamma D_{iO} + D_{iA}), \tag{9.1}$$

where δ_i represents consumption for those who always purchase chicken type i; χ and γ are adjustment coefficients for consumers who purchase the types sometimes and often, respectively; and D_{iS}, D_{iO}, and D_{iA} are the dummies for consumption of chicken type i with a frequency of sometimes, often, or always, respectively. The adjustment coefficients, χ and γ, are identical for all types of chicken, thereby assuring a consistent proportional relationship between the qualitative responses for all chicken types. Equation (9.1) was estimated using maximum likelihood. Theoretically, all of the parameters should be positive and $0 < \chi < \gamma < 1$. The estimated parameters were of the proper sign and relative magnitude without imposing these restrictions. Since this analysis is not the major focus of the study, the regression estimates and a short discussion of the results are contained in the Appendix.

The second stage of the analysis is the primary focus of this study. The estimated coefficients from equation (9.1) provide a basis for transforming the relative frequency variables (sometimes, often, and always) into quantitative weekly consumption, and share of consumption, for each chicken type. In fact, the share of consumption of each chicken type is the dependent variable for the second-stage estimation.

We will estimate the effects of household characteristics on the demand for convenience features. The household characteristics that we use are the age of the respondent; whether they have young children; their religion, defined as a combination of the religious affiliation (Jewish, Muslim, or Christian), and devoutness (secular, Conservative, or Orthodox); and the respondent's personal assessment of leisure time. The effect of leisure time is differentiated by religion. The regression equation may be summarized for a representative consumer as

$$\begin{aligned}\text{Share of chicken}_i = {} & v_{0i} + v_{1i} \cdot \text{age} + v_{2i} \cdot \text{young children} + v_{3i} \cdot \text{income} \\ & + v_{4ij} \cdot \text{religion} + v_{5ij} \cdot \text{leisure(religion)},\end{aligned} \tag{9.2}$$

where the Share of $\text{chicken}_i \equiv \frac{\text{consumption of chicken type } i}{\text{total chicken consumption}}$, v is the coefficient, i is the five types of chicken available, j is different religion types. The five "share of chicken type" equations are estimated simultaneously using an iterative three-stage, least-squares procedure.

The Survey Data

The data for this research come from a survey conducted in peoples' homes in four of the largest Israeli cities as well as in several rural villages. Within the cities, survey locations were selected according to common stratification methods to be representative of income, religious affiliation, and degree of observance within Israel. The interview was face to face, and responders were the adults in the home. When two adults were present, the person who shops most frequently was asked to respond. The responders were asked about family consumption patterns and, in particular, the total quantity of chicken consumed. Shoppers were asked to respond to four-point scale questions about their frequency of buying chicken with different convenience features. The reason for using a scale rather than quantitative measures is that in our pretest it was found that the questionnaire was too long and individuals did not have the patience to recall specific quantities purchased for each category. But they were willing to provide qualitative answers on the scaled question. The survey took approximately 15 minutes to complete, with a range of 12–18 minutes, and the response rate was a very high at 93%.

The survey consisted of 405 households. Seventeen of the responses were dropped because of insufficient responses to the questions, leaving 388 usable responses. Table 9.1 summarizes the statistics of the socioeconomic variables used in this analysis. The average age of respondents was 40 years with the youngest shopper interviewed being 15 and the oldest being 81. Fifty-five percent of the respondents had children at or under 14-years old, and 24% were in the low-income category.

Seventy-eight percent of the respondents were Jewish, where 60% were secular, 23% were Conservative, and 16% were Orthodox. Nineteen percent of the respondents were Muslim, where 36% were secular, 43% were Conservative, and 21% were Orthodox. There were 10 Christians, 1 Druze, and 1 other religion. Forty-two percent of the responders indicated that they had very little leisure time.

Table 9.2 provides information on the number of people who indicated that they had positive purchases of various types of chicken. The chicken type with the highest positive response rate was fresh chicken parts, where 85% of the 388 respondents indicated that they purchase them at least sometimes. Cut and uncut fresh whole chicken were second and third at 70% and 67%, respectively. Frozen parts and frozen uncut whole chicken had the lowest response rates at 60% and 59% of the population, respectively. The high response rates

Table 9.1 Summary statistics of socioeconomic variables

Variable	Number	Percent*
Age of respondent		40 years old
Children under 14-years old	213	55%
Low income	94	24%
Religion, Jewish	304	78%
Secular	183	60% of Jews
Conservative	71	23% of Jews
Orthodox	50	16% of Jews
Religion, Muslim	72	19%
Secular	26	36% of Muslims
Conservative	31	43% of Muslims
Orthodox	15	21% of Muslims
Religion, Christian	10	3%
Secular	6	60% of Christians
Conservative	4	40% of Christians
Low leisure time	163	42%

* Percents given are percent of total sample unless otherwise indicated. Age is mean age.

Table 9.2 Number of people in sample with positive purchases of chicken types

	Sometimes		Often		Always	
Chicken type	Number	Percent	Number	Percent	Number	Percent
Fresh uncut whole	57	14.7	80	20.6	121	31.2
Fresh precut whole	78	20.1	90	23.2	103	26.5
Frozen uncut whole	84	21.6	67	17.3	77	19.8
Fresh parts	47	12.1	108	27.8	175	45.1
Frozen parts	100	25.8	70	18.0	61	15.7

in each of the different purchase categories suggest that consumers buy a variety of chicken types.

Results

In this section, we discuss the results from the second stage of the analysis since this is the focus of the study. A discussion of the statistical results for the first stage of the analysis is contained in the Appendix.

One concern in this analysis is that, since religion enters more than once in the estimation, there is the potential for multicollinearity to cause problems. In order to test whether multicollinearity was affecting the results, we began with the most basic model in which religion and time constraints (low leisure) were variables that did not interact. We then introduced religion and time-constraint interaction variables. If multicollinearity were an issue, the addition of these interacting variables should have affected the coefficients and increased the variance of the estimated parameters relative to the basic model, reducing the estimated significance. In fact, adding the interacting terms had very little effect on the previously estimated coefficients and variances. Moreover, the new interacting variables had reasonable coefficients and variances, all of which suggests that multicollinearity is not a problem in our empirical model.

Our estimated model differentiates secular, Conservative, and Orthodox devotees. Statistical analyses revealed important consumption differences between Conservative and Orthodox Jews while Conservative and Orthodox Muslims were not different. In addition, secular and Conservative Christians did not differ in their consumption behavior. Therefore, the following published results combine Conservative and Orthodox Muslims into one group called "religious" Muslims and combine secular and Conservative Christians into one "Christian" group.

Five equations, one for each chicken type, were estimated simultaneously. Table 9.3 gives the estimated determinants of the share of purchases for fresh uncut whole, fresh precut whole, frozen uncut whole, fresh parts, and frozen parts. The base group is the secular Jewish family with average-to-upper income, without children under 14 years of age, and with average-to-sufficient leisure time.

We find that the age of the respondent affects the type of chicken purchased. Older people tend to purchase less frozen chicken, especially less frozen parts, and buy more of all fresh types, especially precut whole chicken. The H_3 is, therefore, rejected. The analysis suggests a somewhat more traditional pattern in consumption for older people since they consume less frozen chicken and less parts. However, the reduction in chicken parts is compensated by increased precut chicken, which does not increase the amount of work and effort in preparation.

When a family has at least one child under 14 years of age, this slightly reduces the amount of frozen chicken purchased, both whole and parts, and increases the purchases of fresh chicken types although this effect is only mildly significant. This is somewhat consistent with H_4, since frozen chicken may be considered an inferior and less-healthy product that parents do not want to feed young children who are at a "vulnerable" age. However, the support for H_4 is not strong.

Low-income families tend to consume less fresh precut and more uncut whole chicken, both frozen and fresh. The pattern is somewhat consistent with our expectations since low-income families purchase slightly more whole uncut and less cut. However, support for H_5 is weak, particularly in light of the fact that there is insignificant change in frozen and parts consumption. The slight impact of income on the consumption of parts may reflect the fact that we do not distinguish between the types of chicken parts purchased, most of which may be more expensive than uncut whole chicken, but some are cheap. If the analyzed data distinguished between different chicken parts, we might observe low-income families purchasing the parts of the chicken that are cheaper.

We now address how religion affects consumption choice. Jewish Conservatives consume significantly more fresh whole uncut chicken than do secular Jews

Table 9.3 Determinants of the demand for convenience features in chicken

Variable	Fresh uncut whole	Fresh precut whole	Frozen uncut whole	Fresh parts	Frozen parts
Age of respondent	0.0010	0.0018	−0.0006	0.0010	−0.0031
	(0.971)	(**2.507**)	(−0.875)	(1.154)	(**−3.608**)
Children under 14-years old	0.0132	0.0073	−0.0189	0.0111	−0.0128
	(0.542)	(0.439)	(−1.161)	(0.572)	(−0.625)
Low income	0.0174	−0.0273	0.0144	−0.0041	0.0005
	(0.630)	(−1.451)	(0.786)	(−0.185)	(0.020)
Religion, Jewish Conservative	0.1034	−0.0027	−0.0246	−0.0262	−0.0552
	(**2.555**)	(−0.097)	(−0.917)	(−0.816)	(−1.634)
Jewish Orthodox	0.0189	−0.0991	0.1390	−0.1391	0.0803
	(0.395)	(**−3.022**)	(**4.345**)	(**−3.641**)	(**2.002**)
Muslim secular	0.4521	−0.1082	−0.0539	−0.1020	−0.1880
	(**7.413**)	(**−2.594**)	(−1.325)	(**−2.093**)	(**−3.677**)
Muslim religious	0.3203	−0.0665	−0.0392	−0.0574	−0.1572
	(**6.555**)	(**−1.993**)	(−1.205)	(−1.477)	(**−3.838**)
Christian	0.1338	−0.1260	0.0299	−0.0602	0.0225
	(1.427)	(**−1.989**)	(0.482)	(−0.814)	(0.289)
Low leisure time, Jewish secular	−0.0056	−0.0300	0.0358	0.0207	−0.0209
	(−0.170)	(−1.333)	(1.634)	(0.790)	(−0.758)
Jewish Conservative	−0.0971	0.0419	0.0169	0.0433	−0.0049
	(*−1.726*)	(1.092)	(0.452)	(0.970)	(−0.103)
Jewish Orthodox	−0.1127	−0.0462	0.1029	−0.0453	0.1013
	(*−1.784*)	(−1.071)	(**2.447**)	(−0.904)	(*1.918*)
Muslim secular	−0.2452	−0.0243	0.1157	0.0010	0.1547
	(**−2.744**)	(−0.399)	(*1.946*)	(0.014)	(**2.069**)
Muslim religious	0.0422	−0.0314	−0.0389	0.0027	0.0254
	(0.612)	(−0.669)	(−0.849)	(0.049)	(0.440)
Christians	−0.0625	0.0518	−0.0536	0.0443	0.1087
	(−0.434)	(0.532)	(−0.564)	(0.390)	(0.908)
Constant	0.1428	0.1404	0.1189	0.2462	0.3518
	(**2.727**)	(**3.918**)	(**3.403**)	(**5.888**)	(**8.012**)
Equation R-squares	0.295	0.154	0.201	0.115	0.175
System R-square = 0.514, Chi-square = 279.7 with 70 DF					

Note: T-statistics are in parentheses, where **bold** = significant at the 5% level and *italics* = significant at the 10% level.

(the base group of comparison), consuming less of all other products, particularly less frozen parts. While this pattern is similar to that of both the secular and religious Muslims, the coefficients for the secular and religious Muslims are much larger than the coefficients for the Conservative Jews, suggesting that the purchasing behavior of Jewish Conservatives is somewhere between that of the secular Jews and the Muslims. Christians appear to buy less fresh precut and more fresh whole chicken than secular Jews. However, they are the most similar to the secular Jews in their consumption of convenience features.

Jewish Orthodox households have an altogether different consumption pattern, consuming significantly less fresh precut and fresh parts and more frozen chicken, both whole and parts. We attribute this surprising result to institutional factors particularly to the Jewish Orthodox population. Orthodox families typically have access to cheap institutional purchases of frozen chicken through charity organizations that support families of religious scholars studying in "Yeshivas." The large Orthodox families tend to have lower per-capita incomes, thereby contributing to their acceptance of these charity products available through their religious institutions.

These findings suggest mixed support for H_1—the hypothesis that more devout households will purchase fewer time-saving features. The hypothesis holds among the Jews, where the Conservative Jews consume more of the traditional goods and less of the time-saving goods than do secular Jews. We do not consider Orthodox Jew purchases of frozen products to reflect a violation of H_1 but, rather, to reflect distinct institutional factors unavailable to other religious groups. However, H_1 is not supported among the Muslims since the secular Muslims appear to have equal or even greater traditional consumption patterns compared to their religious brethren.

The results also reveal that religious affiliation does affect consumption, rejecting H_2. Muslims of all devoutness consume in a significantly more traditional manner, eating more fresh uncut chicken than do the secular or Conservative Jews or the Christians.

We also examine how the lack of leisure time (i.e., time constraints) affects the purchasing of convenience features. The results show that the effect of time constraints differs by religious category. Secular Jews, Christians, and religious Muslims do not significantly increase their consumption of time-saving convenience features when faced with time constraints (there is a small, but insignificant, trend away from whole uncut toward cut chicken and chicken parts among Christians), which appears to reject H_6—that families with time constraints will buy more time-saving features. In contrast, time-constrained Conservative and Orthodox Jews as well as secular Muslims consume significantly less traditional products and more time-saving products thus supporting H_6. Conservative Jews switch to fresh precut and fresh parts, saving cooking time, whereas Orthodox Jews and secular Muslims switch to frozen products, both whole and parts, thereby saving shopping time.

These findings imply important differences between Muslims, Jews, and Christians, again, rejecting H_2. In the Jewish population, it is the Conservatives who change their buying behavior when time constrained while the seculars do not. We propose that the reason why secular Jews and Christians do not significantly increase their purchases of time-saving features when they are time constrained is that the typical secular Jew and Christian household already consumes a high-level of time-saving features. There is not much room for adjustment in these groups.

By contrast, in the Muslim population, where the typical family, regardless of religious devotion, consumes a very high level of traditional non-time-saving products, it is the secular family that responds to time constraints while religious households are impervious to time constraints and maintain traditional patterns of consumption thus supporting H_1.

Conclusion

This study explains the effect of religion on the demand for convenience foods. Using data from Israel, we analyze how religion, defined in terms of both the general religious affiliation as well as devoutness, interacts with time constraints to determine the demand for chicken that is differentiated in the type of convenience characteristics offered (precut or uncut, whole or in parts, and fresh or frozen). The analysis shows that the effect of religious affiliation and devoutness on consumption decisions are more complex than expected.

Several findings from the empirical analysis deserve reiteration. We found that all Muslims, regardless of religious devoutness, have a more traditional consumption pattern than Jews in general, purchasing less of all chickens that had time-saving features. However, when faced with time constraints, secular Muslims are willing to shift to time-saving products, whereas religious Muslims maintain their traditional consumption patterns.

Among Jews, Conservative households have a consumption pattern that is significantly more traditional than that of secular households although their consumption is much less traditional than the Muslims. When time constrained, secular Jews do not significantly change their purchasing patterns while Conservative Jews adjust by purchasing time-saving features Christians have similar purchasing patterns to the secular Jews in that they purchase a lot of time-saving features and do not respond to time constraints. We suggest that the reason these two groups do not respond to time constraints is that the typical household in these groups is already consuming a high level of time-saving foods.

These findings on time constraints provide interesting conformity and contrast with previous research. Our research may have revealed an explanation for the fact that previous studies have generally found little or no responsiveness of families with working wives to purchases of convenience foods. Similar to the lack of response to time constraints by secular Jews and Christians, these previous findings may arise because the typical household in North America, where most of the studies were conducted, already purchases a high level of time-saving food products and there is little room for modifying consumption when time constrained. However, our study shows that time constraints can affect the consumption decision when the typical household in the group is consuming a high level of traditional (non-time-saving) foods. In our case, the responsiveness occurs among the religious Jews and secular Muslims who are time constrained. The religious Muslims do not change their traditional consumption patterns even when time constrained. These findings demonstrate the powerful effect of religion and devoutness on consumption behavior.

Suppliers of food products need to be aware of the factors that affect the demand for different types of products when deciding to market and distribute products to a particular population. This study provides detailed examination of the effect of religion on demand in a particular product category in which the convenience features are differentiated. This analysis can be extended to a variety of products, not only those that are differentiated in the convenience features but also those differentiated by other characteristics where consumption may be affected by lifestyle and values. For example, the empirical analysis can be extended to examining the choice of clothing, which may be differentiated by the material, color, design, or price. Alternatively, cars differ in their safety features, sportiness, brand, and country of origin. Future research may reveal that religion influences the features chosen by consumers.

Appendix: Disaggregation of Total Chicken Demand into the Demand for Chicken Types

Table 9.4 shows the decomposition of total chicken demand estimated from (equation 9.1). Although we did not place any restrictions on the parameters, parameters (they are all positive), the coefficient for consuming sometimes is smaller than the coefficient for consuming often, and both are smaller than one as they should be. All of the coefficients are significant at the 10% level except the coefficient for fresh whole cut and fresh parts of chicken. The *R*-square (0.83) is quite high for this type of estimation.

People who buy fresh uncut whole chicken have the largest coefficients associated with their purchases, suggesting that, when it is purchased, it is consumed more frequently than other types of chicken. People who buy chicken types sometimes consume about half (0.48) of the amount consumed by those who always buy them while those who buy often consume slightly over half (0.55) of that amount.

The coefficients in Table 9.4 transform the relative measures of purchase (sometimes, often, and always) into estimated weekly consumption of each type of chicken. For example, a respondent whose family is identified as consuming fresh uncut whole chicken always is assigned a consumption of 1.613 times per week for that chicken type while a respondent indicating sometimes in that category is assigned a weekly consumption of fresh whole uncut chicken of 0.77 times per week (=1.613 * 0.48).

Table 9.4 Decomposition of total chicken demand into demand for chicken types

Variable	Coefficient	Standard error	*t*-Ratio
Coefficients for those who always consume types			
Fresh uncut whole	1.613	0.784	**2.059**
Fresh precut whole	1.013	0.784	1.293
Frozen uncut whole	1.265	0.736	*1.719*
Fresh parts	1.429	0.869	1.644
Frozen parts	1.194	0.723	*1.651*
Adjustment coefficients for those who consume			
Sometimes	0.480	0.145	**3.301**
Often	0.553	0.070	**7.869**
R-square	0.831		

Note: Numbers in **bold** = significant at the 5% level, and numbers in *italics* = significant at the 10% level.

References

Bell, Frederick W. "The Pope and the Price of Fish." *The American Economic Review* (December, 1968), 58(5): 1346–1350.

Bellante, Don, and Ann C. Foster."Working Wives and Expenditure on Services." *Journal of Consumer Research* (September, 1984), 11(2): 700–707.

Delener, Nejdet. "Religious Contrasts in Consumer Decision Behaviour Patterns: Their Dimensions and Marketing Implications." *European Journal of Marketing* (1994), 28(5): 36–53.

Delener, Nejdet. "The Effects of Religious Factors on Perceived Risk in Durable Goods Purchase Decisions." *Journal of Consumer Marketing* (Summer, 1990), 7(3): 27–38.

Douglas, Susan P. "Cross-National Comparisons and Consumer Stereotypes: A Case Study of Working and Non-Working Wives in the U.S. and France." *Journal of Consumer Research* (June, 1976), 3(1): 12–20.

Hirschman, Elizabeth C. "American Jewish Ethnicity: Its Relationship to Some Selected Aspects of Consumer Behavior." *Journal of Marketing* (Summer, 1981), 45(3): 102–110.

Jacobs, Eva, Stephanie Shipp, and Gregory Brown."Families of Working Wives Spending More on Services and Nondurables." *Monthly Labor Review* (February, 1989), 112(2): 15–23.

Kim, Chankon. "Working Wives' Time-Saving Tendencies: Durable Ownership, Convenience Food Consumption, and Meal Purchases." *Journal of Economic Psychology* (November, 1989), 10(3): 391–409.

Lavin, Marilyn. "Wives' Employment, Time Pressure, and Mail/Phone Order Shopping – An Exploratory Study." *Journal of Direct Marketing* (Winter, 1993), 7(1): 42–49.

Nickols, Sahron Y., and Karen D. Fox. "Buying Time and Saving Time: Strategies for Managing Household Production." *Journal of Consumer Research* (September, 1983), 10(2): 197–208.

Oropesa, R.S. "Female Labor Force Participation and Time-Saving Household Technology: A Case Study of the Microwave from 1978 to 1989." *Journal of Consumer Research* (March, 1993), 19(4): 567–579.

Reilly, Michael D. "Working Wives and Convenience Consumption." *Journal of Consumer Research* (1982), 8(4): 407–418.

Schaninger, Charles M., and Chris T. Allen. "Wife's Occupational Status as a Consumer Behavior Construct." *Journal of Consumer Research* (1981), 8(3): 189–196.

Schaninger, Charles M., Margaret Nelson, and William D. Danko. "An Empirical Evaluation of the Bartos Model of Wife's Work Involvement." *Journal of Advertising Research* (May/June, 1993), 33(3): 49–63.

Schiffman, Leon G., William R. Dillon, and Festus E. Ngumah."The Influence of Subcultural and Personality Factors on Consumer Acculturation." *Journal of International Business Studies* (Fall, 1981), 12(2): 137–143.

Sood, James, and Yukio Nasu. "Religiosity and Nationality: An Explanatory Study of Their Effect on Consumer Behavior in Japan and the United States." *Journal of Business Research* (September, 1995), 34(1): 1–9.

Strober, Myra H., and Charles B. Weinberg. "Strategies Used by Working and Nonworking Wives to Reduce Time Pressures." *Journal of Consumer Research* (March, 1980), 6(4): 338–348.

Weinberg, Charles B., and Russel S. Winer. "Working Wives and Major Family Expenditures: Replication and Extension." *Journal of Consumer Research* (September, 1983), 10(2): 259–263.

Wilkes, Robert E., John J. Burnett, and Roy D. Howell."On the Meaning and Measurement of Religiosity in Consumer Research." *Journal of the Academy of Marketing Science* (Spring, 1986), 14(1): 47–56.

Part III
The Spread and Control of Avian Influenza

Chapter 10
Epidemiology of Highly Pathogenic Avian Influenza Virus Strain Type H5N1

Guillaume Fournié, Will de Glanville, and Dirk Pfeiffer

Introduction

Highly pathogenic avian influenza (HPAI) is a severe disease of poultry. It is highly transmissible with a flock mortality rate approaching 100% in vulnerable species (Capua et al. 2007a). Due to the potentially disastrous impact the disease can have on affected poultry sectors, HPAI has received huge attention and is classified as a notifiable disease by the World Organisation for Animal Health (OIE).

Among the family *Orthomyxoviridae*, only viruses of the influenza A genus are known to infect birds. Most virus strain types have been isolated in water birds which are considered to be the main hosts (Webster et al. 1992). When avian influenza viruses spill over into susceptible domestic poultry species (e.g. chickens and turkeys) they can cause a mild disease described as low pathogenic avian influenza (LPAI) (Capua et al. 2004; Osterhaus et al. 2008); however, some subtypes (namely H5 and H7) can undergo mutations into a highly pathogenic form (Alexander et al. 1993; Garcia et al. 1996; Rohm et al. 1995). Only 25 HPAI epidemics have been recorded in poultry since 1959. Despite the sporadic nature of outbreaks, there appears to be a trend for increasing frequency over the past two decades, as well as a trend for increasing economic impact per outbreak (Alexander et al. 2009).

Since 2003, 63 countries have reported outbreaks of HPAI strain type H5N1 in domestic and wild birds (OIE 2010a), and the disease is now endemic in poultry populations in several countries in Asia and in Egypt. HPAI virus (HPAIV) H5N1 has also shown potential for cross-species transmission, including infection of humans. HPAI therefore remains of concern for public health, particularly with respect to its pandemic potential (Ferguson et al. 2004).

G. Fournié (✉) • W. de Glanville • D. Pfeiffer
Veterinary Epidemiology and Public Health Group, Department of Veterinary Clinical Sciences, Royal Veterinary College, University of London, UK
e-mail: gfournie@rvc.ac.uk

D. Zilberman et al. (eds.), *Health and Animal Agriculture in Developing Countries*, Natural Resource Management and Policy 36, DOI 10.1007/978-1-4419-7077-0_10,

In this review, we present the main epidemiological characteristics of the ongoing HPAI H5N1 pandemic in poultry.

Worldwide Spread and Continuing Evolution of HPAIV H5N1

1996–2003: Emergence of HPAIV H5N1

The first HPAI H5N1 outbreak is thought to have occurred in a commercial goose farm in Guangdong province, China, in 1996 (Chen et al. 2004; Xu et al. 1999; Alexander 2007a). This strain was likely to have been introduced from wild birds as a LPAI virus and undergone subsequent mutation (Vijaykrishna et al. 2008). The resulting HPAIV was the first of a lineage that has generated multiple genetic re-assortants and is the precursor of all subsequent HPAIV H5N1. A H5N1 variant caused an outbreak in a Hong Kong farm in April 1997, as well as demonstrating potential for human infection and resulting in the first public attention of the potential importance of this strain (de Jong et al. 1997; Subbarao et al. 1998). The same virus re-emerged in Hong Kong in December 1997 and was found to be highly prevalent in live bird markets (LBMs) (Shortridge 1999), where it may have circulated for several months in the absence of effective surveillance (Guan et al. 2009). The first outbreaks in Hong Kong were controlled through the slaughter of all poultry on the island, although additional outbreaks were reported in 2001 and 2002 (Sims et al. 2003a, b).

From 1999 to 2003, multiple genotypes of H5N1 viruses were isolated in domestic waterfowl in Southern China and in birds imported from China to Hong Kong for slaughter (Chen et al. 2004; Sims et al. 2003a, b; Cauthen et al. 2000; Guan et al. 2002; Li et al. 2004; Martin et al. 2006; Wang et al. 2008a). H5N1 viruses were also isolated from duck meat imported from China to South Korea (Tumpey et al. 2002) and Japan (Mase et al. 2005). In Viet Nam, two Ha Noi LBMs were found to be silently infected with H5N1 viruses in 2001 (Nguyen et al. 2005). Moreover, two human cases were reported in Hong Kong in 2003 from a family with recent travel history to China (Peiris et al. 2004).

These findings suggest that H5N1 viruses circulated extensively in southern China following emergence (Sims et al. 2003a; Guan et al. 2002; Martin et al. 2006; Duan et al. 2008). Moreover, the wide genetic variation displayed by isolates indicates that the pool of AI viruses in the region is large and that a high number of H5N1 genotypes have emerged over a relatively short time period. Southern China, and more specifically Guangdong province (Wallace et al. 2007), therefore appears to be the prime source of emergent HPAIV H5N1, which have subsequently spread at both a regional and an international scale (Wallace et al. 2007; Chen et al. 2006). Among the multiple re-assortants, the genotype *Z* has become the dominant genotype since 2002 (Li et al. 2004; Duan et al. 2008). This virus, first detected in Guangxi, China, in 2001 (Guan et al. 2009), is more virulent than its predecessors and infects a wider range of species (Eagles et al. 2009).

2003: Major Outbreak Waves in Southeast Asia

From November 2003 to February 2004, eight countries (Viet Nam, Thailand, Indonesia, South Korea, Japan, Cambodia, Laos and China) reported outbreaks of HPAI H5N1 to the OIE (Li et al. 2004). However, the virus is very likely to have circulated in several of these countries, including Viet Nam and Indonesia, for several months before official notification (Vijaykrishna et al. 2008). Thailand and Viet Nam were particularly affected, with the spread of infection to most provinces by the end of January 2004. A second wave of outbreaks started in June 2004 and culminated in Thailand in late 2004 and in Viet Nam in early 2005 (Guan et al. 2009). Malaysia reported its first cases in August 2004, but subsequently eradicated the disease through mass culling (Martin et al. 2006).

The viruses circulating in Viet Nam, Thailand, Malaysia, Cambodia and Laos at this time belonged to the clade 1 and were derived from viruses previously identified in Yunnan province, China. In contrast, viruses circulating in Indonesia belonged to clade 2.1, believed to have originated in Hunan province, China (Guan et al. 2009; Wang et al. 2008b). China thus appears to have been the epicentre of disease spread, with countries in Indochina acting as a sink (Wallace et al. 2007). However, within these countries, geographic diversity was already emerging with viruses circulating in northern Viet Nam more closely related to those in Thailand, and viruses in the Mekong region related to those in Cambodia (Smith et al. 2006a).

2005–2006: Westerly Virus Spread

The death of more than 6,000 wild birds due to HPAIV H5N1 in May 2005 in Qinghai Lake, Western China, prompted fears of the possibility of wider spread of the virus through bird migration (Liu et al. 2005; Chen et al. 2005). This was the first occurrence of the clade 2.2 H5N1 variant, whose lineage rapidly spread from China to Europe, Africa and the Middle East (Guan et al. 2009).

Soon after the first detection of the virus in wild birds in China, infection was found in wild birds in Mongolia (Gilbert et al. 2006a). In July 2005, several outbreaks were reported across Siberia (Feare 2007), and from October wild bird deaths were detected around the Caspian and Black seas. Poultry outbreaks of HPAI H5N1 were subsequently reported in Crimea (Feare 2007). Although HPAI H5N1 viruses had been detected in Europe before 2006, previous occurrences had been scarce and isolated (Van Borm et al. 2005; Alexander 2007b). From February 2006, dead birds, mainly mute swans, were found in several countries, including Austria, Croatia, Denmark, France, Germany, Greece, Scotland, Sweden and Switzerland. Outbreaks in domestic poultry were also detected in some of these countries but were generally rapidly controlled, with transmission between premises remaining limited (Brown 2010).

Whilst limited disease spread was occurring in Europe, the clade 2.2 variant was also spreading within domestic poultry populations in the Middle East, particularly Iran, Azerbaijan, Afghanistan, Pakistan, as well as into Africa (Feare 2007; Brown 2010), first affecting Nigeria in January 2006 (Cattoli et al. 2009), and Egypt, Niger, Cameroon, Burkina Faso, Sudan, Cote d'Ivoire and Djibouti shortly after. A total of 1,024 outbreaks were reported in Egypt from February to December 2006, either in commercial or backyard flocks (Aly et al. 2008). The disease spread further in Africa in 2007, with Ghana, Togo and Benin all reporting outbreaks (Cattoli et al. 2009).

Phylogenetic studies suggest that there have been several independent introductions of H5N1 viruses into Europe (Salzberg et al. 2007; Gall-Recule et al. 2008; Starick et al. 2008) and Africa (Ducatez et al. 2006, 2007; Fasina et al. 2009). However, all the viruses have been closely related, and viruses circulating in Russia in 2005 have been proposed as putative progenitors of this Euro-African lineage (Cattoli et al. 2009; Ducatez et al. 2007).

From 2007: Virus Maintenance and Genetic Diversification

Although outbreaks are now reported less frequently than during the first outbreak waves in 2003–2006, HPAI H5N1 has become endemic in several regions of Asia and Africa. LBM surveys in southern China from January 2004 to June 2006 (Chen et al. 2006; Smith et al. 2006a) and from 2007 to 2009 (Jiang et al. 2010) demonstrated that H5N1 viruses continue to circulate in a variety of poultry species. A new variant has also emerged during this period: the Fujian-like variant (clade 2.3.4), related to genotype V. Since 2005, this variant has gradually replaced the previously prevailing sub-lineage in China (Duan et al. 2008; Smith et al. 2006a; Li et al. 2010a), as well as in northern Viet Nam (Wan et al. 2008; Dung Nguyen et al. 2008) where the disease also appears to be endemic. This variant had also been detected in Laos, Malaysia and Thailand (Smith et al. 2006a; Saito et al. 2008). As well as 2.3.4, several clades continue to circulate in China, including clade 7 and clade 2.3.2 (Jiang et al. 2010). Introduction of clade 7 has also occurred in Myanmar (Saito et al. 2008) and northern Viet Nam (Nguyen et al. 2009). Moreover, clade 2.3.2 has been isolated in several other Asian countries as well as in Europe, in both poultry and wild birds, as a result of a new wave of cross-continental spread from Asia to Europe (Jiang et al. 2010; Boltz et al. 2010; Kim et al. 2010). Since 2001, nine distinct genotypes have been detected in Viet Nam, at least four of which appear to have emerged in the country, others having been introduced (Wan et al. 2008). New variants in Viet Nam appear to be first detected in northern parts, and to then spread to the south (Wan et al. 2008), although clade 1 is still prevalent in southern Viet Nam (Wan et al. 2008; Dung Nguyen et al. 2008). In Indonesia, the disease has been officially declared as endemic since 2006 (OIE 2010a), and outbreaks in poultry are frequent, particularly on the islands of Java, Bali, Sulawesi and Sumatra (Henning et al. 2010). Clade 2.1 is still the predominant variant in that country (Eagles et al. 2009), although new reassortants with different transmission and evolutionary dynamics appear to continually emerge on Java, the main endemic focus, and subsequently spread to other regions (Lam et al. 2008; Takano et al. 2009). Hence, the co-circulation of multiple sub-lineages and their continuing evolution both in China and in Southeast Asia has led to the generation of new variants that are able to spread widely across the region.

It has been suggested that HPAI H5N1 is unlikely to be endemic in Cambodia, or in Laos (Buchy et al. 2009; Boltz et al. 2006), where outbreaks seem to result from virus reintroduction from neighbouring endemic areas, rather than through perpetuated transmission. In Thailand, the disease appears to have been effectively controlled, with interventions in place to control sporadic emergence. This is also the case in South Korea and Japan (Eagles et al. 2009).

The genetic diversity that exists within African virus isolates appears to be due to the prolonged circulation and evolution of viruses in a segregated area rather than due to the reintroduction of new variants (Cattoli et al. 2009; Salzberg et al. 2007). In Nigeria, for example, the co-circulation of multiple sub-lineages led to the emergence of new variants which gradually replaced introduced virus strains (Owoade et al. 2008; Fusaro et al. 2010; Monne et al. 2008). Some new introductions are likely to have occurred, however, as genotypes closely related to those circulating in Europe and the Middle East in 2007 were detected in July 2008 (Fusaro et al. 2009). No outbreak has been reported in Nigeria since 2009 (OIE 2010b). The disease has been formally declared endemic in Egypt (Aly et al. 2008) and surveillance campaigns have highlighted high prevalence in farms and LBMs (Abdelwhab et al. 2010; Hafez et al. 2010). Several sub-lineages have become established, co-circulate and continue to evolve in the country (Arafa et al. 2010; Abdel-Moneim et al. 2009) to the extent that they were reclassified as a new third-order clade, 2.2.1 (Balish et al. 2010).

Outbreaks have regularly occurred in Bangladesh since 2007, suggesting that the disease may now be endemic in the country (Ahmed et al. 2010; Biswas et al. 2008; ProMED-mail 2010a, b). Moreover, several outbreak waves have been reported in India since 2006. The viruses associated with these waves have been of clade 2.2, and outbreaks are considered to be the result of new introductions, as interventions were reported to mitigate the successive outbreaks (Chakrabarti et al. 2009; ProMED-mail 2009, 2010c; Ray et al. 2008; Mishra et al. 2009; Murugkar et al. 2008).

Wild Birds: Putative Disease Spreaders and Reservoir

From 2002 to 2005, reported H5N1 outbreaks in wild birds in Asia (Hong Kong, Japan, South Korea, Thailand, China, Cambodia) (Feare 2007; Ellis et al. 2004; Kwon et al. 2005a; Desvaux et al. 2009) tended to be isolated and limited. They generally involved a small number of fatalities among non-migrant species, and occurred in the vicinity of poultry outbreaks or among captive or semi-captive wild bird species. However, the emergence of new variants responsible for mass mortality in wild birds in Qinghai Lake, China, in 2005 (Liu et al. 2005; Chen et al. 2005) led to the notion that migratory birds could spread H5N1 viruses beyond Southeast Asia (Olsen et al. 2006).

Virus Spread to Disease-Free Areas

Most virus introductions into Europe in 2005–2006 were probably caused by wild birds (Kilpatrick et al. 2006; Pfeiffer et al. 2006). Indeed, the virus spread to the Caspian and Black sea occurred with the autumn Anatidae migration (Gilbert et al. 2006a), and the subsequent spread to western Europe was very likely due to unusual cold weather that caused wild birds to leave the Caspian and Black sea (Reperant et al. 2010). Moreover, wild birds have been implicated as the cause of disease introduction or reintroduction to several other countries, including Russia (Sharshov et al. 2010), Mongolia (Spackman et al. 2009), Nigeria (Owoade et al. 2008; Fusaro et al. 2010; Gaidet et al. 2008), Egypt (Saad et al. 2007), India (Chakrabarti et al. 2009; Murugkar et al. 2008), Japan (Uchida et al. 2008), and South Korea (Kang et al. 2010). However, the evidence is generally scarce and these conclusions are based on the fact that the timing of poultry outbreaks was associated with bird migration, or in some cases from outbreak investigations. However, and in general, the lack of information and the weakness of surveillance systems in some of the countries involved means it is difficult to rule out other possible causes, such as live bird trade. As such, the relative importance of wild birds in the introduction of disease remains hypothetical and the subject of continued debate. For example, assuming that wild birds are a major virus spreader, outbreaks would be expected in the Philippines, New Zealand, and Australia which are on the flyways of several Asian migratory waterfowl species (Gilbert et al. 2006a; Krauss et al. 2010): this has not yet been the case. However, it was shown experimentally that species migrating to Australia shed lower quantities of viruses than those migrating westward (East et al. 2008). Most surveillance campaigns have either failed to isolate the virus in wild birds or found it only on very rare occasions (e.g. Thailand (Siengsanan et al. 2009), Egypt (Saad et al. 2007), Switzerland (Baumer et al. 2010)), and even more rarely during migration periods (Feare 2010).

An additional argument against the role of wild birds as long distance transporters of H5N1 is that, to date, the majority of infected wild birds have been found either sick or dead (Olsen et al. 2006). In order to carry a virus over long distances, infected wild birds would need to show few or no adverse clinical symptoms (Weber et al. 2007). There have been reports of H5N1 infection in apparently healthy wild birds, particularly ducks (Chen et al. 2006; Starick et al. 2008; Saad et al. 2007; Siengsanan et al. 2009; Feare and Yasue 2006; Globig et al. 2009), as well as some terrestrial birds, such as sparrows (*Passer montanus*) (Kou et al. 2005). Experimental studies have demonstrated that some species, such as mallards (*Anas platyrhynchos*) and pochards (*Aythya ferina*)

(Keawcharoen et al. 2008; Brown et al. 2006), can shed the virus without or with very limited disease signs. Moreover, immunity induced by prior LPAI infection was shown experimentally to prevent overt clinical disease (Fereidouni et al. 2009). However, such studies do not account for the effect of migration on the immunological state of the birds involved. Indeed, the physiological cost of migration is high and the impact of avian influenza viruses on bird fitness will condition their long-distance spread (Weber et al. 2007). Although known to induce no or mild symptoms, LPAI infection was shown to delay migration and increase the frequency of stopovers in free-living mallards (*A. platyrhynchos*) and Bewick's swans (*Cygnus columbianus bewickii*) (van Gils et al. 2007; Latorre-Margalef et al. 2009).

The length of the asymptomatic infectious period may allow wild bird species to take part in short-distance disease spread (Kalthoff et al. 2008; Brown et al. 2008) as was probably the case for mute swans in Europe. Hence, whilst some birds, and in particular the dabbling ducks, may have a putative role in long-distance disease spread, there remains considerable uncertainty in the role of wild birds in all but the short-distance spread of HPAIV H5N1.

Farm-to-Farm Virus Spread and Virus Maintenance in Wild Birds

Opportunities for free-range poultry and wild birds to mix are numerous: terrestrial wild birds are likely to mix with scavenging poultry and the transformation of wetland areas into rice fields may have increased the rate of contacts between wild and domestic waterfowl (Artois et al. 2009). Access of wild birds to food and watering sources for poultry may also allow indirect transmission. On a local scale, wild birds could therefore transmit the infection to domestic birds and play a role in the spread of virus between farms. Indeed, presence of wild birds in feed troughs or in poultry confinement areas was identified as a risk factor for farm infection in case-control studies in Hong Kong, Viet Nam and Bangladesh (Henning et al. 2009a; Kung et al. 2007; Biswas et al. 2009a). Moreover, the risk of infection was higher in the vicinity of wetlands or water bodies in China, Bangladesh, Thailand and Romania (Biswas et al. 2009a; Ward et al. 2008; Paul et al. 2010a; Fang et al. 2008).

Outbreaks in wild birds are often associated with outbreaks on poultry farms (e.g. South Korea (Lee et al. 2008), Japan (Uchida et al. 2008), Russia (Feare 2007), Pakistan, India, Czech Republic, Poland and Ukraine (Feare 2010)). Virus isolates from both poultry and wild bird populations are often closely related phylogenetically, indicating that viruses were transmitted from one population to another, although the direction of transmission (i.e. from wild birds to poultry or vice versa) cannot be determined with certitude (Lee et al. 2008). In Thailand (Siengsanan et al. 2009), infected wild birds, mostly peri-domestic and commensal species, were rarely found in the areas where the disease was reported in poultry and are therefore unlikely to play a significant role in the epidemiology of the disease in that country. Moreover, experimental studies have shown that for terrestrial (i.e. non-aquatic) wild birds (e.g. sparrows), the intra-species transmission rates or rate of transmission to chickens is relatively low (Boon et al. 2007; Forrest et al. 2010).

Although the isolation of H5N1 viruses in wild birds is generally rare, it has been suggested that sample sizes used are often insufficiently large to detect very low prevalence (Fereidouni et al. 2009). It has been assumed that H5N1 viruses could be maintained at low prevalence levels in small subpopulation pockets of certain wild bird species (Haase et al. 2010), particularly ducks (Krauss et al. 2010), and such populations could then act as a virus reservoir for poultry. With regard to LPAI viruses, environmental contamination is very likely to play a role in virus maintenance, allowing transmission between wild bird populations that do not share the same site temporally (Brown et al. 2007, 2008). These viruses are mainly transmitted via the faecal-oral route (Webster et al. 1992) and high virus titres are released into the environment where they remain for long periods, particularly in surface water (Brown et al. 2007; Stallknecht et al. 2010). Indeed, viruses shed in the Arctic during one breeding season may remain infectious at the return of migrating birds for the following season (Ito et al. 1995). The isolation of LPAI viruses in wintering sites (Gaidet et al. 2007; Stallknecht and Shane 1988) supports the hypothesis that viruses could be also perpetuated in migratory bird populations throughout the year. In contrast, the maintenance of H5N1 viruses among wild bird populations is uncertain. H5N1 viruses do not persist as long as LPAI viruses in water (Brown et al. 2007). Moreover, these viruses are predominantly shed by the respiratory tract by Anseriformes species (Keawcharoen et al. 2008; Brown et al. 2006), and hence the transmission of H5N1 viruses requires high contact rates, and could thus be maladapted to natural ecosystems, where the contact rates vary between seasons and species.

Trade and Live Bird Markets

Disease Introduction into Disease-Free Areas

In the last 20 years, poultry production has increased at a huge rate in Asia. For example, between 1985 and 2005, the production of chicken and duck meat in China increased by almost 7 times (Gilbert et al. 2007). Increases in poultry production have also led to increases in both local and international trade in poultry products, including both legal and illegal activities. The commercial movement of live birds and poultry products may therefore have played a major role in the virus spread within and beyond Asia (Sims et al. 2005). Indeed, cross-border trade is very likely to have been responsible for the initial spread of the virus from southern China to Southeast Asia, as well as the continued introduction of new variants into the latter region (Wang et al. 2008b; Wan et al. 2008). A large number of spent hens and ducklings are known to move daily from China to Viet Nam (personal observations). The isolation of clade 7 viruses from poultry seized at the border between China and Viet Nam, and subsequently in Vietnamese LBMs, demonstrates how cross-border trade can lead to the introduction of new variants into Viet Nam, and how the local market chain can spread introduced viruses locally (Nguyen et al. 2009; Davis et al. 2010). The close similarity between viruses isolated in northern Viet Nam, Thailand and Malaysia is thought to be due to commercial movements of birds, as legal and illegal trade is well developed in the region (Smith et al. 2006b). Likewise, trade into and across Laos is also the most likely cause of virus introduction into that country. The well-established poultry trade, particularly of ducks, from southern Viet Nam to Cambodia (Van Kerkhove et al. 2009) may have caused multiple introductions of H5N1 viruses into Cambodia, and therefore explain the high degree of homology between viruses isolated in both regions (Buchy et al. 2009; Smith et al. 2006b). Virus spread from northern to southern Viet Nam is also believed to occur via the poultry trade (Wan et al. 2008).

Although migratory birds have been implicated, illegal cross-border trade from infected neighbouring countries, such as Bangladesh, is also a probable route for multiple virus introductions to India (Chakrabarti et al. 2009; Murugkar et al. 2008).

Larger scale poultry movements can also occur, for example legal and illegal trade of live poultry from China (ProMED-mail 2006) to Nigeria were known to be frequent (Cecchi et al. 2008). Phylogenetic studies indicate that the patterns of virus evolution and geographical strain distribution in Africa are coherent with poultry trade patterns (Cattoli et al. 2009). Hence, although the movement of infected wild birds cannot be ruled out, poultry trade is very likely to have had a role in the introduction of the virus into Africa.

Wild bird migration is considered to have been the main route for the introduction of the virus to Europe (Kilpatrick et al. 2006); however, the first two reports of H5N1 infection in Europe (Alexander 2007b; van den Berg et al. 2008), and a subsequent report in 2007 (Irvine et al. 2007), were associated with trade. Trade, rather than bird migration, is thought to have caused the westerly virus spread across Russia in 2005.

Farm-to-Farm Virus Transmission and Maintenance

Trade is also likely to be an important mechanism by which HPAI can spread from farm to farm (Sims 2007). Five out of seven outbreak waves in Viet Nam occurred around the celebration of the Tet, during which poultry trade activities increase drastically (Pfeiffer et al. 2007; Minh et al. 2009). Traders, or poultry collectors, have been particularly implicated in the farm-to-farm virus spread given that they may move between a potentially large number of farms in the course of a single day (Van Kerkhove et al. 2009), and come into direct contact with birds on each of these. In the absence of effective sanitation and disinfection, traders themselves, their equipment and their vehicles may act as important mechanical transmitters of infection. Indeed, farms visited by traders were at higher risk of infection in Hong Kong (Kung et al. 2007) and Thailand (Paul et al. 2010a) whilst farms that always used the same trader or who prevented the entry of traders were at lower risk in Bangladesh (Biswas et al. 2009b) and Nigeria (Métras et al. 2009), respectively. In northern Viet Nam, it was observed that traders tended to link communes with similar infection status, suggesting that they may have had a role in the spread of the virus between them (Soares Magalhaes et al. 2010).

LBMs are likely to play a particularly important role in the spread of HPAI. Retail marketing of live poultry was the main source of exposure to infection on chicken farms in Hong Kong (Kung et al. 2007). During the 2008 epidemic in South Korea, the virus was suspected to have spread throughout the country via LBMs and then to poultry farms (Kim et al. 2010). Investigations in Bangladesh identified egg trays and contaminated vehicles from LBMs as the cause of 47% of farm outbreaks (Biswas et al. 2008).

As well as trade activities that allow transmission at a local scale, the movement of poultry and poultry products along major transport routes is also likely to contribute to virus spread. Proximity to major roads, highways or big cities, and density of roads have been identified as risk factors for HPAI H5N1 in China (Fang et al. 2008), Thailand (Paul et al. 2010b), Bangladesh (Loth et al. 2010), Viet Nam (Pfeiffer et al. 2007), Indonesia (Yupiana et al. 2010) and Romania (Ward et al. 2008). In Nigeria, a high proportion of cases were located in proximity to main roads (Rivas et al. 2010). For countries where the road network is poorly developed, most commercial poultry movements occur on a small number of main roads, with LBMs located along these, with big cities, like Phnom Penh in Cambodia (Van Kerkhove et al. 2009), attracting a huge proportion of the commercial flow, and as such a large number of LBMs.

Farms supplying LBMs are typically either backyard or small-scale commercial farms, with a low level of biosecurity (Van Kerkhove et al. 2009; Tiensin et al. 2005; Soares Magalhaes et al. 2007). Importantly, it has been observed in Viet Nam, Bangladesh and Indonesia that farmers facing an outbreak may attempt to sell apparently healthy or even sick poultry in order to minimise economic losses. Such practices will inevitably increase the probability that infected birds will be introduced into the market chain (Biswas et al. 2009a; Yupiana et al. 2010; Phan Dang et al. 2007).

The major impact of live animal markets in virus spread has been described for severe acute respiratory syndrome (SARS) (Guan et al. 2003) and foot and mouth disease (FMD) (Ferguson et al. 2001; Ortiz-Pelaez et al. 2006). Due to the high density of hosts, LBMs offer conditions for virus amplification, re-assortment and cross-species transmission (Webster 2004). The diversity and abundance of LPAI viruses within LBMs has long been known (Shortridge et al. 1977; Senne et al. 1992), and is important in East Asian markets (Guan et al. 2002; Chen et al. 2006, 2009; Amonsin et al. 2008; Choi et al. 2005; Liu et al. 2003; Ge et al. 2009; Lee et al. 2010). H5N1 viruses have also been identified in these markets where they may circulate silently: whilst HPAI outbreaks were not reported in Viet Nam until 2003, H5N1 virus was identified in two LBMs around Ha Noi in 2001 (Nguyen et al. 2005). Moreover, during the H5N1 epidemics which affected

Hong Kong in 1997, the prevalence of the infection in chickens in LBMs reached 19.5% (Shortridge 1999). Since 2003, H5N1 viruses have been isolated from LBMs in both epidemic and endemic areas, such as China (Jiang et al. 2010; Chen et al. 2009), Thailand (Amonsin et al. 2008), Indonesia (Santhia et al. 2009), South Korea (Kang et al. 2009), Bangladesh (Biswas et al. 2008) and Nigeria (Joannis et al. 2008). A survey in Egypt in 2009 found that 12.4% (71/573) of sampled LBMs were infected (Abdelwhab et al. 2010). The environment of 47% markets sampled in 2007–2008 in Indonesia was found to be contaminated (Indriani et al. 2010).

Following multiple outbreaks of HPAI H5N1 in Hong Kong between 1997 and 2003, control strategies were implemented across the LBM chain (Sims et al. 2003a, b; Guan et al. 2007). These interventions appear to have been successful as only one outbreak has been reported on the island since 2003 (ProMED-mail 2008). Among control measures, the provision of rest days that allow markets to be emptied and disinfected, have been associated with a significant decrease in the rate of isolation of LPAI viruses in Hong Kong retail markets (Kung et al. 2003; Lau et al. 2007). Similar observations have been noted in the United States where surveys highlighted that rest days, frequent cleaning and disinfection decrease the risk that the market is positive for LPAI (Bulaga et al. 2003; Garber et al. 2007; Yee et al. 2008; Trock et al. 2008). These observations suggest that the level of infection in markets is not simply the result of multiple introductions of infected birds, but the consequence of virus re-circulation and amplification within them. In Indonesia, the H5N1 virus isolation rate was higher among poultry sampled in LBMs than in farms, also suggesting that the virus may be amplified in the market chain (Santhia et al. 2009).

Thus, LBMs may play a key role in the epidemiology of avian influenza viruses; acting as a network "hub", they may be responsible for sustaining endemic infection within the poultry sector. Poultry that are purchased alive, birds that return unsold, or the movement of people or equipment contaminated with virus at such markets may play an important role in the onward spread of the disease.

Ducks: Silent Viral Vectors and Potential Reservoir

Asymptomatic Infection

It has been noted that the susceptibility of ducks to H5N1 has varied since these viruses first emerged. Until 2002, infected ducks tended to show mild or no clinical signs following infection (Shortridge et al. 1998; Alexander 2000; Perkins et al. 2002). In contrast, H5N1 viruses isolated in Hong Kong in 2002 were pathogenic for wild waterfowl (Ellis et al. 2004; Sturm-Ramirez et al. 2004) whilst new variants that have emerged since 2003 have tended to have lower pathogenicity.

Such viruses are excreted in high titres for extended periods of up to 17 days, often in the absence of clinical signs (Hulse-Post et al. 2005). Indeed, viruses have been isolated from healthy ducks in Laos (Boltz et al. 2006), and serological surveys in Viet Nam (Takakuwa et al. 2010) found H5 and N1 inhibiting antibodies in unvaccinated birds suggesting that ducks had been infected and survived, although no outbreak was reported. In Thailand in 2005, free-range duck flocks shed viruses for 5–10 days before being culled with few or no disease signs (Songserm et al. 2006a). Moreover, H5N1 viruses were reported to have circulated silently in a commercial duck flock in Germany (Harder et al. 2009).

There appears to be some variability in the pathogenicity of H5N1 viruses for ducks by H5N1 strain (Londt et al. 2008; Saito et al. 2009; Sturm-Ramirez et al. 2005; Tian et al. 2005; Middleton et al. 2007; Vascellari et al. 2007; Bingham et al. 2009). In an experimental setting, Saito et al. (2009) found that the mortality rate could vary from 50 to 75% according to the strain. During outbreaks in South Korea in 2008, the morbidity and mortality rates of an infected duck farm were 60 and 50%, respectively (Kim et al. 2010). In contrast, mortality in waterfowl in Egypt appeared to be lower than 30% (Abdel-Moneim et al. 2009). Even when the infection is lethal, virus shedding tends to persist longer in ducks than in chickens. Saito et al. (2009) showed that the mean death time (MDT) varied between 4.8 and 6.3 days in ducks. Moreover, susceptibility may vary with duck breed (Saito et al. 2009) and age (Londt et al. 2010): infection that was always lethal for 8-week-old ducks was mild in 12-week-old ducks. During the 2003–2004 epidemic in South Korea, morbidity and mortality were lower in adult birds than in younger ones, and infection in a duck breeder farm was only detected due to the identification of symptoms in ducklings (Kwon et al. 2005b). This suggests that long-life duck flocks, such as breeder and layer flocks, are at highest risk of amplifying and silently spreading the virus.

Domestic waterfowl may act as asymptomatic virus carriers, and therefore act as a potential virus reservoir for more susceptible species, such as chickens. The presence of these so-called "Trojan horses" complicates the control of the disease.

Farm-to-Farm Spread and Maintenance

The density of duck flocks in a region has been shown to be a risk factor for infection with HPAI H5N1 (Pfeiffer et al. 2007; Paul et al. 2010b; Gilbert et al. 2006b; Tiensin et al. 2009). At the farm level, the number of ducks present and interactions with ducks from other flocks have been also identified as risk factors for disease (Henning et al. 2009a; Biswas et al. 2009a, b; Paul et al. 2010a). The presence of ponds and water bodies, which may act as an interface between domestic and wild waterfowl and between neighbouring waterfowl flocks, also increases the risk of infection (Biswas et al. 2009a; Ward et al. 2008; Paul et al. 2010a; Fang et al. 2008). Water bodies may therefore act as a meeting point where H5N1 viruses can be transmitted directly or indirectly between ducks from different flocks when they congregate at these places.

With the exception of Thailand, which has restructured its duck raising system since 2004, duck flocks are rarely kept in strict confinement in Southeast Asia (Songserm et al. 2006a; Burgos et al. 2008a, b). Flocks are generally free-ranging or have open access to ponds. Biosecurity measures are difficult to implement in such systems, and the same inadequate biosecurity measures that allow HPAIV H5N1 to enter a flock may allow the virus to spread onwards in the event of an outbreak.

The practice of grazing ducks on rice paddies may be a critical factor in the maintenance and spread of H5N1 viruses in Southeast Asia (Henning et al. 2009a, b; Paul et al. 2010a, b; Pfeiffer et al. 2007; Yupiana et al. 2010; Gilbert et al. 2006b, 2008). Young ducks may scavenge for insects and snails during the rice growing period (Minh et al. 2010), while adult ducks are allowed to scavenge on the fields for periods ranging from 2 weeks to 2 months after harvest (Henning et al. 2009a). This husbandry practice involves the frequent movements of flocks from one field to another (Gilbert et al. 2006a), sometimes over long distances (Songserm et al. 2006a; Minh et al. 2010). Hence, rice paddies offer an opportunity for domestic ducks to infect ducks from other flocks, either through direct contacts or indirectly by contaminating the field. Moreover, rice paddies may also provide an interface with wild bird populations. At night, several duck flocks may be housed in common shelters within villages, which may contribute to virus spread between free-grazing duck flocks, as well as wider spread to village poultry. In Thailand, both the first and second waves of outbreaks affected areas with a high density of free-grazing ducks, but these were followed by outbreaks in high chicken density areas (Songserm et al. 2006a) suggesting free-grazing ducks were involved in the dissemination of the virus to the chicken population. The occurrence of most Vietnamese outbreaks around the Tet festival coincides with an increase in poultry trade, as well as the period during which ducks are brought to rice paddies (Pfeiffer et al. 2007).

Impact of Farming Systems and Practices on Virus Spread

Farming Systems

Some studies have found that high poultry density areas or areas with commercial farms were at lower risk of HPAI outbreaks (Biswas et al. 2009a; Yupiana et al. 2010; Henning et al. 2009b). Indeed, small-scale farms have generally appeared to be more susceptible to infection than large-scale industrial farms, probably due to the fact that larger farms applied better husbandry practices, better biosecurity, and were more likely to vaccinate (Sims et al. 2005).

The high proportion of poultry kept in backyard flocks in H5N1 endemic countries such as Egypt, Viet Nam and Cambodia (Burgos et al. 2008a, b; Hosny et al. 2006) has raised the concern that this type of poultry husbandry may contribute to virus maintenance (Peiris et al. 2007; Iqbal 2009). Backyard flocks were also suspected to have played a role in the spread of the virus in Nigeria (Joannis et al. 2008).

Backyard flocks are typically maintained as low input systems, and are allowed to range freely for most of the day. Although they are primarily produced for household consumption, backyard birds may be sold to mobile traders or directly at LBMs. Levels of biosecurity in this sector are extremely low, or non-existent, and as such the risk of infection may be very high (FAO et al. 2004). In Egypt, for example, a survey carried out in 2007 found H5N1 viruses in 30% of sampled backyard flocks (Hafez et al. 2010). Moreover, access to veterinary services is often limited in this sector and background disease burdens are often high, even in the absence of HPAI. Hence backyard poultry owners may not recognise HPAI H5N1 as an immediate threat (Cardona et al. 2010) and outbreaks caused by H5N1 viruses may remain unreported to veterinary services and interventions subsequently delayed.

In Laos and Cambodia, where almost all poultry are reared in backyards, the mechanisms contributing to virus maintenance are uncertain. It is thought that the poultry density is too low to enable the virus to be maintained and therefore that the sporadic outbreaks that occur in these countries are a consequence of repeated virus introductions, in particular from southern Viet Nam (Buchy et al. 2009). In Thailand, subdistricts with backyard flocks were at lower risk of infection than subdistricts with commercial flocks (Tiensin et al. 2009).

Small-scale commercial farms are very likely to play an important role in the spread of HPAI within affected areas. Such farms may contain several hundred or even thousands of birds within a single shed. Levels of biosecurity are generally low and contacts with poultry production stakeholders may be numerous (e.g. traders, feed sellers) (Sims et al. 2005). Hence, the risk of virus introduction, amplification and then spread to other farms is high. Most outbreaks notified in Viet Nam from 2004 to 2007 occurred in farms with 50–3,000 birds (Burgos et al. 2008a). Henning et al. (2009b) found that medium poultry density, which probably represents small commercial farms, was associated with an increase in the risk of infection in Viet Nam.

Fighting cocks were associated with a higher risk of infection in Thailand, but this association was weak (Paul et al. 2010b; Gilbert et al. 2006b; Tiensin et al. 2009). The presence of quail flocks in an area has also been identified as a risk factor for HPAI, although the epidemiological significance of this finding is unknown (Tiensin et al. 2009).

Husbandry Practices

Several husbandry practices that are not associated with trade have been identified as risk factors for HPAI disease outbreaks. Owners living off the farm or having visitors entering the premises were associated with higher risk of infection (Henning et al. 2009a; Kung et al. 2007), and highlight the potential role of humans as mechanical transmitters of virus. Increased risks associated with such indirect contact are also likely to occur through shared equipment, as shown for other avian influenza viruses (Capua et al. 2000, 2007b; Wee et al. 2006; Nishiguchi et al. 2007; Thomas et al. 2005). Likewise, contact with wild animals, rodents and even flies may allow H5N1 to spread from farm to farm (Biswas et al. 2009a, b; Barbazan et al. 2008; Butler 2006; Kuiken et al. 2006; Sawabe et al. 2006). Thai subdistricts with poultry slaughterhouses were also at higher risk, probably due to the regular movement of vehicles and cages to and from slaughterhouses, which may have acted as a virus dissemination point (Tiensin et al. 2009).

Vaccinated flocks had a lower risk of infection in Viet Nam (Henning et al. 2009a). However, in the absence of strict sanitation and disinfection, the movement of vaccinators from farm to farm may actually allow the spread of the virus through mechanical transmission. Indeed, activities associated with the first Vietnamese vaccination campaign may have been responsible for an outbreak wave (Pfeiffer et al. 2007), and vaccinators are likely to play a role in virus dissemination in Egypt (Peyre et al. 2009).

Extension of the Mammalian Host Range

Mammals, Other Than Humans

Although reports of LPAI transmission from birds to mammals are rare, H5N1 viruses have shown a great capacity for xenospecific transmission (Reperant et al. 2009). Among carnivores, various field species have been infected, including tigers (*Panthera tigris*) and leopards (*Panthera pardus*) in Thailand (Keawcharoen et al. 2004) and probably in Cambodia (Desvaux et al. 2009). The susceptibility of domestic dogs and cats has been highlighted by isolation of viruses (Songserm et al. 2006b–c; Leschnik et al. 2007), as well as through serological surveys (Butler 2006). Other carnivore species are also susceptible (Reperant et al. 2009). The role any mammalian species has in the epidemiology of this disease is uncertain. Cats coming into contact with domestic birds, and potentially their droppings, may develop severe disease and excrete virus from the respiratory and digestive tracts (Kuiken et al. 2006). Hence, as well as allowing transmission of H5N1 between cats (Kuiken et al. 2004; Ayyalasomayajula et al. 2008; Rimmelzwaan et al. 2006), such active secretion might suggest a role for these animals in the spread of disease between domestic poultry, although this remains highly speculative (Kuiken et al. 2006).

Although rodents are likely to be exposed to the H5N1 virus through contact with poultry, no cases of natural infection have been reported so far. Mice inoculated experimentally are susceptible and widely used as a model for human infection; however, experimentally inoculated rats appeared to be resistant to the infection (Perkins and Swayne 2003).

Pigs are susceptible to both human and avian influenza viruses and may therefore act as a "mixing vessel" (Ito et al. 1998) for the generation of pandemic viruses through re-assortment, as was recently observed with the emergence of the H1N1 virus (Smith et al. 2009). Surveys carried out in Viet Nam (Choi et al. 2004) suggested that pig susceptibility to H5N1 viruses is low, and this has been confirmed through experimental infection (Lipatov et al. 2008).

Humans are the only primates for which outbreaks have been reported. Macaques have been infected experimentally to serve as a primate model and were found to be susceptible (Rimmelzwaan et al. 2001, 2003).

Humans

To our knowledge, only three HPAIV subtypes have been transmitted to humans: one individual was found infected by the Canadian H7N3 isolate in 2003; in The Netherlands, 89 people were infected by H7N7 viruses, of which one person died (Katz et al. 2009); and as of 31 August 2010, 505 humans have been found to be infected with H5N1, of which 300 have died (WHO 2010). Transmission of H5N1 viruses to humans was first identified in 1997 when a 3-year-old boy died in Hong Kong (Claas et al. 1998). In the following months, 17 additional human cases were reported (Sims et al. 2003a). Viet Nam, Indonesia and Egypt have 79% of all human cases reported to the World Health Organisation. Although some human cases may be the result of human-to-human transmission, especially between family members (Kandun et al. 2006; Gilsdorf et al. 2006; Olsen et al. 2005; Ungchusak et al. 2005; Brankston et al. 2007), this transmission route appears to be very rare. The majority of human cases are thought to have arisen from direct or indirect contact with infected poultry or their products (Wang et al. 2008a; Gambotto et al. 2008).

Pandemic viruses that arose in the last century have all shown an avian origin. Indeed, the H1N1 strain responsible for the 1918–1919 pandemic which caused between 40 and 50 million human deaths (Webster et al. 1992) was entirely derived from an avian virus that adapted to humans. The 1957 and 1968 pandemics, caused by H2N2 and H3N2 viruses, respectively, resulted from re-assortment between viruses of human and avian origins (Kawaoka et al. 1989), and the 2009 H1N1 pandemic was caused by the re-assortment between human, avian and swine viruses (Smith et al. 2009). Moreover, recent re-assortants between H5N1 and human H3N2 viruses have shown high virulence and demonstrated the potential for H5N1 viruses to recombine with strains circulating in the human population (Li et al. 2010b). Therefore, although H5N1 viruses do not seem to transmit easily to humans, the continuing circulation of H5N1 viruses in the poultry population and its high case fatality rate in humans still raises a great concern about the potential emergence of a highly lethal pandemic strain.

Conclusion

Since its emergence in 1996, HPAIV H5N1 has spread across three continents. This sporadic, large-scale spread appears to originate from a small number of areas of virus persistence, particularly in Asia. Such pockets of infection exist within specific agro-ecological niches in which virus persistence and intense genetic diversification allows the continuous emergence of new strains. Trade of live birds and poultry products is likely to be a major pathway for virus dissemination within and beyond these endemic areas, whilst LBMs are likely to contribute to virus persistence. Duck rearing and associated practices create conditions for undetected virus amplification within flocks, and the high level of contacts between duck flocks facilitates the spread of the virus. Moreover, although they are likely to play only a limited role in disease dynamics, wild birds can potentially take part in virus dissemination at both a local and a continental scale.

Due to the multiple features that allow HPAI viruses to persist in some agro-ecosystems, the design of control strategies needs to take into account the local epidemiological patterns and characteristics of these production systems. These policies need to target the pocket of infection as a whole, and thus have to be coordinated at a regional level, as agro-ecosystems generally overlap several countries. Otherwise, attempts to control the virus circulation are likely to fail. This is particularly important as the continuous emergence of new HPAIV strains, and their co-circulation with swine and human influenza viruses, is a major concern for public health.

References

A. S. Abdel-Moneim *et al.*, *Arch Virol* **154**, 1559 (2009).
E. M. Abdelwhab *et al.*, *Avian Dis* **54**, 911 (Jun, 2010).
S. S. Ahmed, A. K. Ersboll, P. K. Biswas, J. P. Christensen, *Epidemiol Infect* **138**, 843 (Jun, 2010).
D. J. Alexander, *Vet Microbiol* **74**, 3 (May 22, 2000).
D. J. Alexander, *Vaccine* **25**, 5637 (Jul 26, 2007a).
D. J. Alexander, *Avian Dis* **51**, 161 (Mar, 2007b).
D. J. Alexander, I. H. Brown, *Rev Sci Tech* **28**, 19 (Apr, 2009).
D. J. Alexander, S. A. Lister, M. J. Johnson, C. J. Randall, P. J. Thomas, *Vet Rec* **132**, 535 (May 22, 1993).
M. M. Aly, A. Arafa, M. K. Hassan, *Avian Dis* **52**, 269 (Jun, 2008).
A. Amonsin *et al.*, *Emerg Infect Dis* **14**, 1739 (Nov, 2008).
A. Arafa, D. L. Suarez, M. K. Hassan, M. M. Aly, *Avian Dis* **54**, 345 (Mar, 2010).
M. Artois *et al.*, *Rev Sci Tech* **28**, 69 (Apr, 2009).
S. Ayyalasomayajula, D. A. DeLaurentis, G. E. Moore, L. T. Glickman, *Zoonoses Public Health* **55**, 497 (Oct, 2008).
A. L. Balish *et al.*, *Avian Dis* **54**, 329 (Mar, 2010).
P. Barbazan *et al.*, *Vector Borne Zoonotic Dis* **8**, 105 (Spring, 2008).
A. Baumer *et al.*, *Avian Dis* **54**, 875 (Jun, 2010).
J. Bingham *et al.*, *Avian Pathol* **38**, 267 (Aug, 2009).

P. K. Biswas *et al.*, *Emerg Infect Dis* **14**, 1909 (Dec, 2008).
P. K. Biswas *et al.*, *Emerg Infect Dis* **15**, 1931 (Dec, 2009a).
P. K. Biswas *et al.*, *Vet Rec* **164**, 743 (Jun 13, 2009b).
D. A. Boltz *et al.*, *Emerg Infect Dis* **12**, 1593 (Oct, 2006).
D. A. Boltz *et al.*, *J Gen Virol* **91**, 949 (Apr, 2010).
A. C. Boon *et al.*, *Emerg Infect Dis* **13**, 1720 (Nov, 2007).
G. Brankston, L. Gitterman, Z. Hirji, C. Lemieux, M. Gardam, *Lancet Infect Dis* **7**, 257 (Apr, 2007).
I. H. Brown, *Avian Dis* **54**, 187 (Mar, 2010).
J. D. Brown, D. E. Stallknecht, J. R. Beck, D. L. Suarez, D. E. Swayne, *Emerg Infect Dis* **12**, 1663 (Nov, 2006).
J. D. Brown, D. E. Swayne, R. J. Cooper, R. E. Burns, D. E. Stallknecht, *Avian Dis* **51**, 285 (Mar, 2007).
J. D. Brown, D. E. Stallknecht, D. E. Swayne, *Emerg Infect Dis* **14**, 136 (Jan, 2008).
P. Buchy *et al.*, *Emerg Infect Dis* **15**, 1641 (Oct, 2009).
L. L. Bulaga *et al.*, *Avian Dis* **47**, 996 (2003).
S. Burgos, J. Hinrichs, J. Otte, D. Pfeiffer, D. Roland-Holst, "Poultry, HPAI and Livelihoods in Viet Nam – A Review" (Rome, 2008a).
S. Burgos *et al.*, "Poultry, HPAI and Livelihoods in Cambodia – A Review" (Rome, 2008b).
D. Butler, *Nature* **439**, 773 (Feb 16, 2006).
I. Capua, D. J. Alexander, *Avian Pathol* **33**, 393 (Aug, 2004).
I. Capua, D. J. Alexander, *Influenza Other Respi Viruses* **1**, 11 (Jan, 2007).
I. Capua, S. Marangon, *Avian Pathol* **29**, 289 (Aug, 2000).
I. Capua, S. Marangon, *Vaccine* **25**, 5645 (Jul 26, 2007).
C. J. Cardona *et al.*, *Avian Dis* **54**, 754 (Mar, 2010).
G. Cattoli *et al.*, *PLoS One* **4**, e4842 (2009).
A. N. Cauthen, D. E. Swayne, S. Schultz-Cherry, M. L. Perdue, D. L. Suarez, *J Virol* **74**, 6592 (Jul, 2000).
G. Cecchi, A. Ilemobade, Y. Le Brun, L. Hogerwerf, J. Slingenbergh, *Geospat Health* **3**, 7 (Nov, 2008).
A. K. Chakrabarti *et al.*, *PLoS One* **4**, e7846 (2009).
H. Chen *et al.*, *Proc Natl Acad Sci U S A* **101**, 10452 (Jul 13, 2004).
H. Chen *et al.*, *Nature* **436**, 191 (Jul 14, 2005).
H. Chen *et al.*, *Proc Natl Acad Sci U S A* **103**, 2845 (Feb 21, 2006).
J. Chen *et al.*, *Virus Res* **146**, 19 (Dec, 2009).
Y. K. Choi, S. M. Goyal, H. S. Joo, *Am J Vet Res* **65**, 303 (Mar, 2004).
Y. K. Choi, S. H. Seo, J. A. Kim, R. J. Webby, R. G. Webster, *Virology* **332**, 529 (Feb 20, 2005).
E. C. Claas *et al.*, *Lancet* **351**, 472 (Feb 14, 1998).
C. T. Davis *et al.*, *Avian Dis* **54**, 307 (Mar, 2010).
J. C. de Jong, E. C. Claas, A. D. Osterhaus, R. G. Webster, W. L. Lim, *Nature* **389**, 554 (Oct 9, 1997).
S. Desvaux *et al.*, *Emerg Infect Dis* **15**, 475 (Mar, 2009).
L. Duan *et al.*, *Virology* **380**, 243 (Oct 25, 2008).
M. F. Ducatez *et al.*, *Nature* **442**, 37 (Jul 6, 2006).
M. F. Ducatez *et al.*, *J Gen Virol* **88**, 2297 (Aug, 2007).
T. Dung Nguyen *et al.*, *Emerg Infect Dis* **14**, 632 (Apr, 2008).
D. Eagles *et al.*, *Rev Sci Tech* **28**, 341 (Apr, 2009).
I. J. East, S. Hamilton, G. Garner, *Geospat Health* **2**, 203 (May, 2008).
T. M. Ellis *et al.*, *Avian Pathol* **33**, 492 (Oct, 2004).
L. Q. Fang *et al.*, *PLoS One* **3**, e2268 (2008).
FAO, "Recommendations on the Prevention, Control and Eradication of Highly Pathogenic" (Rome, 2004).

F. O. Fasina, S. P. Bisschop, T. M. Joannis, L. H. Lombin, C. Abolnik, *Epidemiol Infect* **137**, 456 (Apr, 2009).
C. J. Feare, *Avian Dis* **51**, 440 (Mar, 2007).
C. J. Feare, *Avian Dis* **54**, 201 (Mar, 2010).
C. J. Feare, M. Yasue, *Virol J* **3**, 96 (2006).
S. R. Fereidouni *et al.*, *PLoS One* **4**, e6706 (2009).
N. M. Ferguson, C. A. Donnelly, R. M. Anderson, *Science* **292**, 1155 (May 11, 2001).
N. M. Ferguson, C. Fraser, C. A. Donnelly, A. C. Ghani, R. M. Anderson, *Science* **304**, 968 (May 14, 2004).
H. L. Forrest, J. K. Kim, R. G. Webster, *J Virol* **84**, 3718 (Apr, 2010).
A. Fusaro *et al.*, *Emerg Infect Dis* **15**, 445 (Mar, 2009).
A. Fusaro *et al.*, *J Virol* **84**, 3239 (Apr, 2010).
N. Gaidet *et al.*, *Emerg Infect Dis* **13**, 626 (Apr, 2007).
N. Gaidet *et al.*, *Emerg Infect Dis* **14**, 1164 (Jul, 2008).
G. L. Gall-Recule *et al.*, *Avian Pathol* **37**, 15 (Feb, 2008).
A. Gambotto, S. M. Barratt-Boyes, M. D. de Jong, G. Neumann, Y. Kawaoka, *Lancet* **371**, 1464 (Apr 26, 2008).
L. Garber, L. Voelker, G. Hill, J. Rodriguez, *Avian Dis* **51**, 417 (Mar, 2007).
M. Garcia, J. M. Crawford, J. W. Latimer, E. Rivera-Cruz, M. L. Perdue, *J Gen Virol* **77** (**Pt 7**), 1493 (Jul, 1996).
F. F. Ge *et al.*, *J Clin Microbiol* **47**, 3294 (Oct, 2009).
M. Gilbert *et al.*, *Emerg Infect Dis* **12**, 1650 (Nov, 2006a).
M. Gilbert *et al.*, *Emerg Infect Dis* **12**, 227 (Feb, 2006b).
M. Gilbert *et al.*, *Agric Ecosyst Environ* **119**, 409 (2007).
M. Gilbert *et al.*, *Proc Natl Acad Sci U S A* **105**, 4769 (Mar 25, 2008).
A. Gilsdorf *et al.*, *Euro Surveill* **11**, 122 (2006).
A. Globig *et al.*, *Transbound Emerg Dis* **56**, 57 (Apr, 2009).
Y. Guan *et al.*, *Proc Natl Acad Sci U S A* **99**, 8950 (Jun 25, 2002).
Y. Guan *et al.*, *Science* **302**, 276 (Oct 10, 2003).
Y. Guan *et al.*, *BMC Infect Dis* **7**, 132 (2007).
Y. Guan, G. J. Smith, R. Webby, R. G. Webster, *Rev Sci Tech* **28**, 39 (Apr, 2009).
M. Haase *et al.*, *Infect Genet Evol* **10**, 1075 (Oct, 2010).
M. H. Hafez *et al.*, *Poult Sci* **89**, 1609 (Aug, 2010).
T. C. Harder *et al.*, *Emerg Infect Dis* **15**, 272 (Feb, 2009).
K. A. Henning *et al.*, *Prev Vet Med* **91**, 179 (Oct1, 2009a).
J. Henning, D. U. Pfeiffer, T. Vu le, *Vet Res* **40**, 15 (May-Jun, 2009b).
J. Henning *et al.*, *Emerg Infect Dis* **16**, 1244 (Aug, 2010).
F. Hosny, "The Structure and Importance of the Commercial and Village Based Poultry Systems in Egypt, FAO/ECTAD/AGAP report" (2006).
D. J. Hulse-Post *et al.*, *Proc Natl Acad Sci U S A* **102**, 10682 (Jul 26, 2005).
R. Indriani *et al.*, *Emerg Infect Dis* **16**, 1889 (Dec, 2010).
M. Iqbal, *J Mol Genet Med* **3**, 119 (2009).
R. M. Irvine *et al.*, *Vet Rec* **161**, 100 (Jul 21, 2007).
T. Ito *et al.*, *Arch Virol* **140**, 1163 (1995).
T. Ito *et al.*, *J Virol* **72**, 7367 (Sep, 1998).
W. M. Jiang *et al.*, *J Gen Virol* **91**, 2491 (Oct, 2010).
T. M. Joannis *et al.*, *Euro Surveill* **13**, (Oct 16, 2008).
D. Kalthoff *et al.*, *Emerg Infect Dis* **14**, 1267 (Aug, 2008).
I. N. Kandun *et al.*, *N Engl J Med* **355**, 2186 (Nov 23, 2006).
S. J. Kang *et al.*, *Virus Genes* **38**, 80 (Feb, 2009).
H. M. Kang *et al.*, *J Wildl Dis* **46**, 878 (Jul, 2010).
J. M. Katz *et al.*, *Poult Sci* **88**, 872 (Apr, 2009).
Y. Kawaoka, S. Krauss, R. G. Webster, *J Virol* **63**, 4603 (Nov, 1989).

J. Keawcharoen *et al.*, *Emerg Infect Dis* **10**, 2189 (Dec, 2004).
J. Keawcharoen *et al.*, *Emerg Infect Dis* **14**, 600 (Apr, 2008).
A. M. Kilpatrick *et al.*, *Proc Natl Acad Sci U S A* **103**, 19368 (Dec 19, 2006).
H. R. Kim *et al.*, *Vet Microbiol* **141**, 362 (Mar 24, 2010).
Z. Kou *et al.*, *J Virol* **79**, 15460 (Dec, 2005).
S. Krauss, R. G. Webster, *Avian Dis* **54**, 394 (Mar, 2010).
T. Kuiken *et al.*, *Science* **306**, 241 (Oct 8, 2004).
T. Kuiken, R. Fouchier, G. Rimmelzwaan, A. Osterhaus, P. Roeder, *Nature* **440**, 741 (Apr 6, 2006).
N. Y. Kung *et al.*, *Avian Dis* **47**, 1037 (2003).
N. Y. Kung *et al.*, *Emerg Infect Dis* **13**, 412 (Mar, 2007).
Y. K. Kwon *et al.*, *J Wildl Dis* **41**, 618 (Jul, 2005a).
Y. K. Kwon *et al.*, *Avian Pathol* **34**, 367 (Aug, 2005b).
T. T. Lam *et al.*, *PLoS Pathog* **4**, e1000130 (2008).
N. Latorre-Margalef *et al.*, *Proc Biol Sci* **276**, 1029 (Mar 22, 2009).
E. H. Y. Lau *et al.*, *Emerging Infectious diseases* **13**, 1340 (2007).
Y. J. Lee *et al.*, *Emerg Infect Dis* **14**, 487 (Mar, 2008).
H. J. Lee *et al.*, *Avian Dis* **54**, 738 (Mar, 2010).
M. Leschnik *et al.*, *Emerg Infect Dis* **13**, 243 (Feb, 2007).
K. S. Li *et al.*, *Nature* **430**, 209 (Jul 8, 2004).
Y. Li *et al.*, *J Virol* **84**, 8389 (Sep, 2010a).
C. Li *et al.*, *Proc Natl Acad Sci U S A* **107**, 4687 (Mar 9, 2010b).
A. S. Lipatov *et al.*, *PLoS Pathog* **4**, e1000102 (Jul, 2008).
M. Liu *et al.*, *Virology* **305**, 267 (Jan 20, 2003).
J. Liu *et al.*, *Science* **309**, 1206 (Aug 19, 2005).
B. Z. Londt *et al.*, *Avian Pathol* **37**, 619 (Dec, 2008).
B. Z. Londt *et al.*, *Influenza Other Respi Viruses* **4**, 17 (Jan, 2010).
L. Loth, M. Gilbert, M. G. Osmani, A. M. Kalam, X. Xiao, *Prev Vet Med* **96**, 104 (Aug 1, 2010).
V. Martin *et al.*, *Dev Biol (Basel)* **124**, 23 (2006).
M. Mase *et al.*, *Virology* **339**, 101 (Aug 15, 2005).
R. Métras *et al.*, paper presented at the Proceedings of the 12th Symposium of the International Society for Veterinary Epidemiology and Economics, Durban, 2009.
D. Middleton *et al.*, *Virology* **359**, 66 (Mar 1, 2007).
P. Q. Minh *et al.*, *Prev Vet Med* **89**, 16 (May 1, 2009).
P. Q. Minh, M. A. Stevenson, B. Schauer, R. S. Morris, T. D. Quy, *Prev Vet Med* **94**, 101 (Apr 1, 2010).
A. C. Mishra *et al.*, *Virol J* **6**, 26 (2009).
I. Monne *et al.*, *Emerg Infect Dis* **14**, 637 (Apr, 2008).
H. V. Murugkar *et al.*, *Vet Rec* **162**, 255 (Feb 23, 2008).
D. C. Nguyen *et al.*, *Journal of Virology* **79**, 4201 (Apr, 2005).
T. Nguyen *et al.*, *Virology* **387**, 250 (May 10, 2009).
A. Nishiguchi *et al.*, *Zoonoses Public Health* **54**, 337 (2007).
OIE. (2010a) H5N1 Notified in Domestic Poultry 2003-2009. Paris, France. See http://www.oie.int/animal-health-in-the-world/web-portal-on-avian-influenza/about-ai/h5n1-notified-in-domestic-poultry-2003-2009/
OIE. (2010b) Facts & Figures: Avian Influenza (ed. OIE). Paris, France. See http://www.oie.int/eng/info_ev/en_AI_factoids_2.htm
S. J. Olsen *et al.*, *Emerg Infect Dis* **11**, 1799 (Nov, 2005).
B. Olsen *et al.*, *Science* **312**, 384 (Apr 21, 2006).
A. Ortiz-Pelaez, D. U. Pfeiffer, R. J. Soares-Magalhaes, F. J. Guitian, *Prev Vet Med* **76**, 40 (Sep 15, 2006).
A. D. Osterhaus, G. A. Poland, *Vaccine* **26 Suppl 4**, D1 (Sep 12, 2008).
A. A. Owoade *et al.*, *Emerg Infect Dis* **14**, 1731 (Nov, 2008).
M. Paul *et al.*, in *Society for Veterinary Epidemiology and Preventive Medicine* L. A. K. a. t. S. E. C. L. Alban, Ed. (Nantes, 2010a), pp. 249–257.

M. Paul *et al.*, *Vet Res* **41**, 28 (May-Jun, 2010b).
J. S. Peiris *et al.*, *Lancet* **363**, 617 (Feb 21, 2004).
J. S. Peiris, M. D. de Jong, Y. Guan, *Clin Microbiol Rev* **20**, 243 (Apr, 2007).
L. E. Perkins, D. E. Swayne, *Avian Dis* **46**, 53 (Jan-Mar, 2002).
L. E. Perkins, D. E. Swayne, *Avian Dis* **47**, 956 (2003).
M. Peyre *et al.*, *J Mol Genet Med* **3**, 198 (2009).
D. U. Pfeiffer *et al.*, "Migratory Birds and their Possible Role in the Spread of Highly Pathogenic Avian Influenza" (2006).
D. U. Pfeiffer, P. Q. Minh, V. Martin, M. Epprecht, M. J. Otte, *Vet J* **174**, 302 (Sep, 2007).
T. Phan Dang *et al.*, paper presented at the 12th International Conference of the Association of Institutions for Tropical Veterinary Medicine, Frontignan, 2007.
ProMED-mail. (2006) Avian Influenza (33): Avian influenza, poultry vs migratory birds. See http://www.promedmail.org/pls/apex/f?p=2400:1202:1137555469103831::NO::F2400_P1202_CHECK_DISPLAY,F2400_P1202_PUB_MAIL_ID:X,33790
ProMED-mail. (2008) Avian Influenza (75): China (Hong Kong S.A.R.) Bird flu virus found in market stalls. See http://www.promedmail.org/pls/otn/f?p=2400:1001:4465822385385527::NO::F2400_P1001_BACK_PAGE,F2400_P1001_PUB_MAIL_ID:10001,72755
ProMED-mail. (2009) Avian Inlfuenza (41): India (West Bengal). See http://www.promedmail.org/pls/apex/f?p=2400:1202:3528901951843742::NO::F2400_P1202_CHECK_DISPLAY,F2400_P1202_PUB_MAIL_ID:X,77772
ProMED-mail. (2010a) Avian Influenza (06): Bangladesh (Rajshahi) correction. See http://www.promedmail.org/pls/apex/f?p=2400:1202:3528901951843742::NO::F2400_P1202_CHECK_DISPLAY,F2400_P1202_PUB_MAIL_ID:X,81088
ProMED-mail. (2010b) Avian Influenza (25): Bangladesh, OIE. See http://www.promedmail.org/pls/apex/f?p=2400:1202:3528901951843742::NO::F2400_P1202_CHECK_DISPLAY,F2400_P1202_PUB_MAIL_ID:X,82513
ProMED-mail. (2010c) Avian Infuenza (03): India (West Bengal) OIE. See http://www.promedmail.org/pls/apex/f?p=2400:1202:3528901951843742::NO::F2400_P1202_CHECK_DISPLAY,F2400_P1202_PUB_MAIL_ID:X,80946
K. Ray *et al.*, *Virus Genes* **36**, 345 (Apr, 2008).
L. A. Reperant, G. F. Rimmelzwaan, T. Kuiken, *Rev Sci Tech* **28**, 137 (Apr, 2009).
L. A. Reperant, N. S. Fuckar, A. D. Osterhaus, A. P. Dobson, T. Kuiken, *PLoS Pathog* **6**, e1000854 (Apr, 2010).
G. F. Rimmelzwaan *et al.*, *J Virol* **75**, 6687 (Jul, 2001).
G. F. Rimmelzwaan *et al.*, *Avian Dis* **47**, 931 (2003).
G. F. Rimmelzwaan *et al.*, *Am J Pathol* **168**, 176 (Jan, 2006).
A. L. Rivas *et al.*, *Epidemiol Infect* **138**, 192 (Feb, 2010).
C. Rohm, T. Horimoto, Y. Kawaoka, J. Suss, R. G. Webster, *Virology* **209**, 664 (Jun 1, 1995).
M. D. Saad *et al.*, *Emerg Infect Dis* **13**, 1120 (Jul, 2007).
T. Saito *et al.*, *Vet Rec* **163**, 722 (Dec 13, 2008).
T. Saito *et al.*, *Vet Microbiol* **133**, 65 (Jan 1, 2009).
S. L. Salzberg *et al.*, *Emerg Infect Dis* **13**, 713 (May, 2007).
K. Santhia *et al.*, *Influenza Other Respi Viruses* **3**, 81 (May, 2009).
K. Sawabe *et al.*, *Am J Trop Med Hyg* **75**, 327 (Aug, 2006).
D. A. Senne, J. E. Pearson, B. Panigrahy, in *Proceeding of Third International Symposium on Avian Influenza,* C. W. Beard, B. C. Easterday, Eds. (United States Animal Health Association, Richmond, VA, 1992), pp. 50–58.
K. Sharshov *et al.*, *Emerg Infect Dis* **16**, 349 (Feb, 2010).
K. F. Shortridge, *Vaccine* **17 Suppl 1**, S26 (Jul 30, 1999).
K. F. Shortridge, W. K. Butterfield, R. G. Webster, C. H. Campbell, *Bull World Health Organ* **55**, 15 (1977).
K. F. Shortridge *et al.*, *Virology* **252**, 331 (Dec 20, 1998).
J. Siengsanan *et al.*, *J Wildl Dis* **45**, 740 (Jul, 2009).
L. D. Sims, *Avian Dis* **51**, 174 (Mar, 2007).
L. D. Sims *et al.*, *Avian Dis* **47**, 832 (2003a).

L. D. Sims *et al.*, *Avian Dis* **47**, 1083 (2003b).
L. D. Sims *et al.*, *Vet Rec* **157**, 159 (Aug 6, 2005).
G. J. Smith *et al.*, *Proc Natl Acad Sci U S A* **103**, 16936 (Nov 7, 2006a).
G. J. Smith *et al.*, *Virology* **350**, 258 (Jul 5, 2006b).
G. J. Smith *et al.*, *Nature* **459**, 1122 (Jun 25, 2009).
R. Soares Magalhaes, D. Q. Hoang, K. L. Lai Thi, "Farm Gate Trade Patterns and Trade at Live Poultry Markets Supplying Ha Noi: Results of a Rapid Rural Appraisal" (Rome, 2007).
R. J. Soares Magalhaes *et al.*, *BMC Vet Res* **6**, 10 (2010).
T. Songserm *et al.*, *Emerg Infect Dis* **12**, 575 (Apr, 2006a).
T. Songserm *et al.*, *Emerg Infect Dis* **12**, 681 (Apr, 2006b).
T. Songserm *et al.*, *Emerg Infect Dis* **12**, 1744 (Nov, 2006c).
E. Spackman *et al.*, *Virol J* **6**, 190 (2009).
D. E. Stallknecht, S. M. Shane, *Vet Res Commun* **12**, 125 (1988).
D. E. Stallknecht, V. H. Goekjian, B. R. Wilcox, R. L. Poulson, J. D. Brown, *Avian Dis* **54**, 461 (Mar, 2010).
E. Starick *et al.*, *Vet Microbiol* **128**, 243 (Apr 30, 2008).
K. M. Sturm-Ramirez *et al.*, *J Virol* **78**, 4892 (May, 2004).
K. M. Sturm-Ramirez *et al.*, *J Virol* **79**, 11269 (Sep, 2005).
K. Subbarao *et al.*, *Science* **279**, 393 (Jan 16, 1998).
H. Takakuwa *et al.*, *Microbiol Immunol* **54**, 58 (2010).
R. Takano *et al.*, *Virology* **390**, 13 (Jul 20, 2009).
M. E. Thomas *et al.*, *Prev Vet Med* **69**, 1 (Jun 10, 2005).
G. Tian *et al.*, *Virology* **341**, 153 (Oct 10, 2005).
T. Tiensin *et al.*, *Emerg Infect Dis* **11**, 1664 (Nov, 2005).
T. Tiensin *et al.*, *J Infect Dis* **199**, 1735 (Jun 15, 2009).
S. C. Trock, M. Gaeta, A. Gonzalez, J. C. Pederson, D. A. Senne, *Avian Dis* **52**, 160 (Mar, 2008).
T. M. Tumpey *et al.*, *J Virol* **76**, 6344 (Jun, 2002).
Y. Uchida *et al.*, *Emerg Infect Dis* **14**, 1427 (Sep, 2008).
K. Ungchusak *et al.*, *N Engl J Med* **352**, 333 (Jan 27, 2005).
S. Van Borm *et al.*, *Emerg Infect Dis* **11**, 702 (May, 2005).
T. van den Berg *et al.*, *Comp Immunol Microbiol Infect Dis* **31**, 121 (Mar, 2008).
J. A. van Gils *et al.*, *PLoS One* **2**, e184 (2007).
M. D. Van Kerkhove *et al.*, *Vaccine* **27**, 6345 (Oct 23, 2009).
M. Vascellari *et al.*, *Vet Pathol* **44**, 635 (Sep, 2007).
D. Vijaykrishna *et al.*, *PLoS Pathog* **4**, e1000161 (2008).
R. G. Wallace, H. Hodac, R. H. Lathrop, W. M. Fitch, *Proc Natl Acad Sci U S A* **104**, 4473 (Mar 13, 2007).
X. F. Wan *et al.*, *PLoS One* **3**, e3462 (2008).
H. Wang *et al.*, *Lancet* **371**, 1427 (Apr 26, 2008a).
J. Wang *et al.*, *J Virol* **82**, 3405 (Apr, 2008b).
M. P. Ward, D. Maftei, C. Apostu, A. Suru, *Vet Res Commun* **32**, 627 (Dec, 2008).
T. P. Weber, N. I. Stilianakis, *Emerg Infect Dis* **13**, 1139 (Aug, 2007).
R. G. Webster, *Lancet* **363**, 234 (Jan 17, 2004).
R. G. Webster, W. J. Bean, O. T. Gorman, T. M. Chambers, Y. Kawaoka, *Microbiol Rev* **56**, 152 (Mar, 1992).
S. H. Wee *et al.*, *Vet Rec* **158**, 341 (Mar 11, 2006).
WHO. (2010) Cumulative Number of Confirmed Human Cases of Avian Influenza A/(H5N1) Reported to WHO (ed. WHO). Geneva, Switzerland. See http://www.who.int/influenza/human_animal_interface/EN_GIP_LatestCumulativeNumberH5N1cases.pdf
X. Xu, K. Subbarao, N. J. Cox, Y. Guo, *Virology* **261**, 15 (Aug 15, 1999).
K. S. Yee, T. E. Carpenter, S. Mize, C. J. Cardona, *Avian Dis* **52**, 348 (Jun, 2008).
Y. Yupiana, S. J. de Vlas, N. M. Adnan, J. H. Richardus, *Int J Infect Dis* **14**, e800 (Sep, 2010).

Chapter 11
Mathematical Models of Infectious Diseases in Livestock: Concepts and Application to the Spread of Highly Pathogenic Avian Influenza Virus Strain Type H5N1

Guillaume Fournié, Patrick Walker, Thibaud Porphyre, Raphaëlle Métras, and Dirk Pfeiffer

Introduction

Animal health governance faces new challenges as the ecology of infectious livestock diseases is changing (Tomley and Shirley 2009). Environmental and climate changes, intensification of livestock production, modification in land-use and agricultural practices, globalization of human travel, the development of the trade of livestock and livestock products have created conditions for an increase in the emergence and re-emergence of infectious agents in the last decades (Weiss and McMichael 2004; Randolph and Rogers 2010; Jones et al. 2008; Gibbs 2005). The frequency of emergence of new highly pathogenic avian influenza viruses (HPAIV) has increased over the past 20 years, as well as the economic impact of associated outbreaks (Alexander and Brown 2009). Bluetongue virus serotypes have continuously increased their spatial distribution, specifically in a northern direction. Treatment-resistant strains, such as methicillin-resistant *Staphylococcus aureus* (MRSA), have appeared. Numerous infectious diseases such as foot-and-mouth disease (FMD) are endemic in many parts of the world, and may have a high impact on animal health and farmer livelihood. Moreover, they constrain the ability of affected countries to trade livestock and livestock-derived products. Production systems in developed countries are also vulnerable. For example, outbreaks of FMD in United Kingdom in 2001, classical swine fever in Holland in 1997/1998, and highly pathogenic avian influenza H7N7 in Holland in 2003 resulted in the loss of millions of animals, mainly as a result of culling of affected and exposed animals. Finally,

G. Fournié (✉) • T. Porphyre • R. Métras • D. Pfeiffer
Veterinary Epidemiology and Public Health Group, Department of Veterinary Clinical Sciences, Royal Veterinary College, University of London, UK
e-mail: gfournie@rvc.ac.uk

P. Walker
MRC Centre for Outbreak Analysis Modelling, Department of Infectious Disease Epidemiology, Imperial College, London, UK

D. Zilberman et al. (eds.), *Health and Animal Agriculture in Developing Countries*, Natural Resource Management and Policy 36, DOI 10.1007/978-1-4419-7077-0_11,

infectious livestock diseases are a threat for public health: about 75% of human infectious agents that emerged in the last 25 years had an animal origin (King et al. 2006).

To meet these challenges, transparent scientific approaches are needed to obtain a better understanding of epidemiological patterns and to inform policy-development. Epidemiological systems are composed of multiple processes interacting nonlinearly. They are thus too complex to be represented using purely mental thought processes. The ability of humans to mentally conceptualize a system seems to be reached when more than three variables and six transitions from a state to another are involved (Klein 1998). Therefore, the study of epidemiological systems can benefit significantly from the use of quantitative modeling tools. By integrating individual animal-level knowledge of epidemiological, biological and behavioral factors, mathematical models of infectious diseases can provide insights into disease dynamics at the population-level and predictions of the impact of control strategies on outbreak outcome. Their mathematical formulation makes the underlying assumptions explicit, thereby ensuring the transparency of the approach, at least to those with the relevant understanding of the underlying mathematical methodologies. According to the Royal Society's Infectious diseases in Livestock report (Society 2002), "Quantitative modeling is one of the essential tools both for developing strategies in preparation for an outbreak and for predicting and evaluating the effectiveness of control policies during an outbreak."

The aim of this chapter is to present the basic concepts of mathematical modeling of infectious diseases using examples of models developed for the spread of avian influenza viruses.

Use of Mathematical Models to Study the Spread of Infectious Diseases

Mathematical models of infectious disease transmission can be defined as a set of equations conceptualizing the spread of an infectious agent in a host population. They are a simplification of a complex phenomenon, but the simplification should have a limited effect on the disease dynamics properties, which are under study (Britton and Lindenstrand 2009). Models can be used either to simulate disease spread, they are then referred to as simulation models, or to estimate epidemiological parameters. Simulation models use existing knowledge regarding the spread of infection and host population dynamics to investigate mechanisms underlying disease spread, or to predict the future trajectory of an epidemic and the impact of control strategies. Alternatively, given assumptions as to the nature of transmission, models can quantitatively estimate key transmission parameters retrospectively from existing outbreak or experimental infection data.

Simulating Infectious Disease Spread

Models can be used as an explanatory tool to improve our understanding of the dynamics of infectious diseases, and to obtain insight into the impact of interventions. They provide a rigorous framework where the complexity of disease transmission is disentangled and each aspect of the disease spread can be monitored. Allowing identification of the underlying factors that drive disease dynamics, these models can thus be used to generate and to test hypotheses. They can also help in defining priorities for epidemiological data collection by identifying which elements of the disease dynamics are most important for meaningful model experimentation.

Models can also be used to predict the outcome of a disease outbreak and to develop practical control strategies that will minimize its impact. Models then provide quantitative outputs that can be used to inform decision-making. Also, to accurately reproduce the behavior of an epidemic and to improve the reliability of the predictions, those models should ideally include all biological mechanisms and heterogeneities known to influence the disease dynamics.

To date mathematical models have played a major role in elucidating fundamental principles of infectious disease spread. This has included the description of the properties of the dynamics of an epidemic within a population (Dietz 1993; Ross and Hudson 1917; Anderson and May 1991; Kermack and McKendrick 1991; Fine 1993), the demonstration of the role of heterogeneity in driving an epidemic (Yorke et al. 1978; Woolhouse et al. 1997), and the assessment of the possible effects of different control strategies for a wide range of diseases (Ross et al. 2008; Velasco-Hernandez et al. 2002; Hallett et al. 2008; Woolhouse 1992; Grassly et al. 2006).

In recent times, inspired by events such as human disease outbreaks including variant Creutzfeldt-Jakob disease and severe acute respiratory syndrome (SARS), the spread of resistant bacteria such as MRSA and the 2009 H1N1 influenza pandemic, modeling has increasingly become a standard component of public health decision making (Temime et al. 2008; Fraser et al. 2009; Austin and Anderson 1999; Li et al. 2004; Lipsitch et al. 2003; Ghani et al. 2000). In the field of livestock diseases, models of the spread of diseases such as bovine spongiform encephalopathy, scrapie and classical swine fever have all been developed to address pertinent issues regarding ongoing epidemics such as optimal culling strategies, farming restrictions, and the impact upon human health (Anderson et al. 1996; Ferguson et al. 1998; Stringer et al. 1998; Woolhouse et al. 1998; Boender et al. 2008). The 2001 FMD epidemic in Great Britain further served to demonstrate that, following advances in computing power and given access to the best available data (Levin et al. 1997), mathematical models can play an active role in providing policy makers with vital information as to the optimum disease control strategy during a rapidly progressing epidemic (Ferguson et al. 2001; Keeling et al. 2001). However, how useful this information was is still debated, since modeling outputs might have led to an excessive slaughtering of uninfected animals (Kitching et al. 2006).

The spread of avian influenza viruses in poultry has also led to a surge in modeling associated scientific publications over the last decade. Simulation models have been developed to get a better understanding of factors driving the disease spread, such as the potential role played by environmental transmission in the maintenance of avian influenza viruses in natural ecosystems (Lebarbenchon et al. 2009; Breban et al. 2009; Rohani et al. 2009; Guberti et al. 2007; Roche et al. 2009); and the impact of poultry population management on the silent spread of HPAIV strain type H5N1 viruses in live bird markets (Fournié et al. 2010). Further models have been developed to investigate direct and indirect effects of control measures, including the hypothesis that large-scale culling campaigns may impede the evolution of avian host resistance to the virus (Shim et al. 2009), and the evolution and spread of resistance following vaccination campaigns (Iwami et al. 2009a, b). At the farm level, models allowed evaluating the impact of different mortality and morbidity thresholds for the detection of infection (Savill et al. 2006; Carpenter et al. 2004) and investigating the use of sentinel birds for detecting the emergence of HPAIV (Verdugo et al. 2009). The potential for infection to spread silently in vaccinated populations, as a result of lower levels of clinical signs and flock mortality (Savill et al. 2006) within a commercial flock, and the dynamics of waning immunity in backyard systems (Lesnoff et al. 2009) was also evaluated using models. Predictive models have been developed to investigate the potential spread of HPAIV H5N1 in the United Kingdom and to develop control strategies that could efficiently mitigate an outbreak (Dent et al. 2008; Sharkey et al. 2008; Truscott et al. 2007). A simulation model was used to explore the different pathways of infection from a commercial broiler farm in the United States (Dorea et al. 2010).

Quantifying Epidemiological Parameters

Such simulation models, whilst providing useful insight into the infection process, do, however, often require the specification of a range of model parameters. In the absence of context-specific outbreak data, sensitivity analyses can, to a certain extent, address issues associated with misspecification of parameter values. However, modelers are generally reliant upon parameter estimates available in the scientific literature or expert opinion to obtain a reasonable indication of the ranges of values such inputs may take.

In contrast, transmission models fitted to outbreak data are generally a lot simpler, requiring less parameters. This is partially because fewer parameters are easier to fit, but also because this provides parsimonious and easily interpretable measures by which to quantify the spread of infection and the incremental effect of interventions. Fitting such models allows estimation of important epidemiological parameters which, due to the nonlinearity of the transmission process, would not otherwise be possible.

In the case of avian influenza, at the level of transmission between birds, such models are often fitted to data obtained through laboratory experiments. This is because such controlled environments allow relatively detailed data collection, and often offer the opportunity to quantify differences in transmissibility between host species, viral strains, or subtypes and are straightforward to model. Such analyses have been used to estimate the transmissibility of avian influenza viruses in various host species within experimental settings, such as HPAI H7N7 viruses in chickens (van der Goot et al. 2005, 2007a), HPAI H5N1 in chickens (Bouma et al. 2009) and ducks (van der Goot et al. 2007b).

In a field setting, heterogeneity in the distribution of various risk factors such as environmental conditions, host species, poultry housing and the provision of preventative control measures complicates the dynamics of infection, and it is rarely possible to collect detailed data on these factors, or even only on the incidence of infection. As such it is more difficult to ensure that the underlying model dynamics are truly representative of those of the outbreak. However, inferring the transmissibility of infection from such "real-life" settings provides important information as to the effectiveness of ongoing interventions and the incremental effectiveness of alternative strategies necessary for achieving control (Tiensin et al. 2007; Soares Magalhaes et al. 2010; Bos et al. 2010). Moreover, such analyses can provide insight into the dynamics of infection at a larger population scale than that possible in an experiment setting. This is particularly relevant for quantifying between-flock transmission (Stegeman et al. 2004; Garske et al. 2007; Mannelli et al. 2007; Ward et al. 2009) and the geographical spread of infection (Boender et al. 2007; Le Menach et al. 2006; Walker et al. 2010).

Infectious Agents, Transmission, and Infection

Microparasites and Macroparasites

Based on the characteristics of their population biology, infectious agents can be grouped into two categories: microparasites, including bacteria, viruses, prions, protozoa, and macroparasites, including helminths, arthropods. In this chapter, we focus on microparasites. With some exceptions, the duration of microparasitic infections, such as avian influenza, is generally much lower than the host life span, and they reproduce to very high numbers within hosts (Anderson and May 1992).

Transmission

Transmission can be defined as the passing of an infectious agent from an infected host to a susceptible one. Susceptibility refers to the level of vulnerability of a host to an infectious agent. It can be reduced by prior immunization, due to former infection or vaccination. Transmission can be direct, where an infectious agent

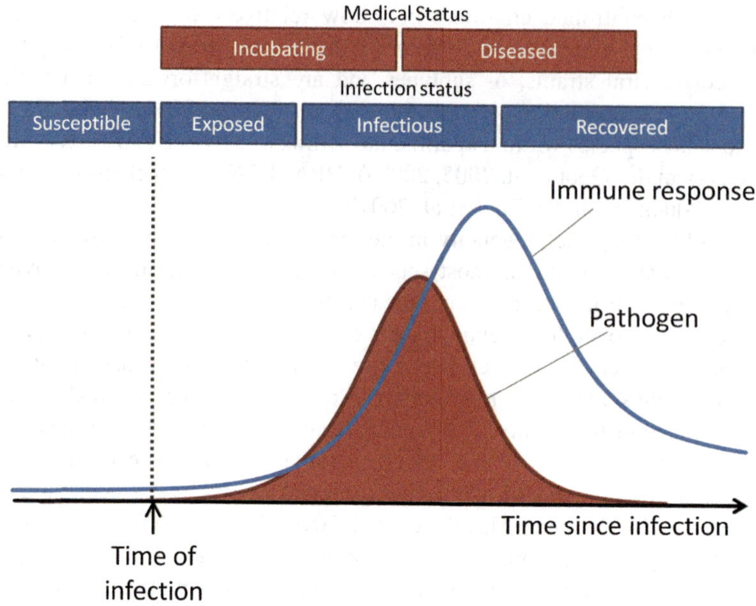

Fig. 11.1 Infection states and disease progress adapted from Keeling and Rohani (2007)

passes to a susceptible host due to close contact with an infected host, or indirect, where transmission is mediated by the environment or by animate or inanimate vectors. In the case of avian influenza viruses, while direct bird-to-bird contact is likely to be the predominant mode of within-flock transmission, indirect transmission may be crucial for transmitting viruses between farms, and for virus sustainability in natural ecosystems.

Transmission efficacy depends on the infectiousness of infectious hosts and the susceptibility of individuals yet to be infected. Infectiousness is defined by Grassly and Fraser (2008) as the characteristics of an infected host that determine the rate at which susceptible individuals become infected. Both, biological factors, such as the within-host dynamics of pathogenic agents which affect the level of shedding, and behavioral factors, which influence the rate at which contacts are made between individuals, determine the extent to which direct transmission contributes to infectiousness. In contrast, indirect transmissibility is modulated by factors such as the survival of infectious agents in the environment and the ecology of vectors.

Infection

The abundance dynamics of microparasites within hosts are often simplified, and the infection is modeled as the passing of a host through several infection states. One widely used mathematical model of the infection process is known as the SEIR model where a host may move through four states, or compartments (Fig. 11.1): Susceptible, Latent (or exposed), Infectious, and Recovered (or Removed). Initially, susceptible hosts may become infected after contact with an infected host or the infected environment. In the early stages of infection, infectious agents may not be shed in sufficient quantity to transmit infection to other hosts. Individuals are then said to be in the latent (or exposed) state. Once they are able to transmit the infection, they are classified as infectious and they remain infectious for a period of time. Then, depending on the pathogenicity of the parasite and the immune system of the infected individual, the host either dies or the infection is cleared. Following clearance, the host may become immune to infection for a duration dependent upon the parasite in question.

An infection is thus described according to the host's capacity to transmit the infectious agent, the number of infection states a host can move through, and by the time individuals remain in each state. The average time infected hosts spent in the latent and infectious states are known as the latent and infectious periods, respectively. Such a model may need to be further developed to incorporate other important characteristics of a particular disease that may have a bearing on disease dynamics. For example, as in the case of HPAIV H5N1 in ducks, an infection may not always lead to clinical symptoms. It may also be the case that expression of disease signs leads to a reduction in host activity, or to its isolation, and thus lowers the contact rate and subsequently the infectiousness.

Moreover, symptomatic and infectious periods may not be synchronous. If the time from infection to the onset of symptoms, referred to as the incubation period, is longer than the latent period, an infectious host will be able to transmit an infectious agent before the onset of symptoms (Fig. 11.1). This feature may have important implications for disease control (Fraser et al. 2004).

A Simple Approach to Model Disease Spread

Modeling Transmission

In many models, transmission is assumed to follow a "mass-action" principle. This means that the rate of infection is proportional to the product of the density of infectious and susceptible individuals within a population. This makes the assumption of homogenous mixing within the population: each individual has the same probability to contact any other individual in the population. This mass-action principle can be modulated depending on selected transmission characteristics, specifically how contact probability varies with changes in population size. The two main approaches are either to define the transmission as density-dependent (pseudo mass-action), where the number of contacts increases with population size, or as frequency-dependent (true mass action), where the number of contacts remains independent of population size. The rate at which individuals get infected per unit of time, the so-called force of infection, is then equal to:

$$\text{Force of infection} = \frac{\beta}{N} I_t$$

where β is the rate of transmission. I_t is the number of infectious individuals at time t. For density-dependent transmission, N is equal to N_0, the initial number of susceptible individuals. For frequency-dependent transmission, N is equal to N_t, the population size at time t.

The rate of transmission β refers to the infectiousness previously described. It is often defined as the product of the contact rate C, the average number of contacts made by an individual per unit of time, and the probability p_c that such a contact leads to an infection (Vynnycky and White 2010):

$$\beta = Cp_c.$$

Characterizing an Epidemic

An epidemic can then be defined as a chain of transmission, and is principally governed by two key individual-level transmission parameters: the basic reproduction number R_0 and the generation time T_g (Grassly and Fraser 2008; Ferguson et al. 2003). T_g is the period between the infection of a given individual and the infection of another one as a result of transmission from this individual (=secondary case), and R_0 is defined as the expected number of secondary cases of a typical infected case in a fully susceptible population (Diekmann et al. 2000). R_0 assesses the intrinsic transmissibility of an infectious agent and thus its potential to be sustained in a given population. If R_0 is greater than one, an infected host will transmit the infection to more than one individual, the epidemic can spread throughout the population. In contrast, if it is lower than one, an infected host will transmit the infection, on average, to less than one individual, and the epidemic will fade out. R_0 influences both the intensity of the peak and overall final size of the epidemic. Together, R_0 and T_g determine the rate of growth in the number of infected individuals during the initial stages of an epidemic.

For simple models, R_0 can be easily derived analytically. In the case of the SEIR model presented in Box 11.1, R_0 is the product between the rate of transmission β and the duration of the infectiousness D:

$$R_0 = \beta D = Cp_c D.$$

During the early stages of the spread of an infectious agent in a susceptible population, the number of infectious individuals is much lower than the number of susceptible ones, and the incidence rate increases exponentially. For a closed population in which there is no supply of susceptible individuals, either through new individuals entering the population or through individuals becoming

Box 11.1 Formulation of a SEIR Model

The progression of avian influenza virus infection in many contexts can be represented with a SEIR model, with birds either recovering or being removed from the population depending on the lethality of a particular strain within a given species. It is not each individual that is explicitly modeled, but the flow of individuals from an infection state, or compartment, to another. This can be expressed using the following differential equations:

$$\frac{dS_t}{dt} = -\frac{\beta}{N_0} S_t I_t,$$

$$\frac{dE_t}{dt} = \frac{\beta}{N_0} S_t I_t - \alpha E_t,$$

$$\frac{dI_t}{dt} = \alpha E_t - \gamma I_t,$$

$$\frac{dR_t}{dt} = \gamma I_t.$$

Here, the transmission process is density-dependent and the infectiousness is assumed not to vary during the infectious period of an individual. α and γ are the rates at which individuals leave the exposed and infectious compartments, respectively. They can be calculated as the inverse of the latent and infectious period.

susceptible again following recovery from infection or waning immunity, the ongoing transmission leads to the depletion of the number of susceptible individuals. This causes a reduction in the average number of secondary cases infected by an infectious individual, defined as the net reproductive number R_n, also referred to as the effective reproduction number. This can be calculated as the product between R_0 and the fraction s of susceptible individuals in the population, such as:

$$R_n = R_0 s.$$

Contrary to R_0, R_n varies as a function of time and informs on the evolution of the incidence. If R_n is greater than one the incidence increases, if it is less than one the incidence decreases. As a result an epidemic moves to extinction when $R_n < 1$, meaning that the proportion of susceptible individuals is lower than $1/R_0$.

In such closed populations, the average number of secondary infections produced by an infected individual eventually falls below 1 and the epidemic

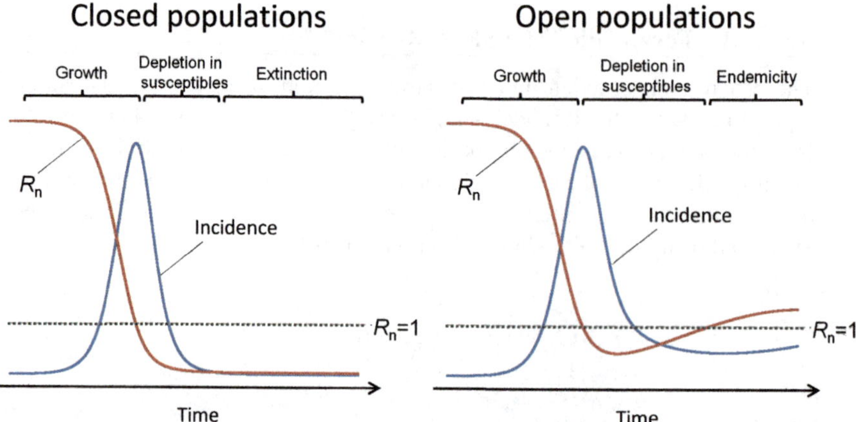

Fig. 11.2 Course of an transmission in a closed and an open population. In a closed population, an epidemic spreads and then fades out. In an open population, the endemic transmission can be sustained

fades out (Fig. 11.2). The final fraction of the number of individuals that experienced infection, F_∞, is given by:

$$F_\infty = 1 - e^{-R_0 F_\infty}.$$

In contrast, if the susceptible fraction of the population is replenished due to demographic processes such as birth and migration, or due to the absence, or short-lived nature, of immunity, the transmission may be sustained in a population and the infection may become endemic (Fig. 11.2). In that case, the final number of infected individuals cannot be deduced from the former equation. Instead the epidemic may converge to an equilibrium prevalence or may be subject to seasonal peaks.

Modeling Control Strategies

As mentioned earlier, one of the essential goals of mathematical models is to assess the efficiency of interventions and possibly to optimize them. When facing an outbreak, a range of strategies can be implemented to control its magnitude. Their effect will depend on the epidemiology of the infectious agent, the host population dynamics, and the scale of the epidemic. Interventions may achieve a reduction in transmissibility by affecting one, or any combination, of the components of R_0: the contact rate, the probability that a contact leads to infection and the infectious period. Therefore, if the values of the components that an intervention acts upon and R_0 are known, the effectiveness of an intervention at reducing these components required to prevent the spread of infection can be obtained. In this section, we describe some of the most frequently used control strategies to mitigate the spread of HPAI outbreaks in poultry, and how they impact on transmission.

Main Strategies to Control Livestock Disease Outbreaks

Vaccination

By inducing a strong immune response through the production of neutralizing antibodies, vaccination aims to protect individuals from disease signs and death and to prevent outbreaks by reducing susceptibility to infection and viral shedding. The vaccination thus impacts transmissibility by lowering the probability that a contact leads to an infection. The following example assumes a fully effective vaccine, meaning that vaccinated individuals are no longer susceptible to the infection. As discussed in Section "Characterizing an Epidemic" transmission is not sustained if the value of R_n remains below its threshold of one. This is achieved in a randomly mixing population if the proportion of susceptible individuals is lower than $1/R_0$. Therefore, in this case, the eradication of a disease does not require all individuals to be vaccinated as long as the fraction of the population that has been vaccinated is higher than the herd immunity threshold (Fine 1993):

$$\text{Herd immunity threshold} = 1 - \frac{1}{R_0}.$$

Stamping Out

Stamping out is widely used to control infectious diseases in livestock. The culling of all birds in infected and exposed premises successfully contributed to the eradication of HPAI H7N7 in the Netherlands (Stegeman et al. 2004) and HPAI H7N1 in Italy (Mannelli et al. 2007). The successive outbreak waves of HPAI H5N1 occurring in Hong Kong since 1997 (Sims et al. 2003), and also during the epidemics that have affected South Korea and Japan since 2003, have also been controlled through the depopulation of poultry farms and live bird markets. As it depletes both the number of infected and susceptible individuals, stamping out operates by reducing the infectious period and bird-to-bird contact rates. The usefulness of this strategy for developed countries facing HPAI outbreaks has also been highlighted by mathematical models (Truscott et al. 2007; Le Menach et al. 2006), and was shown to highly depend on the scale of the disease spread and the early detection of infected flocks. Once disease becomes widespread, the costs of sustaining such a strategy can become prohibitive and may not lead to a sustainable reduction in transmission (Sims 2007).

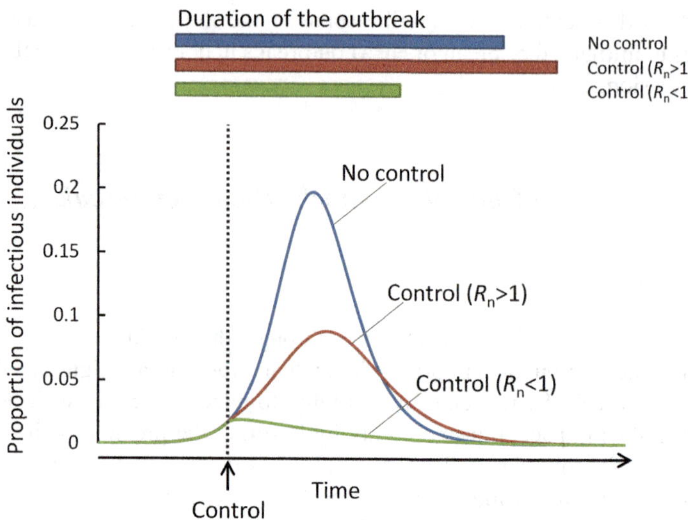

Fig. 11.3 Impact of lowering R_n on the epidemic curve. Three scenarios are shown for a SEIR model with similar initial parameters: No control (blue line), controls but R_n remains higher than one (red line); controls with R_n lower than one (green line). This figure is adapted from Ferguson et al. (2003)

Movement Restrictions

If implemented quickly after the detection of an outbreak, movement restrictions lower disease transmissibility by reducing the number of contacts between farms, and thus prevent disease spread. Importantly, in the context of the spatial spread of an epidemic, when contacts can potentially be made over long distances (such as in the case of live bird trade or indirect transmission by fomites), this may also prevent the disease from accessing regions, which have yet to experience infection. During the 2001 FMD outbreak in United Kingdom, long-range movements of sheep occurring before restrictions were put in place had spread the disease all around the country (Kao 2002).

Impact of Control Strategies on the Course of an Epidemic

As shown in Fig. 11.3, when interventions are able to reduce R_n below its threshold of one, the transmission is not sustained anymore and both the number of individuals becoming infected, and the outbreak duration will be curtailed. However, reducing R_n below one may be neither logistically nor economically possible, particularly for diseases with a high R_0 and/or affecting a fully susceptible host population. In the case that R_n is reduced but remains higher than one, the disease may still spread in the population, but the outbreak size, the total number of individuals that experience infection, will be reduced (Fig. 11.3).

In some cases, interventions may have unintended consequences on the components of R_0. If the vaccination coverage is suboptimal, or if the vaccine is not fully protective at the individual level, it may create conditions for a silent spread of the infectious agent within a host population. It was shown that insufficient vaccination coverage for HPAIV H5N1 may extend the infectious period of a poultry flock. Indeed, as the vaccination coverage rises, fewer birds become infected but outbreaks become harder to detect (Savill et al. 2006). Likewise, Walker et al. (2010) found that the time from infection to reporting was significantly increased in northern Viet Nam following vaccination campaigns. Stamping out may also have unintended effects: failing to implement effective incentive policies leads to underreporting and late detection of outbreaks. In Egypt (Meleigy 2007; Peyre et al. 2009) and Viet Nam, humans are now acting as "sentinels" for the disease in poultry (Minh et al. 2009a).

Increasing Model Complexity

The simplicity of models similar to the SEIR model has helped to elucidate major principles driving the dynamics of numerous infectious agents in a parsimonious manner. However, to extend our understanding of epidemiological patterns and improve the extent to which outcomes reflect reality, it is often necessary to incorporate a higher degree of complexity into the modeled disease dynamics. In this section, we describe various methods to achieve this for livestock diseases.

Stochasticity

So far, all considered models were deterministic: for the same initial conditions and parameter values, all simulations give the same outcome, which is expected to be the average epidemic behavior. However, even in the most controlled settings, for example during laboratory-based transmission experiments, the course of an epidemic would be expected to vary, due to chance. Incorporating stochasticity in mathematical models allow taking into account the random nature of certain epidemiological features, such as the number of transmission events and the latent and infectious periods. Therefore, contrary to deterministic models, several simulations of a stochastic model do not provide a single outcome but a probability distribution of the outcome.

There are several ways to implement stochasticity in a model. Input parameter values can be drawn from probability distributions. This is possible when the extent to which a parameter varies, as a result for instance of interindividual variability or external forces such as climate or vector density, is known. Stochasticity can also concern infection processes. The transition of individuals between infection states, for instance from susceptible to infected, while being associated with an underlying probability, can also be thought of as a random process. Several approaches exist to implement such stochastic processes in a model. One of them is described in Box 11.2. Other algorithms are presented by Keeling and Rohani (2007), and Vynnycky and White (2010).

Accounting for the random nature of disease spread can greatly impact the course of a modeled epidemic. In a deterministic model, if R_0 is higher than one, then an infectious agent will always invade a host population, resulting in a major outbreak, and possibly in its endemicity. In contrast, the infectious agent will not be transmitted if R_0 is lower than one. In reality, for some diseases, an infectious individual can recover before transmitting the infection. As a result, during the early stage of a disease invasion when the number of infectious individuals is low, the disease can become extinct before leading to a major outbreak. The probability that the outbreak avoids this "epidemic fade out" and leads on to a major outbreak following the introduction of a single infectious individual can be calculated as follows:

$$\text{Probability of disease invasion} = 1 - \frac{1}{R_0}.$$

This equation applies to values of R_0 higher than one. If R_0 is lower than one, transmission events may occur, but they will be limited and will not lead to a major outbreak. As R_0 increases above unity, the distribution of the final epidemic size becomes bimodal (Fig. 11.4): either the epidemic fades out before invading the population, or the disease becomes established and spreads to most individuals. Outbreaks involving a large number of infected individuals become more likely with increasing values of R_0.

Box 11.2 Formulation of a Stochastic SEIR Model

Dividing simulated time into steps of arbitrarily defined small length Δt, the number of individuals passing from one compartment to another from time t to $t + \Delta t$ can be generated using a binomial distribution. A binomial distribution, denoted $B(n, p)$, gives the probability distribution of the number of successes in a sequence of n independent trials, each associated with the same probability p of success. The following symbols are the same as the ones used in Box 11.1. If there is one infectious individual, the probability of a given susceptible individual becoming infected will be $q = \frac{C}{N} p_c \Delta t$. If there are I_t infectious individuals, this probability will then be $1 - (1 - q)^{I_t}$. The number of newly infected individuals at time $t + \Delta t$ then follows a binomial distribution $B(S_t, 1 - (1 - q)^{I_t})$. Similarly, the number of individuals becoming infectious and recovering follows $B(E_t, \alpha\Delta t)$ and $B(I_t, \gamma\Delta t)$, respectively. The resulting number of individuals in each infection state is therefore equal to an integer number.

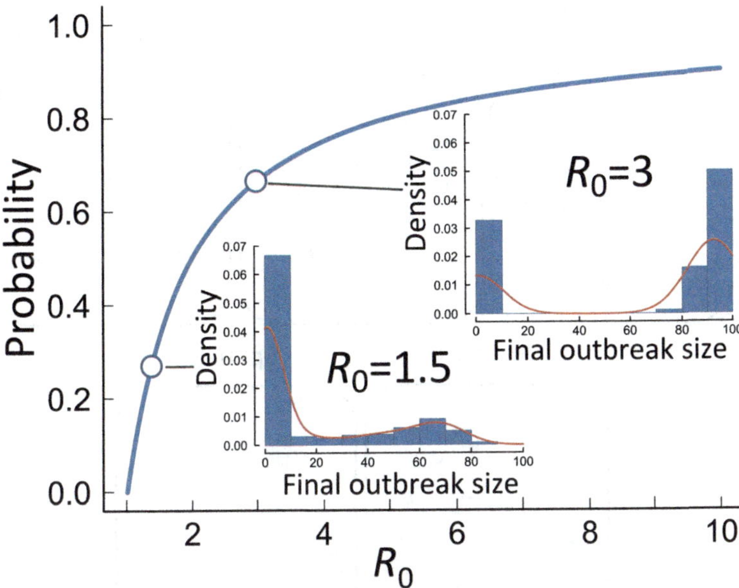

Fig. 11.4 Probability of disease invasion and distribution of the final epidemic size for R_0 equal to 1.5 and 3. The probability of disease invasion refers to the probability that the introduction of a single infected host into a fully susceptible population leads to a major outbreak. The distribution of the final epidemic size is obtained from 1,000 simulations of the stochastic SEIR model described in Box 11.2. The density function describes the probability associated with each possible final epidemic size. The initial population size is 100

Heterogeneity

The strong assumptions involved by models described so far may limit their use when investigating infectious agents or host populations for which the heterogeneity in transmission is thought to be an important characteristic.

The rate of transmission β has been assumed to be constant over the host's infectious period. Such an assumption is often not justifiable, especially for diseases where the infectious period lasts for a significant fraction of the host's lifespan. In this case, the infectiousness may vary considerably with time, as is the case with HIV infection. Therefore, assuming a constant rate of transmission over the infectious period may be misleading. The infectious compartment I may then be divided into N multiple and successive sub-compartments I_n (with $n = 1,.., N$), each being associated with a specific level of infectiousness β_n.

Individuals within a population often differ in their contact patterns and can mix preferentially with a particular subgroup or even with some specific individuals. Therefore, assuming that individuals mix randomly is an oversimplifying assumption. Nonrandom mixing may have an important impact on disease dynamics and on the implementation of control strategies. In the following paragraphs, increasingly complex ways to take into account heterogeneity in mixing patterns are described. They are illustrated in Fig. 11.5. Five ways to represent mixing patterns

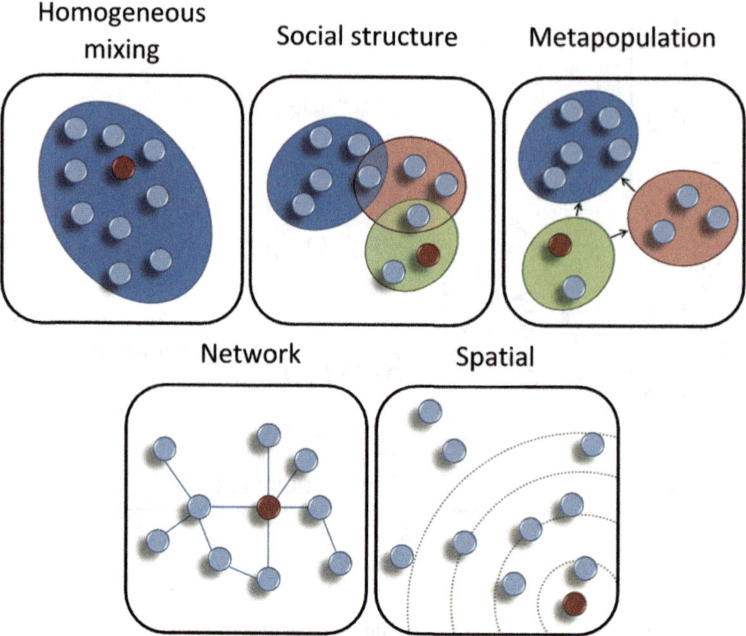

Fig. 11.5 Representing mixing patterns

are shown. Each circle represents an individual. The red circles represent the infectious individuals and the blue circles are the susceptible ones. In randomly mixing populations, each individual has the same probability of making contact with the infectious host. The presence of a particular social structure allows the population to be divided into groups, assuming that the rate of transmission varies within and between these groups. In a metapopulation model, the population is divided into subpopulations that are linked by either contacts, or movements of individuals from one subpopulation to another. In a network, the contact patterns between all individuals are known and explicitly represented. For spatial transmission, each individual is assigned a location and the probability of disease transmission depends on the distance from an infectious individual.

Social Structure

Factors such as infectiousness, susceptibility, and the duration of latent or infectious periods may vary between hosts according to intrinsic characteristics, such as their age, gender, or species. Such heterogeneity can often be incorporated into a model by stratifying the host population into groups. A set of compartments is attributed to each group and different rates of transmission are specified within and between those groups. For example, the course of HPAIV H5N1 infection is known to vary according to poultry species. Ducks show a long infectious period, may not exhibit clinical signs and may even survive the infection, in contrast to chickens for which disease is more pathogenic and almost always causes fatality shortly following infection (Saito et al. 2009; Hulse-Post et al. 2005; Spickler et al. 2008). Therefore, simulating the spread of HPAIV H5N1 within a flock composed by chickens and ducks would require modeling the two species separately, with one set of compartments for ducks, and another for chickens. In that case, the force of infection for chickens can be formulated as follows:

$$\text{Force of infection} = \frac{\beta_{C\to C}}{N_C} I_{C,t} + \frac{\beta_{D\to C}}{N_C} I_{D,t}$$

where $\beta_{C\to C}$ is the rate of transmission between chickens, and $\beta_{D\to C}$ N_C is the rate of transmission from ducks to chickens. N_C is the initial size of the chicken population, $I_{C,t}$ and $I_{D,t}$ are the number of infectious chickens and ducks at time t, respectively.

Accounting for the structure of a population may have a major impact on control strategy design. Indeed, if a group of individuals is responsible for most of the transmission, interventions targeting this group may have a higher impact than interventions applied at random.

Metapopulation

A metapopulation is composed of multiple distinct subpopulations, which are separated spatially and are then coupled only by the movement of individuals between these subpopulations (Colizza et al. 2007) or by indirect contacts. This approach has been recommended when the spatial structure of the population is known to have an important role in the disease dynamics. This is generally the case for infectious diseases affecting livestock where animals are usually aggregated into farms, and the movements of animals and fomites between farms can drive the transmission. Models with such a metapopulation structure can replicate patterns of infection that other models cannot. For infectious diseases of livestock, epidemics which would eventually become extinct in a single farm can become endemic in a production system consisting of a population of farms. This can occur as a consequence of infectious agents becoming amplified within farms and then spreading to others, leading to the sustainability of transmission.

A metapopulation framework has been used to model the spread of HPAIV H5N1 through a market chain (Fournié et al. 2010). At the end of a farm production cycle, birds are sent to a wholesale market and then from a wholesale market to regional markets where they are sold and slaughtered. In addition to this movement of birds from farms to markets, the disease can also be transmitted by indirect contacts between farms or between farms and markets. Such a design was required to evaluate the impact of various interventions that are targeted in subpopulations rather than being indiscriminately applied over the entire population. These included stamping out and vaccination at the farm level and hygiene measures at the market level.

Spatial and Network Transmission

The spatial spread of an infectious agent between livestock premises is generally characterized by a combination of two types of transmission (Riley 2007):

- Short-range transmission, for which the probability that a premise becomes infected is determined by the infectious state of the other premises in proximity. This can reflect air or vector-borne transmission, or the extent to which farms preferentially contact neighboring farms, for example through sharing equipments and movement of people. Such transmission is generally modeled either by categorizing premises into distinct spatial patches, or by assuming that the probability of transmission scales with the radial distance from the infectious source.
- Long-range transmission, which is often determined by an underlying network of contacts. This contact network defines how individuals are connected between them through, for example, movements of animals from farm to farm or to slaughterhouses, or movements or visitors such as feed providers or traders.

During the 2003 HPAI H7N7 outbreak that occurred in Holland, the proximity of poultry farms within densely populated poultry areas is likely to have favored intense local transmission of viruses between farms, requiring the implementation of severe interventions to eradicate the disease (Alexander and Brown 2009). In contrast, during the FMD outbreak that affected the United Kingdom in 2001, in addition to local spatial transmission, the movement of infected animals in the days preceding the implementation of interventions was considered to be a major driver of the spread of infection across the country, and thus a key contributor to the large scale of the outbreak (Kao 2002).

The spread of HPAIV H5N1 between farms is often thought to be driven by these short- and long-range modes of transmission. Outbreak waves that occurred in Viet Nam between 2003 and 2007 appeared to result from a combination of short and long-range transmission (Walker et al. 2010; Minh et al. 2009b). Simulation models developed to simulate the spread of HPAIV H5N1 among poultry farms in the United Kingdom have thus taken into account local spatial spread and long-range transmission between premises, mediated, for instance, by the transport of poultry to slaughterhouses (Truscott et al. 2007). For the former, a spatial kernel function is often used to describe how the probability that any infectious premise infects any susceptible premise decays with distance. Truscott et al. (2007) used the following spatial kernel to describe the spatial spread of HPAIV H5N1 between premises:

$$k_d = \left(1 + \frac{d}{\alpha}\right)^{-\gamma},$$

where d is the distance between two premises, and α and γ are two constant parameters. Long-range transmission was modeled by specifying a network of contacts between premises. Le Menach et al. (2006) chose a different approach, defining three values for the rate of transmission, each one being associated to a range of distances between two premises: short, medium, and long-range.

Impact of Heterogeneity on Disease Transmission and Control

The distribution of factors that determine disease spread, such as the spatial location of farms, contact patterns, production type, and environmental factors, is likely to be heterogeneous. Because of this heterogeneity, the risk for premises to be infected or to transmit the infection follows a distribution that is likely to be highly skewed. Generally, R_0 will be higher when susceptibility and infectiousness are positively correlated (i.e., premises with a high risk of being infected are also at high risk to spread the infection to a high number of other premises) than if the same overall level of infectiousness and susceptibility is evenly distributed within a population (Diekmann et al. 2000). The role of such heterogeneity was highlighted during the SARS epidemic where "superspreading" events were identified as key factor in the spread of infection (Galvani and May 2005), as some infectious hosts infected a much higher number of individuals than the average infectious case. In the cattle movement network formed by Scottish holdings, it was also estimated that 20% holdings were responsible for 80% of the value of R_0 (Woolhouse et al. 2005). From a policy development perspective, heterogeneity can have an important impact when assessing the effectiveness of control measures. Indeed, heterogeneity may cause strong stochastic fluctuations in the early stages of an epidemic, and therefore control policies may need to be readily adaptable to changes in the observed rate of transmission. Moreover, interventions targeting potential high-risk premises may lead to effective disease control with less effort than interventions, which do not consider this level of heterogeneity.

Finally, integrating information relating to the spatial location of premises and their contact patterns increases the level of realism, and, if this can be accurately parameterized with existing data, is likely to improve the reliability of predictions. However, the number of compartments necessary to monitor the number of individuals in each disease state increases rapidly as they are stratified by each additional risk factor. Therefore, when a relatively complex level of heterogeneity is required, individual-based models, where the infection status of each individual in a population is explicitly modeled, are often preferable.

Trade-Off Between Simplicity and Complexity

Simple models, such as the SEIR model, have advantage of being transparent and the way in which the model components interact and drive the disease dynamics can be easily mathematically understood (Keeling and Rohani 2007). Such models often generate analytical solutions (formulae for the general behavior of the model can be obtained directly from the set of model equations) and can thus provide qualitative insights into fundamental principles driving the disease dynamics. However, a higher level of realism in model assumptions is generally needed when seeking quantitative predictions.

Incorporating more detail about host demography and infection processes may result in a model that cannot be solved analytically. Instead, the dynamics of such models have to be simulated in order to obtain numerical approximations of the general model behavior. Although increasing complexity of models improves their realism, it also reduces their transparency, with the mechanisms underlying the disease dynamics becoming increasingly hard to identify, as the number of assumptions and parameters rises. The high number of parameters makes the task of robustly estimating them from data, or finding reliable estimates from existing literature, increasingly arduous. Thus, parameter values or distributions often have to be assumed, subjecting the model outputs to additional and often unquantifiable uncertainty. Moreover, as the desired level of complexity of a model increases, it becomes progressively more difficult to validate any given choice of model assumptions. Therefore, complex models may well become less reliable than simpler ones.

In many cases, over-complexity is just as undesirable as over-simplification (Grassly and Fraser 2008). Models should ideally be as parsimonious as possible, including only mechanisms describing properly the epidemic patterns while ensuring that additional complexity does not alter the outcome. In practice, the level of detail represented in a model needs to be informed by the availability of demographic and disease data. For example in Viet Nam, it has been possible to model the overall dynamics of waves of outbreaks at the commune-level, because this was the geographical resolution at which outbreaks were reported (Walker et al. 2010). Moreover, model parameterization should also be complemented by uncertainty and sensitivity analyses. These involve systematically investigating how potential errors in parameter estimation and changes in parameter values are likely to affect the modeled outputs.

However, whatever their level of detail and the robustness of parameter estimates, models are always a reflection of our current understanding of mechanisms underlying disease transmission, which often remains limited. Thus, models cannot be expected to predict the exact course of an epidemic. In light of this, Medley (2001) recommends their use for relative rather than absolute predictions. Nevertheless, if absolute predictions are made, they should always be associated with a measure of the uncertainty arising from both the stochastic nature of a transmission process and our limited understanding of processes affecting disease spread.

Conclusion

Mathematical models have been widely used to simulate the spread of infectious agents within susceptible host populations, with the principle aims of understanding the fundamental mechanisms underpinning disease dynamics, predicting the course of an epidemic, and assessing the impact of control strategies. Mathematical models can also be used to assess important epidemiological parameters, such as the basic reproduction number and the generation time. Depending upon the scale at which the dynamics of transmission need to be understood, the availability of data and the ability to accurately estimate parameters, models can be developed to incorporate an increasing level of complexity to be able to more faithfully replicate the real-life behavior of epidemics. However, to do so, modelers need to carefully assess the implications this is likely to have for the reliability and transparency of modeled conclusions and predictions. As a result, in many cases, the optimal approach is to develop collections of models that have sequentially increasing complexity. Doing this can provide both general insights into qualitative aspects of disease dynamics and the factors which are most likely to affect transmission, and a robust foundation from which to generate more detailed and realistic predictions.

References

D. J. Alexander, I. H. Brown, *Rev Sci Tech* **28**, 19 (Apr, 2009).

R. M. Anderson, R. M. May, *Infectious diseases of humans: dynamics and control.* (Oxford University Press, UK, 1991).

R. M. Anderson, R. M. May, *Infectious diseases of humans: dynamics and control.* (Oxford University Press, UK, 1992).

R. M. Anderson *et al.*, *Nature* **382**, 779 (Aug 29, 1996).

D. J. Austin, R. M. Anderson, *Philos Trans R Soc Lond B Biol Sci* **354**, 721 (Apr 29, 1999).

G. J. Boender *et al.*, *Plos Comput Biol* **3**, 704 (Apr, 2007).

G. J. Boender, G. Nodelijk, T. J. Hagenaars, A. R. W. Elbers, M. C. M. de Jong, *BMC veterinary research* **4**, (Feb 25, 2008).

M. E. Bos *et al.*, *Prev Vet Med* **95**, 297 (Jul 1, 2010).

A. Bouma *et al.*, *PLoS Pathog* **5**, e1000281 (Jan, 2009).

R. Breban, J. M. Drake, D. E. Stallknecht, P. Rohani, *Plos Comput Biol* **5**, (Apr, 2009).

T. Britton, D. Lindenstrand, *Math Biosci* **222**, 109 (Dec, 2009).

T. E. Carpenter, M. C. Thurmond, T. W. Bates, *J Vet Diagn Invest* **16**, 11 (Jan, 2004).

V. Colizza, M. Barthelemy, A. Barrat, A. Vespignani, *C R Biol* **330**, 364 (Apr, 2007).

J. E. Dent, R. R. Kao, I. Z. Kiss, K. Hyder, M. Arnold, *BMC veterinary research* **4**, (Jul 23, 2008).

O. Diekmann, J. A. Heesterbeek, *Mathematical Epidemiology of Infectious Diseases: Model Building, Analysis and Interpretation.* W. Series, Ed., (Chichester, 2000).

K. Dietz, *Stat Methods Med Res* **2**, 23 (1993).

F. C. Dorea, A. R. Vieira, C. Hofacre, D. Waldrip, D. J. Cole, *Avian Dis* **54**, 713 (Mar, 2010).

N. M. Ferguson, A. C. Ghani, C. A. Donnelly, G. O. Denny, R. M. Anderson, *Proceedings of the Royal Society of London Series B-Biological Sciences* **265**, 545 (Apr 7, 1998).

N. M. Ferguson, C. A. Donnelly, R. M. Anderson, *Science* **292**, 1155 (May 11, 2001).

N. M. Ferguson *et al.*, *Nature* **425**, 681 (Oct 16, 2003).

P. E. Fine, *Epidemiol Rev* **15**, 265 (1993).

G. Fournié, F. J. Guitian, P. Mangtani, A. C. Ghani, *J R Soc Interface*, (Aug 7, 2011).

C. Fraser, S. Riley, R. M. Anderson, N. M. Ferguson, *Proc Natl Acad Sci U S A* **101**, 6146 (Apr 20, 2004).

C. Fraser *et al.*, *Science* **324**, 1557 (Jun 19, 2009).

A. P. Galvani, R. M. May, *Nature* **438**, 293 (Nov 17, 2005).

T. Garske, P. Clarke, A. C. Ghani, *PLoS ONE* **2**, e349 (2007).

A. C. Ghani, N. M. Ferguson, C. A. Donnelly, R. M. Anderson, *Nature* **406**, 583 (Aug 10, 2000).

E. P. Gibbs, *Vet Rec* **157**, 673 (Nov 26, 2005).

N. C. Grassly, C. Fraser, *Nat Rev Microbiol* **6**, 477 (Jun, 2008).

N. C. Grassly *et al.*, *Science* **314**, 1150 (Nov 17, 2006).

V. Guberti, M. Scremin, L. Busani, L. Bonfanti, C. Terregino, *Avian Dis* **51**, 275 (Mar, 2007).

T. B. Hallett *et al.*, *PLoS ONE* **3**, (May 21, 2008).
D. J. Hulse-Post *et al.*, *Proc Natl Acad Sci U S A* **102**, 10682 (Jul 26, 2005).
S. Iwami, T. Suzuki, Y. Takeuchi, *PLoS ONE* **4**, (Mar 18, 2009a).
S. Iwami, Y. Takeuchi, X. N. Liu, S. Nakaoka, *Journal of Theoretical Biology* **259**, 219 (Jul 21, 2009b).
K. E. Jones *et al.*, *Nature* **451**, 990 (Feb 21, 2008).
R. R. Kao, *Trends Microbiol* **10**, 279 (Jun, 2002).
M. J. Keeling, P. Rohani, *Modeling Infectious Diseases in Humans and Animals.* (Princeton University Press, UK, 2007).
M. J. Keeling *et al.*, *Science* **294**, 813 (Oct 26, 2001).
W. O. Kermack, A. G. McKendrick, *Bull Math Biol* **53**, 33 (1991).
D. A. King, C. Peckham, J. K. Waage, J. Brownlie, M. E. Woolhouse, *Science* **313**, 1392 (Sep 8, 2006).
R. P. Kitching, M. V. Thrusfield, N. M. Taylor, *Rev Sci Tech* **25**, 293 (Apr, 2006).
G. Klein, *Sources of power: how people make decisions.* (MIT Press, Cambridge, MA, 1998).
A. Le Menach, E. Vergu, R. F. Grais, D. L. Smith, A. Flahault, *Proceedings of the Royal Society B-Biological Sciences* **273**, 2467 (Oct 7, 2006).
C. Lebarbenchon *et al.*, *PLoS One* **4**, e7289 (2009).
M. Lesnoff, M. Peyre, P. C. Duarte, J. F. Renard, J. C. Mariner, *Epidemiol Infect* **137**, 1405 (Oct, 2009).
S. A. Levin, B. Grenfell, A. Hastings, A. S. Perelson, *Science* **275**, 334 (Jan 17, 1997).
Y. Li *et al.*, *American journal of epidemiology* **160**, 719 (Oct 15, 2004).
M. Lipsitch *et al.*, *Science* **300**, 1966 (Jun 20, 2003).
A. Mannelli, L. Busani, M. Toson, S. Bertolini, S. Marangon, *Prev Vet Med* **81**, 318 (Oct 16, 2007a).
A. Mannelli, L. Busani, M. Toson, S. Bertolini, S. Marangon, *Preventive Veterinary Medicine* **81**, 318 (Oct 16, 2007b).
G. F. Medley, *Science* **294**, 1663 (Nov 23, 2001).
M. Meleigy, *Lancet* **370**, 553 (Aug 18, 2007).
P. Q. Minh *et al.*, *Transbound Emerg Dis* **56**, 311 (Oct, 2009a).
P. Q. Minh *et al.*, *Prev Vet Med* **89**, 16 (May 1, 2009b).
M. Peyre *et al.*, *J Mol Genet Med* **3**, 198 (2009).
S. E. Randolph, D. J. Rogers, *Nat Rev Microbiol* **8**, 361 (May, 2010).
S. Riley, *Science* **316**, 1298 (Jun 1, 2007).
B. Roche *et al.*, *Infect Genet Evol* **9**, 800 (Sep, 2009).
P. Rohani, R. Breban, D. E. Stallknecht, J. M. Drake, *P Natl Acad Sci USA* **106**, 10365 (Jun 23, 2009).
R. Ross, H. P. Hudson, *Proceedings of the Royal Society London A* **43**, 225 (1917).
A. Ross *et al.*, *PLoS ONE* **3**, e2661 (2008).
T. Saito *et al.*, *Vet Microbiol* **133**, 65 (Jan 1, 2009).
N. J. Savill, S. G. St Rose, M. J. Keeling, M. E. J. Woolhouse, *Nature* **442**, 757 (Aug 17, 2006).
K. J. Sharkey, R. G. Bowers, K. L. Morgan, S. E. Robinson, R. M. Christley, *Proceedings of the Royal Society B-Biological Sciences* **275**, 19 (Jan 7, 2008).
E. Shim, A. P. Galvani, *PLoS ONE* **4**, (May 11, 2009).
L. D. Sims, *Avian Dis* **51**, 174 (Mar, 2007).
L. D. Sims *et al.*, *Avian Dis* **47**, 832 (2003).
R. J. Soares Magalhaes, D. U. Pfeiffer, J. Otte, *BMC Vet Res* **6**, 31 (2010).
Royal Society, *Infectious Diseases in Livestock.* (The Royal Society, London, UK, 2002).
A. R. Spickler, D. W. Trampel, J. A. Roth, *Avian Pathol* **37**, 555 (Dec, 2008).
A. Stegeman *et al.*, *Journal of Infectious Diseases* **190**, 2088 (Dec 15, 2004a).
A. Stegeman *et al.*, *J Infect Dis* **190**, 2088 (Dec 15, 2004b).
S. M. Stringer, N. Hunter, M. E. J. Woolhouse, *Math Biosci* **153**, 79 (Nov, 1998).
L. Temime, G. Hejblum, M. Setbon, A. J. Valleron, *Epidemiol Infect* **136**, 289 (Mar, 2008).

T. Tiensin *et al.*, *J Infect Dis* **196**, 1679 (Dec 1, 2007).
F. M. Tomley, M. W. Shirley, *Philos Trans R Soc Lond B Biol Sci* **364**, 2637 (Sep 27, 2009).
J. Truscott *et al.*, *Proc Biol Sci* **274**, 2287 (Sep 22, 2007).
J. A. van der Goot, G. Koch, M. C. M. de Jong, M. van Boven, *P Natl Acad Sci USA* **102**, 18141 (Dec 13, 2005).
J. A. van der Goot, M. van Boven, G. Koch, M. C. M. de Jong, *Vaccine* **25**, 8318 (Nov 28, 2007a).
J. A. van der Goot, M. van Boven, M. C. de Jong, G. Koch, *Avian Dis* **51**, 323 (Mar, 2007b).
J. X. Velasco-Hernandez, H. B. Gershengorn, S. M. Blower, *Lancet Infect Dis* **2**, 487 (Aug, 2002).
C. Verdugo, C. J. Cardona, T. E. Carpenter, *Prev Vet Med* **88**, 109 (Feb 1, 2009).
E. Vynnycky, R. White, *An Introduction to Infectious Disease Modelling*. (Oxford University Press, UK, 2010).
P. G. Walker *et al.*, *PLoS Comput Biol* **6**, e1000683 (2010).
M. P. Ward, D. Maftei, C. Apostu, A. Suru, *Epidemiol Infect* **137**, 219 (Feb, 2009).
R. A. Weiss, A. J. McMichael, *Nat Med* **10**, S70 (Dec, 2004).
M. E. J. Woolhouse, *Acta Tropica* **50**, 189 (Feb, 1992).
M. E. Woolhouse *et al.*, *Proc Natl Acad Sci U S A* **94**, 338 (Jan 7, 1997).
M. E. J. Woolhouse, S. M. Stringer, L. Matthews, N. Hunter, R. M. Anderson, *Proceedings of the Royal Society of London Series B-Biological Sciences* **265**, 1205 (Jul 7, 1998).
M. E. Woolhouse *et al.*, *Biol Lett* **1**, 350 (Sep 22, 2005).
J. A. Yorke, H. W. Hethcote, A. Nold, *Sexually transmitted diseases* **5**, 51 (Apr-Jun, 1978).

Chapter 12
Large-Scale Vaccination for the Control of Avian Influenza: Epidemiological and Financial Implications

Jan Hinrichs and Joachim Otte

Introduction

Since its emergence in 1996 in China, highly pathogenic avian influenza H5N1 virus has infected 61 countries, caused more than 300 human fatalities, and resulted in disease mortality and culling of several hundred million domestic birds. In most of the affected countries, the H5N1 virus could be eliminated through swift and determined interventions of national animal health systems. In some countries, however, the virus appears to have become endemic in specific eco and production systems, leading to resurgence of infection in poultry and humans the moment control efforts are relaxed. The countries in which HPAI H5N1 virus can currently be considered endemic comprise Bangladesh, China, Egypt, Indonesia, and Vietnam as well as the Indian State of West Bengal (FAO in press).

Prior to the massive HPAI H5N1 epidemics of in Southeast Asia, only few attempts to control HPAI in domestic poultry populations using large-scale vaccination campaigns have been carried out. Pre-H5N1 experience in the use of vaccination in HPAI control programs was gained in Mexico (H5N1, 1994), Italy (H7N1, 2000), and Pakistan (H7N3, 2003) (van den Berg et al. 2007). In the course of the current H5N1 avian pandemic, large-scale vaccination campaigns have been implemented by national animal health authorities in China, Hong Kong, Vietnam, Indonesia, and Egypt. Experience has shown that despite the theoretical potential of vaccination to control HPAI epidemics (Swayne et al. 2000; van den Berg et al. 2007), in practice, large-scale vaccination efforts rarely realize their theoretical potential due to

J. Hinrichs (✉)
Animal Health Economist, FAO Regional Office for Asia and the Pacific (FAO-RAP), 39 Phra Atit Road, Bangkok 10200, Thailand
e-mail: jan.hinrichs@fao.org

J. Otte
Food and Agriculture Organization of the United Nations, Rome, Italy

D. Zilberman et al. (eds.), *Health and Animal Agriculture in Developing Countries*, Natural Resource Management and Policy 36, DOI 10.1007/978-1-4419-7077-0_12,

the numerous constraints to delivering and administering vaccine in large and heterogeneous poultry populations (van den Berg et al. 2007).

Large-scale vaccination campaigns require substantial financial and human resources and therefore are unlikely to be sustainable over long periods. This article, therefore, assesses the technical, epidemiological as well as the financial implications of large-scale vaccination campaigns for HPAI control in developing countries, both from a theoretical as well as from a practical perspective, drawing on existing literature on HPAI vaccines, HPAI epidemics, and large-scale HPAI H5N1 vaccination programs as well as on calculations carried out by the authors.

The paper starts by summarizing the literature on various characteristics of commercially available HPAI vaccines, which determine their utility as a tool for the control of HPAI in different poultry species. The following section presents the three HPAI vaccination strategies proposed by the Office International des Epizooties (OIE) and reviews estimates of immunization rates that would be necessary to suppress H5N1 virus transmission to a level where infection dies out. Part IV provides estimates of the maximum vaccination coverage that would be achievable in different poultry production systems through mass vaccination campaigns carried out under ideal conditions and summarizes experiences with the implementation of such programs. Part V assesses of the costs and returns of vaccination from a producer perspective as well as considering the broader public aspects of embarking on vaccination as part of a national HPAI control program. Finally, the last section discusses the advantages and drawbacks of large-scale vaccination programs and their potential contribution to HPAI control.

Characteristics of Commercial HPAI Vaccines

Route and Schedule of Vaccine Administration

All commercially available HPAI vaccines require administration to poultry via injection. Vaccines for administration via other routes, such as aerosols, drinking water, or injection during egg incubation are not currently available (Swayne and Kapczinski 2008). Generally, a two-dose vaccination schedule (14 day interval between injections) is required to achieve satisfactory protection in any poultry species and birds should not be vaccinated before 7 days of age.

Vaccine Efficacy

Vaccination efficacy in individual birds can be quantified using the following three parameters: (1) degree of protection from infection when exposed to a given amount of infectious virus, (2) the degree of reduction of morbidity and mortality given infection occurred, and (3) the level of reduction of virus excretion by infected poultry.

The efficacy of most commercially available vaccines has been determined in studies with chickens or turkeys, since globally they represent the economically most important poultry species. However, in many developing countries, particularly in Southeast Asia, other species such as ducks, muscovy ducks, and quails also represent significant parts of the poultry sector. Thus, vaccines for use in poultry populations with significant shares of species other than chicken and turkeys require efficacy testing in these species to contain viral transmission.

Although almost all commercially available vaccines provide some level of protection from infection with virulent virus and significantly reduce mortality in infected chicken, no HPAI vaccine has so far proved to satisfactorily perform on all three of the above parameters (Swayne 2006; van den Berg et al. 2007).

Onset and Duration of Immunity

The time lag between vaccination and protective immunity and the respective duration of protection depends on the vaccine used, timing of vaccination, number of doses given, species, immunologic condition of the birds, and the challenge virus. The literature on vaccine efficacy is dominated by studies with specific pathogen free (SPF) birds reared under laboratory conditions. The results of these studies cannot safely be extrapolated to field conditions, and in these protection is often assessed by serology, the assumption being that birds with a specified antibody level (e.g., a hemagglutination inhibition [HI] titer $\geq$ 16) are protected. However, under conditions of antigenic variability and diversity of the HPAI viruses circulating in the field, a given titer found against a laboratory strain cannot unequivocally be interpreted as protective against field virus challenge. Published literature on vaccine efficacy and the onset of protection under field and laboratory conditions indicates that protection in chicken is in no circumstances achieved prior to 11–13 days post vaccination (Ellis et al. 2004, 2005; van der Goot et al. 2005). About 20.5 weeks after successful vaccination of layer chicken with two doses of killed LPAI H6N2 vaccine, all birds tested sero-negative again (Cardona et al. 2006).

Vaccine Cost and Storage Requirements

Legok and Harbin Werke vaccines purchased in large quantities for use in vaccination programs in Indonesia and Vietnam cost about USD 0.02–0.03 per dose. Swayne and Kapczinski (2008) provide a wholesale price for HPAI vaccines of USD 0.05–0.15 per dose. Avian influenza vaccines have to be stored within a temperature range of 2–8°C (Amorij et al. 2008) and cold storage and a cold chain is required to maintain the efficacy of all commercially available vaccines (CAST 2007).

Antigenic Drift and Long-Term Vaccine Efficacy

Avian influenza viruses vary antigenically and evolve rapidly, which poses a major challenge for the sustained use of vaccines as HPAI control measure. Although several studies demonstrated cross-protection for HPAI viruses, a correlation between virus shedding and antigenic differences of vaccine and field strains was shown by Lee et al. (2004) for the Mexican lineage H5N2 virus and by Swayne et al. (2000) for a range of H5 HPAI viruses. During and after the extensive use of about two billion doses of H5N2 vaccine in commercial poultry farms in Mexico, molecular drifts with a yearly trend have been shown (Escorcia et al. 2008). Antigenic drift of avian influenza viruses was also observed in the USA after vaccination programs for LPAI in commercial poultry (Suarez et al. 2006). Variant field strains that escaped the protection by the commonly used vaccines emerged in Shanxi China during 2006, in Egypt in late 2006, and in Indonesia early 2007 (Swayne and Kapczinski 2008). Using a deterministic patch-structured model, Iwami et al. (2009) found that an avian influenza vaccination campaign can lead to the prevalence of a vaccine-resistant strain, which could result in the replacement of viral strains in areas without vaccination via migration of asymptomatic birds.

HPAI Vaccination "Strategies" and Effective Immunization Rates

OIE (2007) lists three HPAI vaccination "strategies" with distinct objectives: (1) preventive vaccination, (2) emergency vaccination, and (3) routine vaccination. A summary of the objective, time frame, and critical success factors of these HPAI vaccination strategies is given in Table 12.1.

Preventive Vaccination

Preventive vaccination is proposed as an option to prevent the infection of poultry flocks in a country or region that is free of disease but at "high" risk of virus introduction and in which early detection and elimination of infection may not be feasible or realistic. Incorporation of DIVA[1] is recommended as part of such a strategy. For example, in the Netherlands, vaccination of free-range laying hens and hobby poultry with inactivated H5N9 vaccine was permitted as an alternative risk reduction measure to indoor housing in 2006. In Hong Kong, a killed oil-adjuvanted

[1] DIVA: Differentiating Infected from Vaccinated Animals, i.e., vaccinated birds can be distinguished from (vaccinated and subsequently) infected birds.

Table 12.1 Vaccination strategies, objectives, time frame, critical success factors and alternative/complementary control measures

Vaccination strategy	Objective	Time frame	Critical factor	Alternatives
Preventive	Protect individual/specific flocks/birds in danger of exposure	Variable, depending on risk of exposure to infectious virus	Accuracy of the exposure risk assessment	Improve bio-security, limit contact to secure sources
Emergency	Curtail potential of an acute epidemic after virus introduction	Short-term	Time to achieve immunity	Movement control and pre-emptive depopulation
Routine	Reduction of mortality/production losses in endemic situations; in longer term, may facilitate eradication of HPAI virus in domestic poultry	Medium- to long-term	Effective immunization rate (reduction of between-flock $R_n < 1$)	Passive and active surveillance with rapid stamping out
	Reduction of human health risk		Reduction of viral shedding by infected birds	Behavior change reducing human exposure

H5N2 vaccine is used in broiler chicken farms since the HPAI outbreak in 2002 to reduce the likelihood of outbreaks, if introduction of infection were to occur from mainland China (EFSA 2007).

Emergency Vaccination

This vaccination strategy is considered an option for the control of HPAI introduced into the national flock when the epidemiological situation suggests an immediate and high risk of massive and rapid spread of infection, which cannot be contained by culling and movement restrictions. Emergency vaccination includes "ring" vaccination of flocks located within a predefined (but not further specified) radius around detected outbreaks to create a "buffer zone." This strategy was applied in northern Pakistan within a 3 km ring after H7N3 outbreaks in 2003 when layer and breeder flocks were vaccinated (EFSA 2007).

Routine Vaccination

Routine vaccination is listed as an appropriate measure in "countries and regions where the disease is endemic and where the classical control cannot be effectively implemented to eliminate the virus." It can achieve a reduction in poultry mortality and in the longer term decrease the prevalence of infection to a level where surveillance and stamping out could be applied cost-effectively. Eradication of HPAI virus is not stated in the OIE (2007) document as an objective that is achievable solely through routine vaccination[2] In addition, the contribution of vaccination to reducing the risk of human cases via reducing the virus load is mentioned in OIE (2007) as a potential result of any vaccination strategy.

All three of the above "strategies" can be applied in different "tactics," i.e., either in a "mass/blanket" or a "targeted" manner, targeting specific subpopulations (age, species, location, production systems) and/or times within production cycles.

Effective Immunization Rates

Whichever the applied vaccination strategy, its effectiveness depends on the proportion of poultry which are rendered immune and will therefore not significantly contribute to virus transmission in the case of exposure. As poultry populations are segregated into flocks (or other types of management units), both within-flock and between-flock transmission need to be sufficiently contained to avoid sustained virus transmission and potential development of endemicity.

A number of studies estimated the minimum within-flock immunization rate required to stop virus transmission within a poultry flock. Depending on the assumptions made, the calculated within-flock immunization rate required to avoid disease spread range from 50 to 90% (Bouma et al. 2009; Lesnoff et al. 2009; Savill et al. 2006; Tiensin et al. 2007).

A factor of high importance for disease control is the contact rate between flocks and the level of risk of each contact to transmit infection (Beach et al. 2007). The objective of emergency and routine vaccination would be to interrupt the infection chain between flocks/farms, i.e., achieve a between-flock/farm-to-farm "reproductive number" (R_n) that is below unity. The required proportion of flocks that would have to be immunized within an affected region can be derived from the infection dynamics in that region in the absence of control measures. For HPAI outbreaks in Holland, Canada, and Italy, Garske et al. (2007) estimated mean farm-to-farm reproductive numbers prior to the introduction of control measures to range from 1.1 to 2.4. For Vietnam and Romania, countries with less industrialized poultry

[2] "Routine vaccination" was successfully used for the eradication of other transboundary diseases such as Rinderpest in Africa (Normile 2008) and FMD in parts of South-America (Melo 2002).

sectors, R_n was estimated to have been in the order of 2–3 prior to the introduction of control measures (Walker et al. 2009; Ward et al. 2009).

Under the ideal condition of full protection of vaccinated flocks, immunization of between half and two thirds of flocks would be necessary to stop sustained between-flock transmission under the situation prevailing in Vietnam (based on the fraction of $1 - 1/R_0$, Anderson and May 1992). Under conditions where only partial immunity of vaccinated flocks is achieved, a higher proportion of immunized flocks would be required to interrupt disease transmission to the extent needed to interrupt infection chains.

Poultry Production Systems and Vaccination Coverage

Production system characteristics are essential determinants of the potential vaccination coverage and whether this coverage is capable of preventing, or at least significantly reducing, virus transmission (Alders et al. 2007). Important characteristics are the average lifespan of poultry, origin of replacement birds (home-bred vs. bought), synchronization of flock age, and the disease status and immuno-competency of flocks. Furthermore, poultry owners' incentives to vaccinate against HPAI strongly influence potential vaccination coverage.

Poultry Production Systems

Within the poultry sector, four main production systems can be identified: (1) Breeder flocks, (2) layer flocks, (3) broiler flocks, and (4) backyard multi-purpose flocks. This production system classification roughly applies to both chickens and ducks (Rushton et al. 2010). Subsystems, such as short- and long-lived broiler production and free-range grazing duck production prevail in some countries. The major poultry production systems are briefly described in the following to provide an overview of the main characteristics relevant for the effectiveness of HPAI vaccination campaigns.

Breeder flocks (grand parent and parent) are kept in closed houses and cages, and the birds are bought from specialized poultry genetic supply companies. Several batches of birds of different ages are required to meet the continuous demand for day-old chicks (DOCs). The average lifespan of breeder chicken for the production of DOCs varies between 63 and 65 weeks. The production period of breeder duck flocks varies between 52 and 104 weeks. Breeder ducks are kept in houses with outdoor access (Desvaux and Ton 2008).

Layer chicken are typically kept in cages in open or closed housing without outdoor access. Hens start to lay at 23–25 weeks of age and are kept for 63–74 weeks. Continuous egg production requires a layer flock with chickens of several age groups. Either DOCs or pullets (at 16–25 weeks of age) are bought to replace spent hens, which are either sold for immediate slaughter or for fattening. In Asia, layer ducks are kept between 1 and 3 years and for a varying proportion of this time flocks can be free-ranging in rice fields to use left over rice, weeds, and snails as a feed resource. Semi-confined fishponds with temporary or permanent shelters are used to house the ducks when they are not left to range in rice fields (Desvaux and Ton 2008).

In Southeast Asia, two broiler chicken production systems need to be distinguished, namely long- and short-finish systems. Short-finish, "industrial," broilers are usually kept indoors on the ground, whereas long-finish, crossbred broilers are kept under semiconfined conditions to utilize some feed resources from the natural environment. "Industrial" broilers are kept in batches of the same age for 5–7 weeks while "crossbred" broilers are kept for 9–26 weeks and achieve a premium price in local markets. Similar to broiler chicken production systems, the production cycle length of broiler ducks varies depending on the breed and feeding system used. Scavenging broiler ducks are reared under similar conditions to layer ducks and are usually sold for slaughter after about 80 days. Confined and intensively fed ducks, usually of specific meat type breeds, are finished in about 60 days (Desvaux and Ton 2008; Seng 2007; Songserm et al. 2006).

Mixed backyard poultry production systems are characterized by scavenging indigenous birds that consume left over feed and produce their own replacement chicks with very little cash investment of the owner (Otte 2006). The systems typically have high mortality rates in the early stages of life due to predation, poor diets, and diseases (Dessie 1996; Gunaratne et al. 1993; Rushton and Ngongi 1998; Spradbrow 1993). Birds that die get replaced with newly hatched chicks resulting in a high population turn over, which significantly limits the duration of flock immunity that can be maintained with vaccination campaigns.

Production System Specific Vaccination Coverage Achievable with Vaccination Campaigns

For nonbackyard poultry production systems, which in most countries constitute a significant share of the standing poultry population, no estimates of the coverage achievable through vaccination campaigns was available in the literature. The authors therefore had to make their own estimates, which are based on the simplifying assumption that all birds within a specific flock and production system are of the same age (i.e., all-in/all-out flock management) and that the age of all flocks in a region is uniformly distributed (i.e., no seasonal production). The maximum flock vaccination coverage with a vaccination campaign, in which vaccination teams visit each farm once only (or twice in case of booster application) is then theoretically given by the time a particular flock is eligible for vaccination within its production cycle (including the idle time between batches). The estimated maximum achievable vaccination coverage applies to the area that can be vaccinated within a day. For areas which require more than 1 day for vaccination teams to visit all flocks, the theoretical maximum vaccination coverage is lower due to flock turnover. The results of the calculations are displayed in Table 12.2. The potential coverage of HPAI vaccination

Table 12.2 Estimated flock vaccination coverage and time until flocks are fully susceptible again after an HPAI vaccination campaign

	Length of production cycle (days)	Idle time between batches (days)	Time eligible for first shot (days)	Time eligible for second shot (days)	Proportion of flocks eligible for first shot (%)	Proportion of flocks eligible for second shot 14 days later (%)	Time until all flocks are 100% naïve again after first shot (days)[a]	Time until all flocks are 100% naïve again after second shot (days)[b]
Layer chicken; vaccination during laying period	490	0	476	462	97	94	180	180
Layer chicken; no vaccination during laying period	490	0	140	126	29	26	147	133
Broiler chicken; industrial	32	14	18	4	39	9	25	11
Broiler chicken; crossbred	120	14	106	92	79	69	113	99
Broiler ducks; intensive confined	60	21	46	32	57	40	53	39
Broiler ducks; scavenging	80	–	66	52	83	65	73	59

Source: authors' calculations[a]Protection assumed to be lost 180 days after vaccination due to waning immunity[b]It should be noted that the delivery of a second shot is a challenge due to the difficult accessibility of scavenging ducks in rice paddies and will require significantly higher vaccinator time inputs for traveling and catching ducks

in backyard poultry with uncontrolled but more or less continuous replacement dynamics has been modeled by several authors (e.g., Lesnoff et al. 2009; Taylor 2008; Udo et al. 2006) and field data from countries implementing large scale vaccination campaigns is available for comparison.
Chicken breeder and layer flocks. In theory, high vaccination coverage can be achieved for chicken layer flocks with a production cycle length of 490 days, assuming owners would allow vaccination during the laying period (Table 12.2). Since layer flocks usually comprise birds of several age groups (are not kept under all in/all out management), the coverage of 94% for a single shot campaign and 92% for a double shot campaign is a reflection of the share of birds that could be vaccinated at any point in time within a flock of heterogeneous age composition. It should, however, be recognized that in most layer flocks, birds are vaccinated before point of lay and owners are reluctant to (re)vaccinate during egg laying periods. If vaccination during the laying period is not accepted, vaccination coverage would not exceed 26% in a double shot vaccination campaign.

Scavenging layer duck flocks. The age distribution and the number of layer duck flocks are related to the rice harvest seasons. Vaccination coverage similar to those in layer chicken could be achieved with well-timed vaccination campaigns that take into account the seasonality of rice harvest and laying period. However, the poor accessibility of scavenging duck flocks in rice paddies makes the administration of vaccine, especially the booster shot, very difficult and experience from China shows that scavenging duck breeder and layer flocks usually do not receive a booster vaccination (Chen 2009).

Short-lived industrial chicken broiler flocks. During the average 32-day production cycle of an industrial batch of broiler chicken, vaccine can be administered over a period of 18 days, since vaccination should only be applied at a minimum age of 7 days and requires at least 7 days to confer protection, i.e., needs to be applied more than 7 days prior to slaughter to have any effect. Because of an assumed average idle time of 14 days for cleaning and disinfection between batches, a significant proportion (18/46) of broiler houses will either be empty or not be populated by birds eligible for vaccination on any given day. Hence the maximum achievable vaccination coverage with a single injection in a 1-day visit vaccination campaign is 39% of all industrial broiler flocks within a country or region. As described in Part II, two injections are required to achieve full protection. Therefore, if each broiler flock needs to be given a booster shot 14 days after the initial vaccination, only the flocks that received their first injection during the first 4 days of the 18 day window will be eligible for the booster shot 14 days later. Although some previously noneligible farms will have become available for the first vaccination in the 14-day interval, only 9% of all broiler flocks would have received the two injections required to achieve full protection. All industrial broiler flocks would be naïve again within 25 days after a single injection vaccination campaign since vaccinated birds will have been slaughtered and replaced.

Intensively raised duck broiler flocks. The relatively short production cycle length of about 60 days for duck broilers in intensive closed systems results in a low

maximum coverage of 40% for a two-injection and 14-day interval HPAI vaccination campaign. However, experience from China shows that flock owners usually do not vaccinate their duck broilers, due to their short life span (Chen 2009).
Long-lived crossbred chicken broiler flocks. The maximum achievable vaccination coverage for crossbred broiler flocks would be 79 and 69% for a one- or two-injection campaign, respectively. The higher coverage compared to industrial broiler flocks results from the considerably longer production cycle. All crossbred broiler flocks would be naïve again within 113 days after a single injection vaccination campaign.

Scavenging duck broiler flocks. Seasonal peaks of available rice in harvested paddies are utilized by raising broiler ducks around the time of rice harvests. Therefore, the timing of vaccination campaigns should take into account this seasonality. The maximum achievable vaccination coverage by a vaccination campaign will depend on the age distribution of duck broiler flocks within a region. However, under the assumption that the start of broiler flock raising in a region is uniformly distributed over a time period of at least 80 days, 65% of all flocks can be vaccinated in a two shot campaign.[3]

Backyard flocks. Based on a spreadsheet poultry population model, Taylor (2008) estimates that a maximum immunization rate of 52% of all backyard birds can be achieved with a two-shot vaccination campaign under the assumption that 80% of all chicken older than 4 weeks are caught for an initial vaccination and a booster shot 14 days later. Immunization coverage would fall to 19% within 17 weeks due to replacement. These results assume a vaccine efficacy of 80% and that 50% of the eggs are used for human consumption and the remaining 50% are hatched. Also under a two-shot vaccination scenario, vaccinating all poultry $\geq$14 days and assuming 80% vaccine efficacy, Lesnoff et al. (2009) estimate a maximum achievable population immunity rate of 55%. Seventeen weeks after the vaccination campaign, the population immunity rate is estimated to have dropped to 25%.

Table 12.3 provides a compilation of postvaccination sero-monitoring results obtained in different countries, which have embarked on large-scale HPAI vaccination programs. In individual birds, the prevalence of antibody levels regarded as protective ranges from 16 to 72.1%. Testing of 1,113 chicken sera from Guangdong and Guiyang Provinces in China collected at markets in 2005 and 2006 revealed that only 180 (16%) were positive against Ck/HK/YU22/02 (H5N1) and that 55 of the positive sera had low or no neutralizing antibodies against the predominant FJ-like sublineage (Smith et al. 2006). In the provinces West Java, Yogyakarta, and Central Java about 18,000 blood samples from vaccinated backyard chicken and ducks were taken for sero-monitoring as part of World Bank and USAID funded operational research project, of which 33.1% showed a protective HI titer $\geq$ 16 (McLaws 2009).

[3] It should be noted that the delivery of a second shot is a challenge due to the difficult accessibility of scavenging ducks in rice paddies and will require significantly higher vaccinator time inputs for traveling and catching ducks.

Table 12.3 Prevalence of protective HPAI antibody levels in poultry after vaccination campaigns

Country	Species	"Protected"	Comment(s)	References
China	n.s.	69% (n = n.s.)	In 2004	EFSA (2007)
China	Chicken	16% ($n = 1{,}113$)	Sera collected from Nov. 2005 to April 2006 in Guangdong and Guiyang Provinces; HI titer ≥ 20	Smith et al. (2006)
Egypt	n.s.	25.6% ($n = 160$)		CIRAD (2008)
Hong Kong	Chicken (broilers)	75.8% batches ($n = 248$ batches)	1 month after second vaccination, HI titer ≥ 16	Ellis et al. (2005)
Indonesia	Chicken and ducks	33.1% ($n \approx 18{,}000$)	1–2 months after campaign ended, H5N1 HI titer ≥ 16	McLaws (2009)
Vietnam	n.s.	53.8% ($n = 364$)	1–2 weeks after end of first round of 2005 campaign	Nguyen (2008)
	n.s.	44.9% ($n = 203$)	3–4 weeks after end of first round of 2005 campaign	
	n.s.	56.2% ($n = 269$)	1–2 weeks after end of first round of 2006 campaign	
	n.s.	33.3% ($n = 43$)	3–4 weeks after end of first round of 2006 campaign	
	n.s.	72.1% ($n = 1{,}263$)	1 month after end of first round of 2007 campaign	
Vietnam	Ducks Chicken	55% ($n = 182$) 40% ($n = 30$)	4–6 weeks after end of the first round of the 2007 campaign	Henning et al. (2008)
	Ducks Chicken	63% ($n = 302$) 37% ($n = 57$)	>12 weeks after end of the first round of the 2007 campaign	

N.s. not specified

According to Domenech et al. (2009), the government vaccination campaign had little impact on disease incidence and its contribution to reducing the number of reported human cases is unclear. Only 11% of the HPAI vaccinated native chicken populations in Bali, Indonesia had protective HI titers (Sawitri Siregar et al. 2007). In Vietnam, the proportion of tested backyard and commercial poultry with protective antibody levels against HPAI varied between 33 and 72% (Nguyen 2008; Henning et al. 2008).

Vaccination Coverage of National Poultry Populations

The maximum vaccination coverage of national poultry populations achievable through vaccination campaigns is determined by the mix of flock types in the national poultry industry and the time required to carry out a campaign. The latter is determined by the size of the poultry industry and resources devoted to the vaccination campaign. Given that in most countries, broilers are the most common type of poultry followed by backyard birds, while layer and breeder flocks are comparatively rare, immunization rates needed to break infection chains will be difficult to achieve unless vaccination campaigns are complemented by restocking bans.

In China, vaccination coverage of 20–50% was achieved in poultry populations consisting mainly of backyard poultry (Peyre et al. 2008). Relatively low vaccination rates have been observed in waterfowl and the booster vaccination for layer and breeder ducks is not actually administered while broiler ducks are usually not vaccinated at all (Chen 2009). In Egypt, practical difficulties in vaccinating household poultry in the field resulted in an average vaccination coverage of 20–30% in the best vaccinated Governorates and as low as 1% in some villages (Peyre et al. 2009).

In Indonesia, vaccination coverage of Government vaccination campaigns in backyard and semi-intensive production systems rarely exceeded 30% of the poultry population (Sawitri Siregar et al. 2007). A World Bank and USAID funded operational research project vaccinated 2.9 million backyard poultry in 425 villages in the provinces West Java, Yogyakarta, and Central Java with two doses of vaccine. It took 3 weeks each to administer the first and second shots of Legok 2003 H5N1 vaccine and required 1,088 community vaccinators and 64 community vaccinator coordinators. The campaign was estimated to have covered only 32% of the 9.0 million poultry reported by official livestock statistics to be in the subdistricts where vaccination was implemented (FAO 2009b; McLaws 2009).

Vaccination coverage for the vaccination campaigns in Vietnam was estimated by Taylor (2007) and To et al. (2007). Estimations based on the number of vaccinated domestic poultry in relation to available domestic poultry census figures ranged between 22 and 66% for chicken and between 26 and 79% for duck populations. A decreasing trend in vaccination coverage was also seen, which could be related to increasing vaccination fatigue of vaccinators and reduced vaccination acceptance by poultry owners.

Practical Challenges of Large-Scale HPAI Vaccination Programs

It has been emphasized by several authors that the achievable protection through vaccine use in the field is unlikely to reach the potential shown under experimental studies (Swayne et al. 2000; van den Berg et al. 2007). Since commercially available HPAI vaccines are not thermostable, maintaining the cold chain from production to administration is crucial for obtaining high levels of immunization in vaccinated birds in the field. This represents a significant challenge in many developing countries with high daytime temperatures and shortages of cold storage capacity. Other factors that compromise the effectiveness of avian influenza vaccination campaigns are inadequate hygiene during vaccine administration, administration to inappropriate tissues or the birds, or an infection with immunosuppressive disease agents (Wooldridge 2007; Alders et al. 2007).

Because of the changing nature of HPAI virus strains circulating in the field, the need for postvaccination monitoring of vaccine efficacy is stressed in the pertinent literature. Virus isolation and sequencing should be an essential part of any vaccination strategy for early detection of potential genetic shifts affecting vaccine efficacy. The possibility that the efficacy of the vaccine(s) used in the program decreases over time needs to be considered in the planning process because the frequency with which a new representative master-seed for the production of adapted vaccines needs to be found directly influences the profitability of commercial vaccine production. Vaccine producers need a certain level of security about the potential scale and duration of market demand for their vaccine to embark in vaccine development and production. The recommended number of samples required to monitor vaccine performance represents a major challenge for the capacity of most laboratories in developing countries.

It needs to be emphasized that due to scarce veterinary staff and animal health funds the inevitable effect of large-scale vaccination campaigns is a detraction of public animal health services from other disease control activities, both with respect to HPAI and other diseases (e.g., FMD outbreaks increased in Vietnam during and after HPAI vaccination campaigns).

Costs and Incentives for HPAI Vaccination

Costs of Large-Scale Vaccination Campaigns in Indonesia and Vietnam

Essential cost components of any mass vaccination program comprise: (1) planning, monitoring, and communication; (2) vaccinator labor and equipment; (3) storage and distribution of vaccine and equipment; (4) postvaccination sero-monitoring; and (5) the vaccine itself. The recurrent and investment costs differ between countries due to different economic and (veterinary) infrastructure conditions and due to differences in the structure of the poultry industry. Unfortunately, comprehensive cost estimates vaccination campaigns are only available for Indonesia and Vietnam.

The vaccination costs presented in Table 12.4 are derived from *ex-ante* assessments for a government run mass vaccination campaign in Vietnam and a planned mass vaccination campaign in Western Java, Indonesia. The total costs per vaccination range from USD 0.03 per bird in broiler flocks in Vietnam to USD 0.12 per bird in backyard flocks in Indonesia. HPAI vaccination costs per bird differ

Table 12.4 Mass vaccination campaign costs per 100 birds vaccinated (USD)

	Layer		Broiler		Backyard	
Cost component	Indonesia	Vietnam	Indonesia	Vietnam	Indonesia	Vietnam
Vaccine	3.18 (77%)	4.06 (85%)	3.18 (77%)	2.03 (75%)	3.18 (28%)	2.03 (55%)
Storage and distribution	0.04 (1%)	0.01 (0%)	0.04 (1%)	0.01 (0%)	3.27 (29%)	0.01 (0%)
Vaccination (labor, equipment)	0.84 (21%)	0.36 (8%)	0.84 (21%)	0.36 (13%)	4.88 (43%)	1.36 (36%)
Postvaccination monitoring	0.02 (0%)	0.02 (0%)	0.02 (0%)	0.02 (1%)	0.02 (0%)	0.02 (0%)
Planning and communication	0.03 (1%)	0.31 (6%)	0.03 (1%)	0.31 (11%)	0.03 (0%)	0.31 (8%)
Total	4.12	4.75	4.12	2.72	11.38	3.72

Source: authors' calculations

between production systems due to varying accessibility and flock size related differences in achievable vaccinations per vaccinator and day.[4] Predominantly scavenging extensive backyard flocks in remote areas demand significantly higher labor input for vaccination than confined chicken broiler production systems. The relatively high storage and distribution costs for the vaccination of backyard poultry in Indonesia result from necessary investments in motorbikes to supply vaccinators with vaccine.

Private Incentives for HPAI Vaccination

If poultry owners see an economic advantage in vaccinating their flock against HPAI compared to (or in addition to) applying other control measures, they are more likely to comply with a compulsory national vaccination strategy and a higher level of vaccination coverage can be achieved (McLeod and Rushton 2007). Whether poultry owners regard vaccination as a financially worthwhile risk reduction measure for themselves to finance not only depends on the infection risk and cost of vaccination, but also on the overall profitability of their respective poultry enterprise. Indicative values for production inputs, prices of inputs and outputs, performance indicators, and production margins for the various production systems in Southeast Asia have been estimated by Hinrichs et al. (2010). The vaccination campaign costs in Table 12.4 were used to calculate vaccination costs per production cycle in Table 12.5 while an age-based vaccination scheme to achieve the maximum immunization coverage of the birds during the production cycle is assumed. Especially for short production cycle broiler flocks this is not achievable through

[4] 115 (300) birds per day and vaccinator assumed in backyard systems and 500 (500) birds per day and vaccinator assumed in broiler and layer systems in Indonesia (Vietnam).

Table 12.5 Financial breakeven outbreak risks for HPAI vaccination

Production system	Vaccinations required per production cycle	Vaccination costs per prod. cycle (USD cents)		Vaccination costs per year (USD cents)		HPAI outbreak loss (USD)[a]		Breakeven outbreak risk per year (%)[b]	
		Min.	Max.	Min.	Max.	Min.	Max.	Min.	Max.
Grand parent: DOC to spent hen	4	16	19	13	15	55	182	0.1	0.3
Parent: DOC to spent hen	4	16	19	13	15	1	6	3	20
Layer: DOC to pullet	3	12	14	24	41	1	2	18	84
Layer: Pullet to spent hen	1	4	5	4	5	3	5	1	3
Layer: DOC to spent hen	4	16	19	11	16	2	4	3	9
Broiler (crossbred): DOC to broiler	2	5	8	10	40	2	4	3	28
Broiler (industrial): DOC to broiler	2	5	8	33	65	1	6	7	75
Backyard: Egg to chick	2	7	23	77	394	0.02	1	190	29,987
Backyard: Chick to grower	1	4	11	8	28	3	3	3	14
Backyard: Grower to cock/spent hen	1	4	11	5	15	2	3	2	9
Backyard flock		72	219	453	2,191	23	33	17	120

Source: authors' calculations

[a]A potential outbreak is assumed to occur in the middle of the production cycle which would lead to the loss of 50% of the bird value at the production cycle end and 4 weeks of lost gross margin due to subsequent downtime for cleaning and disinfection

[b]Based on 80% vaccine efficacy

mass vaccination campaigns. A more continuous vaccination delivery system would be required, for which the costs are likely to be higher, than the cost estimated for vaccination campaigns.

From a flock owner's perspective, vaccination represents a protective measure against the economic impact of a HPAI outbreak. The decision on whether to contract the "vaccination insurance policy" depends on the vaccination costs ("insurance premium"), which also include production impacts of vaccination such as decreased egg laying, the expected economic loss in case of an outbreak, and the perceived probability of an outbreak. The ratio of vaccination costs to outbreak losses ("breakeven outbreak risk per year" in Table 12.5), adjusted for an average 80% vaccine efficacy, indicates the probability of flock infection at which expenditure on vaccination would be profitable for a risk neutral flock owner. The expected absolute loss in case of an outbreak varies during the production cycle. The applied absolute losses for the calculations of the break even risks in Table 12.5 are based on the assumption of a 50% loss of the maximum bird value and a 4-week gross margin loss due to production downtime subsequent to an outbreak. In case outbreaks occur when birds are of lower value than assumed in Table 12.5 or a salvage value can be derived from selling sick birds, the calculated breakeven risks are underestimated. The breakeven risks presented in Table 12.5 are systematically underestimated due to several other reasons. In all production systems, the "background" mortality requires vaccinating more birds than will eventually be produced. Vaccination can depress production leading to revenue foregone. In Vietnam, layer chicken reportedly showed a decrease in egg production of 5% over 3 weeks following vaccination (Cristalli 2006). This cost component has not been accounted for in the estimation of vaccination costs for layer flocks.

Grand parent and layer chicken production systems have the relatively lowest estimated breakeven outbreak risk and compared to vaccination in other production systems, vaccination in these production systems could be most cost-effective. Depending on the HPAI situation, flock owners have a relatively high private incentive to adopt vaccination, and HPAI vaccination is commonly used by parent stock keepers in Indonesia (Sawitri Siregar et al. 2007). However, even for relatively valuable laying hens with an estimated potential HPAI outbreak loss of USD 6 per bird the minimum annual breakeven risk of 3% would not economically justify the use of vaccination for a risk neutral poultry keeper under infection risk conditions similar to the peak HPAI H5N1 incidence in 2004 in Thailand and Vietnam, which was around 0.2% (Otte et al. 2008). Nevertheless, HPAI vaccination is reportedly widely used by layer flock owners in Vietnam and Indonesia (Sims, personal communication), which indicates that these poultry producers are risk averse and regard vaccination as an important component of their strategy to reduce the likelihood of the high economic losses an HPAI outbreak would cause.

Although the value of broiler flocks increases over their relatively short lifespan, broiler producers have a very small financial incentive to vaccinate against HPAI with a vaccine that becomes fully effective only about 13–21 days after the first of two injections, since they only keep chicken for a short time thereafter. The financial incentive to vaccinate duck broilers, despite their longer production cycle is also low as they often only show mild clinical signs of disease when infected with HPAI (Hulse-Post et al. 2005).

Vaccination of an average backyard chicken flock would only be "profitable" for a risk neutral flock owner, if the annual risk of HPAI infection were higher than 17% and resulted in a loss of USD 1–3 per bird. Since such a high infection risk is very unlikely, the average benefit of free of charge vaccination for backyard chicken flock owners would be marginal. Notwithstanding the relatively high gross margins of backyard poultry production, poultry keepers have shown little motivation to vaccinate against HPAI or other more prevalent diseases such as Newcastle disease. A survey on the willingness to pay for HPAI vaccination in Vietnam showed that only 32% of 62 surveyed poultry owners used Newcastle vaccine at an average price of USD 0.026 per chicken (FAO 2008). If other vaccines to prevent more financially relevant diseases were delivered in addition to HPAI vaccination and if HPAI-vaccinated flocks would be exempted from culling, subsidized HPAI vaccination might be adopted by backyard poultry producers.

If market access of poultry producers is made conditional on the proven use of vaccination, as is the case in Hong Kong and Ho Chi Minh City, Vietnam, vaccination is likely to be used by producers supplying these markets. However, this only holds, if the capacity to control market access and the diagnostic capacity to test for antibodies is sufficiently high. The higher the difference between the calculated breakeven outbreak risk and the actual outbreak risk, the higher is the incentive for flock owners to "bypass" market access restrictions.

Flock owners may consider other available protection measures such as improvements of production hygiene and investments in cleaning and disinfection equipment. The calculated vaccination costs to achieve the maximum achievable protection through vaccination could be considered as a benchmark for the maximum expenditure on other protection measures. For a flock of 1,000 industrial broilers the annual vaccination costs vary between USD 325 and 651. Annual vaccination costs would amount to between USD 106 and 157 for an integrated layer production system with a flock size of 1,000 birds. Detailed assessments of the specific risk factors for the entry of HPAI virus into these production systems would be essential to estimate the potential feasibility, costs and effectiveness of achieving a higher disease protection level for the respective flock types. Nevertheless, simple improvements, such as cleaning and disinfection of equipment, cages, and work clothes are likely to cost less than vaccination. It is recognized that the ease of applying such measures will differ between systems and the quality of housing. For example, layer units with multi-age flocks may not be in a position to regularly disinfect units and may have difficulty in cleaning egg trays, whereas an all-in/all-out broiler system with concrete flooring may be in a better position to apply cleaning and disinfection measures. Such measures require significant investments in training and then need to be followed by management so that they are continuously applied. Production hygiene improvements are also likely to have additional benefits from reduced mortality and morbidity and subsequently increased productivity.

Public "Returns" to Vaccination

Positive externalities from vaccinating poultry flocks result from the reduced probability of secondary outbreaks and the public health benefits from reduced exposure to HPAI virus, neither of which is taken into account in the estimated financial incentives for vaccination presented in Table 12.5. The main benefit of vaccination for the community of poultry producers stems from the expectation that vaccinated flocks will act as "dead ends" of infection chains and thereby also indirectly "protect" nonvaccinated flocks. This indirect benefit complements the "direct" benefits of vaccination, but, given the uncertainty surrounding prevented infections/outbreaks, the inclusion of prevented outbreaks in public cost-benefit analyses is difficult. For Vietnam, Soares Magalhaes et al. (2006) estimated that vaccination of all commercial smallholder farms would have more than halve "secondary" outbreaks by reducing R_n from 2.24 in the case of culling of infected premises and preemptive cull in a 3-km zone to 1.05 in the case of the same measures complemented by vaccination. Additional immunization of 25% of backyard flocks was estimated to further reduce secondary outbreaks from 1.05 to 0.23 per infected premises. These major reductions in secondary outbreaks would result in much higher public cost-benefit ratios than those that can be derived from Table 12.5 for individual producers, the order of magnitude depending on the generations of cases avoided. Assuming three successive generations of cases resulting from one HPAI H5N1 outbreak, vaccination of commercial smallholder farms (in addition to standard culling of infected and surrounding premises) would theoretically lead to a 90% reduction of the outbreak size, i.e., a tenfold increase in cost-effectiveness. Disincentives are needed to counter the "free-rider" problem that results from the above positive externality of reduced overall disease risk, from which farmers, who do not vaccinate, benefit without incurring respective costs.

To reduce the public health risk, it would be essential to limit the contact of humans with infected poultry. Broilers represent the largest share of poultry, which is marketed through live poultry markets and in contact with a magnitude of consumers in many developing countries. Effective immunization of broilers would reduce the exposure of live bird market customers to HPAI virus. However, high immunization coverage of marketed broilers is not likely to be achieved due to the low economic incentive for flock owners to vaccinate their birds. Subsidized vaccination in these flocks may be justified for public health reasons, but even if the vaccine delivery was entirely free, owners' willingness to participate may be affected by their perception of risk and the impact of a vaccination campaign on production in terms of potential losses of birds and their condition. In addition, these costs need to be assessed against market hygiene interventions. The required vaccination costs to supply a medium size live bird market with a daily trade volume of 1,000 broilers would amount to USD 1,151–1,707 per month. Similar to the situation on broiler farms, a detailed assessment of the costs, effectiveness, and feasibility of other market hygiene improvements and behavior changes need to be considered in order to choose the most cost-effective risk reduction strategy.

Discussion and Conclusions

The available literature on field vaccination experiments with commercially available vaccine indicates that antibody titers considered as protective can develop within 13 days after the first vaccination. However, with the exception of Trovac, two injections at 2-week intervals are required to achieve full protection, and one of the few long-term serologic response studies indicates that immunity is lost in most chicken 20.5 weeks after vaccination (Cardona et al. 2006). In general, vaccinated birds have been shown to shed less amounts of virus than unvaccinated controls at specific times postchallenge. Thus, most commercial vaccines have the potential to reduce the level of circulating virus in infected chicken populations. However, a crucial factor for achieving significant reductions in circulating virus in poultry flocks are sufficiently high vaccination coverage levels (50–90% immunization of at least 50% of all flocks at risk of infection) with a vaccine that protects against the circulating virus(es).

Both theoretical considerations as well as field observations show that such high immunization rates are difficult to attain in large poultry populations through vaccination campaigns and that they are even more difficult to maintain over a longer time period due to the high population turn over in short-lived commercial broiler[5] and mixed-age backyard poultry flocks. There are also problems of maintaining immunity levels in long-lived commercial layer and parent flocks as the currently available vaccines do not lead to lifetime immunity. The short to medium term gains in reducing the virus load with vaccination are not likely to result in a cost-effective long-term control approach, if no additional measures are in place, because infection chains are unlikely to be totally interrupted and virus will not be eliminated from the entire poultry population.

A major drawback of vaccination is that the probability of detecting outbreaks may decrease due to a lack or reduction of clinical signs, which could lead to the silent spread of virus (Savill et al. 2006). In China, Jiang et al. (2010) isolated 55 H5 subtype HPAI viruses out of which 43 were retrieved from swab samples taken from clinically healthy chicken. Incentives for disease reporting are relatively low and masking disease signs through vaccination further depresses an already low level of reporting. For northern Vietnam, Walker et al. (2009) estimated a 45% effective vaccination coverage achieved by mass vaccination campaigns, leading to a greatly reduced transmission of virus between communes but also to a marked increase in the commune-level infectious period due to outbreaks remaining unreported for a longer duration. The same authors estimated that, had detection levels been maintained at prevaccination levels, around two-thirds of outbreaks which occurred in the 2007 wave in northern Vietnam would have been prevented. This highlights the fact that, regardless of the underlying reasons for less rapid reporting of outbreaks, in order to translate the reductions in disease transmission following vaccination into greater gains in disease control, more effective reporting and surveillance strategies are required.

Another drawback of the extensive use of vaccination is the increased likelihood of genetic drift as seen in Mexico (Escorcia et al. 2008; Lee et al. 2004) and the US

[5] Theoretically close to 50% of all vaccinated broilers have been replaced by nonvaccinated birds in the 60 days required by the Vietnamese animal health system to conduct one national vaccination campaign.

(Suarez et al. 2006). Therefore, close virus monitoring of circulating field strains, continuous vaccine testing via challenge trials, and subsequent development of new vaccines that protect from infection with evolving field strains are an inevitable component of any longer-term routine vaccination program. This requires considerable financial resources and supporting activities have to be based on surveillance systems that have a high probability of detecting circulating HPAI viruses even in the absence of significant clinical disease. It also requires the sharing of isolates with laboratories capable of assessing the suitability of the vaccines used. At present these significant "collateral" investments to vaccination are rarely found in countries with problems of HPAI endemicity.

Short-lived broilers, mainly chicken but also ducks, constitute a relatively large share of the standing poultry population of most countries, which, due to their rapid turnover, provide a constant and ample supply of susceptible avian hosts. Campaign-based vaccination programs can only achieve a very low coverage in these systems, particularly if two injections are required to achieve immunity. An age-based vaccination schedule for broilers would be an option to achieve higher vaccination coverage and its maintenance over time, but the logistical requirements for age-based vaccine delivery and associated costs differ significantly from those of vaccination campaigns. The private incentives for owners of broiler flocks to regularly vaccinate replacements are low and even if owners do vaccinate, broiler flocks will remain at least partially susceptible for 2–3 weeks, i.e., most of their lifespan. Broilers thus represent the "Achilles heel" of any HPAI control strategy that relies, at least to some extent, on the use of vaccination.

Although vaccination of more valuable breeder and layer flocks is generally more "profitable" from the flock owners' perspective, the incentives to vaccinate are not constant over the production cycle and immunity of birds might have waned toward the end of their productive life. Also, as breeder and layer flocks have relatively high contact rates with other flocks and as HPAI vaccination is frequently used in these production systems, postponed detection of infection due to potential masking of symptoms by vaccination may undermine the success of a vaccination strategy in these systems. Upgrading of biosecurity is likely to be safer and more cost-effective in breeder and layer systems than vaccination.

From a public health and national health security perspective, the reduction of human cases of avian influenza as a means of reducing the risk of a national panic and global pandemic is most important. Human cases of avian influenza receive high media attention and the political pressure to act is high. The impact of poultry vaccination on human health risk is controversially debated in the scientific community. Human cases of H5N1 infections in China in January 2009 raised concerns about the role of vaccination in increasing the virulence of HPAI virus and masking its symptoms in poultry (FAO 2009a, b). Hygiene practices and awareness of risk factors for poultry to human transmission are possibly as important for preventing human infections as reducing virus shedding by vaccination.

The cost-effectiveness of national vaccination efforts need to be weighed against those of alternative measures to reduce disease spread in the national flock. In Vietnam for example, the culling strategy employed during the first wave of outbreaks led to the destruction of about 44 million birds (20% of the standing poultry population) and caused major losses to poultry owners and costs to the government. However, even this extensive depopulation of poultry flocks was not sufficient to break the chain of infection in all locations (Tuan 2007). As a consequence, the government decided to use vaccination as an additional control measure, which, in combination with a modified culling policy, reduced the number of culled poultry, but added substantial vaccination costs. On the contrary, Thailand managed to very significantly reduce or even eliminate the circulation of H5N1 virus in its domestic poultry population within 2 years without resorting to vaccination, largely through intensive active and passive surveillance combined with, progressively restricted, culling in case of outbreaks.

The high and recurrent costs, technical difficulties, and epidemiological drawbacks of large-scale, open-ended blanket vaccination programs in national efforts to control HPAI call for careful targeting of vaccination in national control strategies, which "intelligently" combine available disease control measures. In principle, vaccination can be targeted spatially, temporally, and/or by production system to maximize its impact and cost-effectiveness. Effective targeting, however, requires sound risk assessments, for which data and expertise are often lacking. Strengthening of the epidemiological capacity of national animal health systems would thus be a major prerequisite for large-scale use of vaccination in the control of HPAI.

References

Alders RG, Bagnol B, Young MP, Ahlers C, Brum E, Rushton J (2007) Challenges and constraints to vaccination in developing countries. Dev Biol (Basel) Vol. 130, 73–82.

Amorij J-P, Huckriede A, Wilschut J, Frijlink HW, Hinrichs WLJ (2008) Development of Stable Influenza Vaccine Powder Formulations: Challenges and Possibilities. Expert Review. Pharmaceutical Research, Vol. 25, No. 6, June 2008.

Anderson RM, May, RM (1992) Infectious diseases of humans: dynamics and control. New York, Oxford University Press.

Beach RH, Poulos C, Pattanayak SK (2007): Farm Economics of Bird Flu. Canadian Journal of Agricultural Economics 55:471–483.

Bouma A, Claassen I, Natih K, Klinkenberg D, Donnelly CA, *et al.* (2009) Estimation of Transmission Parameters of H5N1 Avian Influenza Virus in Chickens. PLoS Pathog 5(1): e1000281. doi:10.1371/journal.ppat.1000281.

Cardona CJ, Charlton BR, Woolcock PR (2006) Persistence of immunity in commercial egg-laying hens following vaccination with a killed H6N2 avian influenza vaccine. Avian Diseases 2006 Sep;50(3):374–9.

CAST (2007): Avian Influenza Vaccines: Focusing on H5N1 High Pathogenic Avian Influenza (HPAI). Council for Agricultural Science and Technology (CAST). Special Publication No. 26, Ames, Iowa. ISBN 1–887383–28-X.

Chen H (2009) Avian Influenza vaccination: the experience in China. Rev. sci. tech. Off. int. Epiz., 2009, 28 (1): 267–274.

CIRAD (2008) EPIAAF survey (EPidemiology of Avian Influenza in AFrica). Unpublished presentation at FAO.

Correa Melo E, López E (2002) Control of foot and mouth disease: the experience of the Americas. Rev. sci. tech. Off. int. Epiz., 2002, 21(3): 695–698.

Cristalli A (2006) The Vietnamese vaccination campaign against H5N1 HPAI virus subtype. Internal consultant report for FAO.

Dessie T (1996) Studies on village poultry production systems in the central highlands of Ethiopia. MSc Thesis, Swedish University of Agricultural Sciences, Department of Animal Nutrition and Management, Uppsala, Sweden.

Desvaux S, Ton VD (2008) A general review and a description of the poultry production in Vietnam. PRISE, Agricultural publishing house. http//www.prise-pcp.org.

Domenech J, Dauphin G, Rushton J, McGrane J, Lubroth J, Tripodi A, Gilbert J, Sims L D (2009) Experiences with vaccination in countries endemically infected with highly pathogenic avian influenza: the Food and Agriculture Organization perspective. Rev. sci. tech. Off. int. Epiz., 2009, 28 (1):293–305.

EFSA (2007) Scientific Opinion on "Vaccination against H5 and H7 subtypes in domestic poultry and captive birds". EFSA-Q-2006–30. The EFSA journal 489. European Food Safety Authority (EFSA).

Ellis T, Leung C, Chow M, Bissett L, Wong W, Guan Y, Peiris J (2004) Vaccination of chickens against H5N1 avian influenza in the face of an outbreak interrupts virus transmission. Avian Pathology (August 2004) 33(4): 405–412.

Ellis TM, Sims LD, Wong HKH, Bissett LA, Dyrting KC, Chow KW, Wong CW (2005) Evaluation of vaccination to support control of H5N1 avian influenza in Hong Kong. In: Schrijver RS, Koch G (eds), Wageningen UR Frontis Series: Avian Influenza, Prevention and Control. Springer, Dordrecht; London.

Escorcia M, Vázquez L, Méndez ST, Rodríguez-Ropón A, Lucio E, Nava GM (2008) Avian influenza: genetic evolution under vaccination pressure. Virology Journal: 5:15.

FAO (2008) Data from a HPAI vaccination willingness to pay survey commissioned by FAO Viet Nam and conducted by Abt. Associates Hanoi.

FAO (2009a) Avian Influenza Disease Emergency (AIDE) news. Situation update 57, 15 January 2009.

FAO (2009b) Vaccination and surveillance costs evaluation for operational research project in Indonesia. Draft consultant report by Mohammed Iqbal Rafani to FAO.

FAO (2011) Approaches to controlling, preventing and eliminating H5N1 Highly Pathogenic Avian Influenza in endemic countries. Animal Production and Health Paper. No. 171. Rome. ISBN 978-92- 5-106837–3.

Garske T, Clarke P, Ghani AC (2007) The transmissibility of highly pathogenic influenza in commercial poultry in industrialized countries. PLoS ONE 2(4): e349. doi:10.1371/journal.pone.0000348.

Gunaratne SP, Chandrasiri, A D N, Mangalika Hemalatha W A P, Roberts J A (1993) Feed resource base for scavenging village chickens in Sri Lanka. Trop. Anim. Health. Prod. 25:249–257.

Henning J, Wimbawa H, Henning K, Morton J, Meers J (2009) Prevalences of highly pathogenic avian influenza antibodies on small-scale commercial and backyard farming free-ranging duck enterprises in Southeast Asia. International Symposia on Veterinary Epidemiology and Economics (ISVEE) proceedings, ISVEE 12: Theme 3 - Zoonoses and emerging diseases: Avian influenza, Emerging diseases, Public health, p 53, Aug 2009.

Hinrichs J, Otte J, Rushton J (2010) Technical, epidemiological and financial implications of large-scale national vaccination campaigns to control HPAI H5N1. CAB Reviews: Perspectives in Agriculture, Veterinary Science, Nutrition and Natural Resources, 2010, 5, 021, 1–20.

Hulse-Post DJ, Sturm-Ramirez KM, Humberd J, Seiler P, Govorkova EA, Krauss S, Scholtissek C, Puthavathana P, Buranathai C, Nguyen TD, Long HT, Naipospos TSP, Chen H, Ellis TM, Guan Y, Peiris JSM, Webster RG (2005) Role of domestic ducks in the propagation and biological evolution of highly pathogenic H5N1 influenza viruses in Asia. Proc. Natl. Acad. Sci. USA 102:10682–10687.

Iwami S, Takeuchi Y, Liu X, Nakaoka S. (2009): A geographical spread of vaccine-resistance in avian influenza epidemics. J Theor Biol. 2009 Jul 21;259(2):219–28.

Jiang WM, Liu S, Chen J, Hou GY, Li JP, Cao YF, Zhuang QY, Li Y, Huang BX, Chen JM (2010) Molecular epidemiological surveys of H5 subtype highly pathogenic avian influenza viruses in poultry in China during 2007–2009. J Gen Virol. 2010 Oct;91(Pt 10):2491–6.

Lee CW, Senne DA, Suarez DL (2004) Effect of Vaccine Use in the Evolution of Mexican Lineage H5N2 Avian Influenza Virus. J Virology, 78:8372–8381.

Lesnoff M, Peyre M, Duarte PC, Mariner JC (2009) A simple model for simulating immunity rate dynamics in a tropical free range poultry population after avian influenza vaccination. Epid. Inf. 137, 1405–1413.

McLaws M (2009) Operational Research in Indonesia for more Effective Control of HPAI – Seromonitoring. Presentation at ILRI closing workshop Bandung 1st December 2009.

McLeod A, Rushton J (2007) Economics of animal vaccination. OIE Rev. Sci. Tech., 26 (2):313–326.

Nguyen VL (2008) Post-vaccination surveillance and monitoring for AI virus circulation in Vietnam. Epidemiology Division, Department of Animal Health. Presentation at the International Avian Influenza Research Workshop in Hanoi 16–18 June 2008.

Normile D (2008) Rinderpest; Driven to extinction. Science 21 March 2008. Vol. 319. no. (5870):1606–1609.

OIE (2007) Avian Influenza Vaccination. OIE information document, Verona Recommendations. http://www.oie.int/eng/info_ev/Other%20Files/A_Guidelines%20on%20AI%20vaccination.pdf.

Otte J (2006) The hen which lays the golden eggs - Or why backyard poultry are so popular. FAO PPLPI: http://www.fao.org/ag/AGAinfo/programmes/en/pplpi/docarc/feature01_backyardpoultry.pdf.

Otte J, Hinrichs J, Rushton J, Roland-Holst D, Zilberman D (2008) Impacts of avian influenza virus on animal production in developing countries. CAB Reviews: Perspectives in Agriculture, Veterinary Science, Nutrition and Natural Resources, 2008, 3, 080:1–18, November 2008.

Peyre M, Fusheng G, Desvaux S, Roger F (2008) Review article Avian influenza vaccines: a practical review in relation to their application in the field with a focus on the Asian experience. Epidemiol. Infect., Cambridge University Press, pp. 21.

Peyre M, Samaha H, Makonnen Y J, Saad A, Abd-Elnabi A, Galal S, Ettel T, Dauphin G, Lubroth J, Roger F, Domenech J (2009): Avian influenza vaccination in Egypt: Limitations of the current strategy. Journal of Molecular and Genetic Medicine. Vol 3, No 2, 198–204.

Rushton J, Ngongi SN (1998) Poultry, Women and Development: Old ideas, new applications and the need for more research. World Animal Review 91 (2):pp 43–49.

Rushton J, Viscarra, RE, Taylor N, Hoffman I, Schwabenbauer K (2010) Poultry Sector Development, Highly Pathogenic Avian Influenza and the Smallholder Production Systems. World Journal Poultry Science.

Savill N, Rose S, Keeling M, Woolhouse M (2006) Silent spread of H5N1 in vaccinated poultry. Nature Vol. 442:757.

Sawitri Siregar E, Darminto, Weaver J, Bouma A (2007): The vaccination programme in Indonesia. Dev Biol (Basel) Vol. 130:151–158.

Seng S (2007) Gender and socio-economic impacts of HPAI and its control. Rural livelihoods and bio-security of smallholder poultry producers and poultry value chain in Cambodia. Case study in 36 villages in 4 provinces. Consultancy report by Cambodian Center for Study and Development in Agriculture (CEDAC) for FAO. July 2007.

Smith GJD, Fan XH, Wang J, Li KS, Qin K, *et al.* (2006) Emergence and predominance of an H5N1 influenza variant in China. Proc. Natl. Acad. Sci. USA, Vol 103 (45), 16936–16941.

Soares Magalhaes R, Pfeiffer D, Wieland B, Dung D, Otte J (2006) Commune-level simulation model of HPAI H5N1 poultry infection and control in Viet Nam. FAO-PPLPI Research Report

06–07, 15pp., available at http://www.fao.org/ag/againfo/programmes/en/pplpi/docarc/rep-hpai_modelupdate.pdf.
Songserm T, Jam-on R, Sae-Heng N, Meemak N, Hulse-Post DJ, Sturm-Ramirez KM, Webster, RG (2006) Domestic ducks and H5N1 influenza epidemic, Thailand. Emerg. Infect. Dis. 12, 575–581.
Spradbrow PB (1993) Newcastle Disease in Village Chickens. Poultry Science Rev 5:57–96.
Suarez DL, Lee CW, Swayne DE (2006) Avian influenza vaccination in North America: strategies and difficulties. Dev Biol (Basel) 2006, 124:117–124.
Swayne DE, Garcia M, Beck JR, Kinney N, Suarez DL (2000) Protection against diverse highly pathogenic avian influenza viruses in chickens immunized with a recombinant fowl pox vaccine containing an H5 avian influenza hemagglutinin gene insert. Vaccine 2000;18:1088–1095.
Swayne DE, Kapczinski D (2008) Vaccines, vaccination and immunology for avian influenza viruses in poultry. In: Avian Influenza. Blackwell Publishers, Aimes, Iowa; p. 407–451.
Swayne DE (2006) Principles for vaccine protection in chickens and domestic waterfowl against avian influenza: emphasis on Asian H5N1 high pathogenicity avian influenza. Ann. N. Y. Acad. Sci. 1081:174–81.
Taylor N (2007) An assessment of post-vaccination sero-monitoring and surveillance activities, and the data generated, following HPAI vaccination in Viet Nam (2005 – 2006). Technical report (1) for FAO. January 2007.
Taylor N (2008) HPAI vaccination strategies in Indonesia – Modelling of alternative strategies. Consultancy report for FAO Rome. July 2008.
Tiensin T, Nielen M, Vernooij H, Songserm T, Kalpravidh W, Chotiprasatintara S, Chaisingh A, Wongkasemjit S, Chanachai K, Thanapongtham W, Srisuvan T, Stegeman A (2007) Transmission of the Highly Pathogenic Avian Influenza Virus H5N1 within Flocks during the 2004 Epidemic in Thailand. The Journal of Infectious Diseases 2007; 196:1679–84.
To TL, Bui, QA, Dau NH, Hoang VN, Van DK, Taylor N, Do HD (2007) Control of Avian Influenza: A Vaccination Approach in Viet Nam. In Dev Biol (Basel) Vol. 130:159–160.
Tuan NA (2007) Avian Influenza, Poultry Culling And Support Policy Of The Viet Nam Government. In: McLeod A. and Dolberg, F. (Editors) (2007). Future of Poultry Farmers in Viet Nam after HPAI. FAO and MARD workshop held at Horison Hotel, Hanoi, March 8–9 2007. 99 pp.
Udo HMJ, Asgedom AH, Viets TC (2006) Modelling the impact of interventions on the dynamics in village poultry systems. Agricultural Systems 2006; 88: 255–269.
van den Berg T, Lambrecht B, Marché S, Steensels M, Van Borm S, Bublot M (2007) Influenza vaccines and vaccination strategies in birds. Comparat. Immunol. Microbiol. Infect. Dis. (2007) doi:10.1016/j.cimid.2007.07.004, 31(2–3):121–65.
van der Goot JA, Koch G, de Jong MCM, van Boven M (2005) Quantification of the effect of vaccination on transmission of avian influenza (H7N7) in chickens. Proceedings of the National Academy of Sciences 102(50):18141–18146.
Walker P, Cauchemez S, Metras R, Dung DH, Pfeiffer D, Ghani A (2009) Modelling the Temporal and Spatial Dynamics of the Spread of HPAI H5N1 in Northern Viet Nam. HPAI Research Brief, No. 19, www.hpai-research.net.
Ward MP, Maftei D, Apostu C, Suru A (2009) Estimation of the basic reproductive number (R0) for epidemic, highly pathogenic avian influenza subtype H5N1 spread. Epidemiology and Infection, 137:pp 219–226 doi:10.1017/S0950268808000885.
Wooldridge M (2007) Risk Modelling for Vaccination: a Risk Assessment Perspective. Dev Biol (Basel). Basel, Karger, 2007, vol 130:87–97.

Chapter 13
Poultry Movement and Sustained HPAI Risk in Cambodia

Maria D. Van Kerkhove

Introduction

The threat posed by highly pathogenic avian influenza (HPAI) H5N1 viruses to humans remains significant, given the continued occurrence of sporadic human cases (518 human cases in 15 countries) with a high case fatality rate (approximately 60%; Table 13.1), the endemicity in poultry populations in several countries, and the potential for reassortment with the newly emerging 2009 H1N1 pandemic strain. Additionally, the connectedness of animal networks can lead to large and widespread epidemics of disease and an understanding of human and animal movement and their contact structures could be used to design more targeted surveillance activities and inform models of disease spread, which could result in more cost-effective disease prevention and control (Dent et al. 2008; Green et al. 2008; Kiss et al. 2008; Truscott et al. 2007). However, despite their likely role in the circulation and spread of HPAI in South East Asia, little is understood about the poultry market chains, legal or illegal trade of poultry or the types, and frequencies of contact that exist between rural people raising poultry, local markets, and large-national poultry markets in the major cities. Because trade of poultry may be responsible for some transmission of H5N1 within countries (Normile 2005; WHO 2008), controlling the movement of live poultry and poultry products could contain or reduce the spread of the virus.

Here, I present a case study of avian influenza (H5N1) in Cambodia; a comprehensive assessment to evaluate poultry movement and the extent of interaction between humans and poultry in Cambodia to better define the risks of sustained transmission of H5N1 in poultry and onward potential transmission to humans (Van Kerkhove 2009). First, I will present the state of the literature evaluating risk factors for H5N1 infection in humans (Van Kerkhove et al. 2011), then describe my

M.D. Van Kerkhove (✉)
MRC Centre for Outbreak Analysis and Modelling, Imperial College London, London, UK
e-mail: m.vankerkhove@imperial.ac.uk

D. Zilberman et al. (eds.), *Health and Animal Agriculture in Developing Countries*, Natural Resource Management and Policy 36, DOI 10.1007/978-1-4419-7077-0_13,

own research aimed to identify populations living in rural Cambodia with the highest H5N1 exposure potential (Van Kerkhove et al. 2008a) and finally, I will describe the current movements of poultry throughout Cambodia and determine how these movements influence the potential spread of HPAI at local, regional, and national levels (Van Kerkhove et al. 2009a, b). All of the research presented in this chapter was conducted in collaboration with the National Veterinary Research Institute (NaVRI), Department of Animal Health and Production, Ministry of Agriculture, Forestry and Fisheries in Cambodia, the Ministry of Health and Institut Pasteur du Cambodge.

Table 13.1 Characteristics of human cases of highly pathogenic avian influenza H5N1 virus infection reported to WHO from 1997 to 16 March 2010 by country

Country	Total		Crude	Clade(s)[a]	Median age of	% Male *n*/
	Cases	Deaths	CFR (%)		cases (range)[b]	total (%)[b]
Azerbaijan	8	5	62.5	2.2	10	9/16 (56)[c]
Turkey	12	4	33.3	2.2	16.5 (5–20)[c]	
Bangladesh	1	0	0.0	2.2	16 mo (–)	1/1 (100)
China	38	25	65.8	2.2, 2.3.4, 7	30 (12–41)[d]	3/8 (38)[d]
Hong Kong, SAR (1997)	18	6	33.3	0, 1	6 (1.5–60)	6/15 (40)
Djibouti	1	0	0.0	2.2	2 (–)	0/1 (0)
Egypt	106	32	30.2	2.2	12.5 (1–75)[e]	12/38 (32)[e]
Indonesia	163	135	82.8	2.1.2, 2.1.3	18.5 (1.5–45)[d]	33/54 (61)[d]
Iraq	3	2	66.7	2.2	15 (3–39)	2/3 (66.7)
Lao People's Democratic Republic	2	2	100.0	2.3.4	28.5 (15–42)	0/2 (0)
Myanmar	1	0	0.0	2.3.4	7 (–)	0/1 (0)
Nigeria	1	1	100.0	2.2	22 (–)	0/1 (0)
Pakistan	3	1	33.3	NR	25 (22–27)	3/3 (100)
Cambodia	9	7	77.8	1	14–22 (2–58)[f]	19/41 (46)[f]
Thailand	25	17	68.0	1		
Vietnam	116	58	50.0	1, 2.3.4		
Total	489	289	59.1	–	–	–

Adapted from WHO (2008), Writing Committee of the Second World Health Organization Consultation on Clinical Aspects of Human Infection with Avian Influenza AV (2008), Biswas et al. (2008), WER (2007), Mounts et al. (1999), Fasina et al. (2010), and Otte et al. (2010)

NR not released

[a]Clade(s) isolated from humans

[b]Data from cases up to 1 Jan 2009

[c]Includes data from 2006 cases only

[d]Includes data from 2005 to 2006 cases only

[e]Includes data from 2006 to 2007 cases only

[f]Includes data from 2004 to 2005 cases only

Overview of Existing Research

This section provides a systematic review of the epidemiology literature on HPAI risk to humans (Van Kerkhove et al. 2011). Several epidemiologic studies have evaluated the risk factors associated with increased risk of H5N1 infection among humans who were exposed to H5N1 viruses. Twenty-four published studies evaluating risk and/or risk factors for human infection conducted in eight countries (Thailand, Vietnam, Indonesia, Cambodia, Nigeria, China, Azerbaijan, and Germany) and Hong Kong were included in the review. Four studies focused on the initial 1997 outbreaks in Hong Kong, while the remaining 20 studies were conducted in Asian, African, and European countries in areas with confirmed outbreaks in human and/or domestic poultry populations from 2003 to 2009. Based on the population under study and principal objective, the 24 studies fall into two categories: case-control studies to evaluate risk factors for human infection among laboratory-confirmed H5N1 cases ($n = 5$; 2 related to the 1997 outbreak and 3 related to outbreaks occurring 2003–2009; Table 13.2); or seroepidemiology studies ($n = 19$; 3 relating to the 1997 outbreak and 16 related to outbreaks

Table 13.2 Risk factors for H5N1 infection: summary of published case-control studies

References	Study population	Risk factors RR, OR, 95% CI
Mounts et al. (1999)	Hong Kong, 15 cases 41 matched controls	Exposure to poultry at live/wet markets was associated with a fourfold increased risk (OR = 4.5, 1.2–21.7)
Dinh et al. (2006)	Viet Nam, 28 cases 106 matched controls	Univariate analysis: preparing/cooking unhealthy poultry (OR = 31, 2.4–1,150), having sick or dead poultry in the household (OR = 7.41, 2.7–59), presence of sick/dead poultry in the neighborhood (OR = 3.9, 1.0–55.7), no indoor water source in the household (OR = 5.0, 1.3–77.0) Multivariate analysis: no water in the household (OR = 6.5, 1.2–34.8), sick or dead poultry in the household (OR = 4.9, 1.2–20.2), prepare and cook sick or dead poultry (OR = 9.0, 0.98–82.0)
Areechokchai et al. (2006)	Thailand, matched case control study of 16 cases and 64 controls	Direct touching of unexpectedly dead poultry OR 29.0 (2.7–308.2)
Zhou et al. (2009)	China, ten urban and 18 rural cases; 134 matched controls	Infection included direct (OR = 506.6, 95% CI 15.7–16,319.6) or indirect (OR = 56.9, 95% CI 4.3–745.6) contact with sick or dead poultry, visiting a LBM (OR = 15.4, 95% CI 3.0–80.2) Urban cases were significantly more likely to have visited a LBM, compared with rural cases ($p = 0.002$)
WER (2006)	Azerbaijan, residents in settlements of confirmed cases	9/52 residents tested positive for H5N1 virus No case-control was initiated, but contact with infected wild birds (defeathering) reported as likely cause of infection

Table 13.3 Results of seroprevalence studies to determine the frequency of asymptomatic or subclinical infection and evaluate risk factors for H5N1 virus infection

References	Study population and year of outbreak	Transmission	Seroprevalence results (% seropositive)	Risk factors RR, OR, 95% CI	Comments
Occupationally exposed persons: poultry workers					
Bridges et al. (2002)	Poultry workers, Hong Kong, 1997	Poultry-to-humans	9/293 (3%) GW were seropositive 10% PW were estimated to be seropositive using MN >80 Nested case-control study conducted among 81 seropositive cases and 1,231 controls	Work in retail vs. wholesale/hatchery/farm/other poultry industry 2.7 (1.5–4.9) >10% mortality among poultry 2.2 (1.3–3.7) Jobs: Butchering poultry 3.1 (1.6–5.9) Handling money 1.6 (1.0–2.5) Preparing poultry for restaurants 1.7 (1.1–2.7)	Limited poultry-to-human transmission among PW and GW involved in poultry culling operations
Wang et al. (2006)	Poultry workers, Guangdong China, 2006	Poultry-to-humans	1/110 PW were seropositive using HI with turkey erythrocytes >320	Specific risk factors not identified, but subject slaughtered poultry for 5 years	Specific risk factors not identified
Ortiz et al. (2007)	Poultry workers, Kano Nigeria, 2006	Poultry-to-humans	0/295 PW with median 14 days exposure to H5N1 0/25 laboratory workers with exposure to H5N1 Seropositivity by MN titers if ≥1:80	None	No evidence of H5N1 infection with subjects with repeated exposure to infected poultry
Lu et al. (2008)	Poultry workers, Guangdong China	Poultry-to-humans	2/231 subjects with "occupational exposure" had HI titers >1:80	Occupational exposure including raising, selling slaughtering chickens and ducks in H5N1 outbreak areas	Specific risk factors not identified

Cai et al. (2009)	Firemen, government workers, vets for collection of dead wild birds on Ruegen Island, Germany 2006	Poultry-to-humans	0/97 workers were seropositive Seropositivity by PN or MN assay if >1:20	None	No evidence of H5N1 infection with subjects with exposure to infected wild birds; use of PPE was widespread
Wang et al. (2009a)	Poultry workers in China 2007–2009	Poultry-to-humans	4/2,191 using HI (no cutoff provided) had anti H5 antibodies	None	Limited evidence
Schultsz et al. (2009)	Poultry workers and cullers living on farms with confirmed H5N1 outbreaks in poultry in Vietnam 04–05	Poultry-to-humans	0/500 (183 PW, 317 cullers) using MN and 3/500 (three cullers) using HI >1:80 had anti H5 antibodies	Not evaluated	Limited evidence of poultry-to-human transmission despite exposure to infected poultry
Wang et al. (2009b)	Poultry workers in LBM in Guangzhou in 2006	Poultry-to-humans	0/68 were seropositive using HI (no cutoff provided)	None	No evidence of H5N1 infection with subjects with repeated exposure to infected poultry
Occupationally exposed persons: health care workers					
Bridges et al. (2000)	Health care workers, Hong Kong, 1997	Human-to-human; poultry-to-human	10/526 (8/21 exposed; 2/309 non exposed HCW) using MN >1:80, confirmed by WB	Bathing patients or changing the bed linen of cases (no OR provided); controlled for poultry exposure	Limited human-to-human transmission
Apisarnthanarak et al. (2005)	Health care workers, Thailand, 2004	Human-to-human; poultry-to-human	0/25 among HCW in direct contact with H5N1 patient; Seropositivity tested using MN >1:80, confirmed by WB	None	No serologic evidence of H5N1 among HCW with direct contact with human H5N1 patient
Thanh Liem et al. (2005)	Health care workers, Vietnam, 2004	Human-to-human; poultry-to-human	0/83 among HCW, 95% of which had direct contact with confirmed H5N1 patients Seropositivity tested using MN >1:40 in two independent assays	None	No serologic evidence of H5N1 among HCW with direct contact with human H5N1 patient

(continued)

Table 13.3 (continued)

References	Study population and year of outbreak	Transmission	Seroprevalence results (% seropositive)	Risk factors RR, OR, 95% CI	Comments
Schultsz et al. (2005)	Health care workers, Vietnam, 2004	Human-to-human; poultry-to-human	0/60 HCW in contact with confirmed H5N1 patients Seropositivity tested using MN >1:80 and ELISA >1:80	None	No serologic evidence of H5N1 among HCW with direct contact with human H5N1 patient
Nonoccupational exposure: household and social contacts					
Katz et al. (1999)	Household and Social contacts of H5N1 patients, Hong Kong, 1997	Human-to-human; poultry-to-human	6/51 (12%) household contacts 0/47 co-workers tested positive for H5 antibodies Seropositivity tested using MN or ELISA >1:80, confirmed by WB	None significant; however 21% of seropositive had contact to poultry vs. 5% of seropositive with no poultry contact, $p = 0.13$	Human-to-human transmission was limited
Vong et al. (2006)	Rural Cambodian villagers living in the same villages as two confirmed H5N1 human cases in 2005	Poultry-to-human	0/351 villagers tested positive for H5N1 antibodies ≥1:80 using MN and WB	None	No evidence of H5N1 infection among subjects living in villages with conformed H5N1 in domestic poultry flocks; poultry-to-human transmission was low in this setting
Lu et al. (2008)	Poultry workers, Guangdong China	Poultry-to-humans	12/983 "general citizens" had HI or MN titers ≥1:20	Subjects were general citizens without direct contact with poultry	Specific risk factors not identified
Hinjoy et al. (2008)	Rural poultry farmers in Thailand, 2004	Poultry-to-human	0/322 farmers tested positive for H5N1 antibodies; using MN >1:80, confirmed by WB or ELISA	None	No evidence of H5N1 infection among subjects living in villages with conformed H5N1 in domestic poultry flocks
Vong et al. (2009)					

	Rural Cambodian villagers living in the same villages as confirmed H5N1 human case, 2006	Poultry-to-human	7/674 (1%) seropositive for H5N1 antibodies using MN ≥1:80 85.7% (6/7) male All ≤18 years old Matched case-control study conducted with seven seropositive cases and 24 controls	Swim/bathe in ponds OR 11.3 (1.25–102.2) Water source 6.8 (0.68–66.4) Gathered poultry and placed in cages or designated areas 5.8 (0.98–34.1) Removed/cleaned feces from cages or poultry areas 5.0 (0.69–36.3)	Poultry-to-human transmission was low; possible transmission from the environment to humans via contaminated water
Dejpichal et al. (2009)	Residents in four Thai villages with human cases in 2005	Poultry-to-human	0/901 tested positive for anti-H5 antibodies using MN confirmed by Immunofluorescence >1:40	None	No evidence of H5N1 infection among subjects living in villages with conformed H5N1 in domestic poultry flocks
Santhia et al. (2009)	Residents in 38 villages and three LBM in Bali, 2005	Poultry-to-human	0/841 tested positive for anti-H5N1 antibodies using MN >1:80	None	Despite H5N1 exposure from poultry outbreaks, no evidence of poultry-to-human transmission
Cavailler et al. (2010)	Rural Cambodian villagers living in the same villages as confirmed H5N1 human case, 2007	Poultry-to-human	18/700 (2.8%) seropositive for H5N1 antibodies using MN ≥1:80	Swam/bathed in pond OR 2.52 (95% CI 0.98–6.51) No other risk factors identified	Poultry-to-human transmission was low; possible transmission from the environment to humans via contaminated water

PPE personal protective equipment including masks, gloves, eye protection; *PW* poultry workers; *GW* government workers; *HCW* health care workers; *MN* microneutralization (MN) assay; *HI* hemagglutination-inhibition assay; *WB* Western Blot assay; *PN* plaque neutralization occurring 2003–2009; Table 13.3) to evaluate the predictors of having H5-specific antibody among health care workers (HCW; $n = 4$), poultry workers (PW; $n = 8$), or household/social contacts ($n = 8$) of laboratory-confirmed infected H5N1 cases (one study evaluated both occupational and domestic exposure to poultry and is therefore counted as both a study among PW and social contacts).

Our review shows that most cases are sporadic while occasional limited human-to-human transmission occurs, and that most H5N1 cases are attributed to exposure to sick poultry. The most commonly identified factors associated with H5N1 virus infection included exposure through contact with infected blood or bodily fluids of infected poultry via food preparation practices; touching and caring for infected poultry; consuming uncooked poultry products; exposure to H5N1 via swimming or bathing in potentially virus laden ponds; and exposure to H5N1 at live bird markets.

Research to date has demonstrated that despite frequent and widespread contact with poultry, transmission of the H5N1 virus from poultry to humans is rare. Available research has identified several risk factors that may be associated with infection including close direct contact with poultry and transmission via the environment. However, several important data gaps remain that limit our understanding of the epidemiology of H5N1 in humans. Although infection in humans with H5N1 remains rare, human cases continue to be reported and H5N1 is now considered endemic among poultry in parts of Asia and in Egypt, providing opportunities for additional human infections and for the acquisition of virus mutations that may lead to more efficient spread among humans and other mammalian species. Collaboration between human and animal health sectors for surveillance, case investigation, virus sharing, and risk assessment is essential to monitor for potential changes in circulating H5N1 viruses and in the epidemiology of H5N1 to provide the best possible chance for effective mitigation of the impact of H5N1 in both poultry and humans.

Research Findings

This section presents present findings of two separate, survey based studies of HPAI risk transmission. Empirical findings are discussed here, with general conclusions offered in the last section.

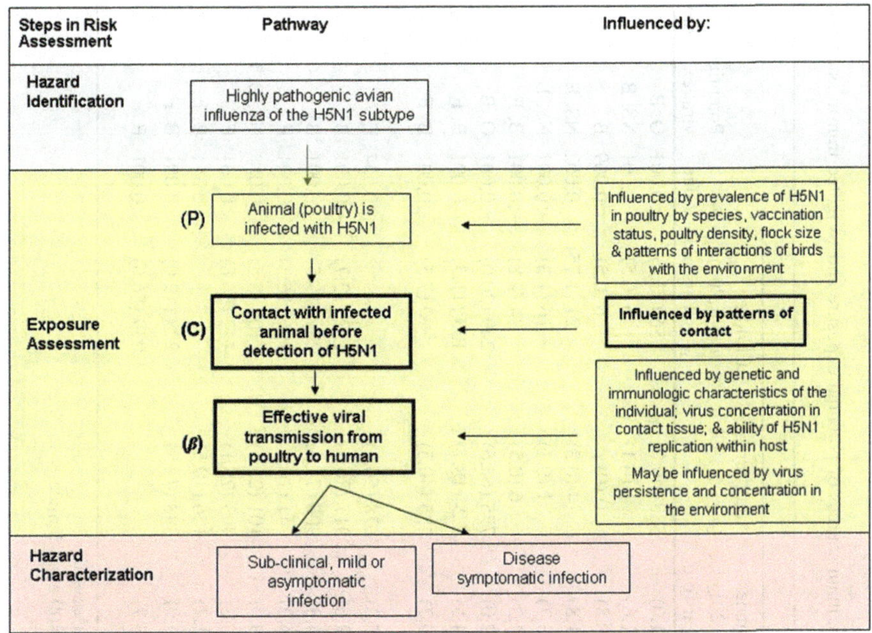

Fig. 13.1 Conceptual Pathway for transmission of H5N1 from poultry to humans via contact with poultry

Study 1: Frequency and Patterns of Contact with Domestic Poultry and Potential Risk of H5N1 Transmission to Humans Living in Rural Cambodia (Van Kerkhove et al. 2008b)

A cross-sectional survey interviewed 3,600 backyard poultry owners from 115 randomly selected villages in six provinces throughout Cambodia. Using risk assessment methods, patterns of contact with poultry as surrogate measures of exposure to H5N1 were used to generate risk indices of potential H5N1 transmission among different populations in contact with poultry (Fig. 13.1). Estimates of human exposure risk for each study participant ($n = 3{,}600$) were obtained by multiplying each reported practice with a transmission risk-weighting factor and summing these over all practices reported by each individual (Table 13.4). Exposure risk estimates were then examined stratified by age and gender (Fig. 13.2a, b). Subjects reported high contact with domestic poultry (chickens and ducks) through the daily care and food preparation practices; however, contact patterns varied by gender and age. Males between the ages 26 and 40 reported practices of contact with poultry that give rise to the highest H5N1 transmission risk potential, followed closely by males between the ages 16 and 25. Overall, males had a higher exposure risk potential than females across all age groups ($p < 0.001$).

Table 13.4 Prevalence of practice associated with poultry in rural Cambodian households, main sources of potential exposure and weighted transmission risk potential (β) ($n = 3{,}600$)

Probability of effective viral transmission (β grouping)	Practice	n (%) Adult males ($n = 1{,}201$) >15 years old	Adult females ($n = 1{,}199$)	Children ($n = 1{,}200$) ≤15 years old	p-Value	Potential viral exposure
High ($\beta 1$)	Remove internal organs (poultry)	733 (61.0)	588 (49.0)	156 (13.0)	<0.001	O, B
	Blow into beak (FC)	19 (1.6)	1 (0.1)	6 (0.5)	0.001	NS, B
	Kiss, suck, lick wounds (FC)	10 (0.8)	0 (0)	6 (0.5)	0.009	B
	Share water from the same bottle (FC)	21 (1.8)	4 (0.3)	21 (1.75)	0.002	NS, B
	Clean trachea (FC)	44 (3.7)	1 (0.1)	16 (1.3)	<0.001	NS, B
	Clean feathers (FC)	52 (4.3)	6 (0.5)	34 (2.8)	<0.001	B, F
	Wash internal organs (poultry)	745 (62.0)	775 (64.6)	249 (20.0)	<0.001	O, B
	Slaughter poultry	655 (54.5)	224(18.7)	138 (11.5)	<0.001	B, F
Moderate ($\beta 2$)	Touch/play with sick poultry or poultry that died from illness	597 (49.7)	485 (40.5)	90 (7.5)	<0.001	B, F
	Use poultry feces as manure	664 (55.3)	678 (56.6)	–	0.534	F
	Cut poultry meat	716 (59.6)	917 (76.5)	152 (12.7)	<0.001	B
	Wash poultry meat	772 (64.3)	906 (75.6)	234 (19.5)	<0.001	B
	Swim/bathe in water source where poultry have access[a]	56 (14.0)	41 (10.3)	196 (16.3)	<0.001	F
	Remove feathers from sick poultry[a]	76 (19.0)	101 (25.3)	102 (8.5)	0.08	NS, B, F
	Cleaning/sweeping poultry areas	843 (70.2)	903 (75.1)	442 (36.8)	<0.001	F
	Shopping at wet/live market for poultry	141 (11.7)	126 (10.5)	–	0.341	B, F
	Boil poultry	673 (56.0)	898 (74.9)	228 (19.0)	<0.001	B, F
Low ($\beta 3$)	Living in a household with poultry (raised chickens or ducks within previous 8 months)	517 (86.7)[b]		1,039 (86.6)	<0.001	F

FC fighting cocks; *B* blood; *F* feces; *NS* nasopharyngeal secretions; *O* organ tissue; – not assessed
[a]This practice was only evaluated in adults from two provinces ; $n = 400$ adult males and 400 adult females
[b]Evaluated from head of household questionnaire only ($n = 600$)

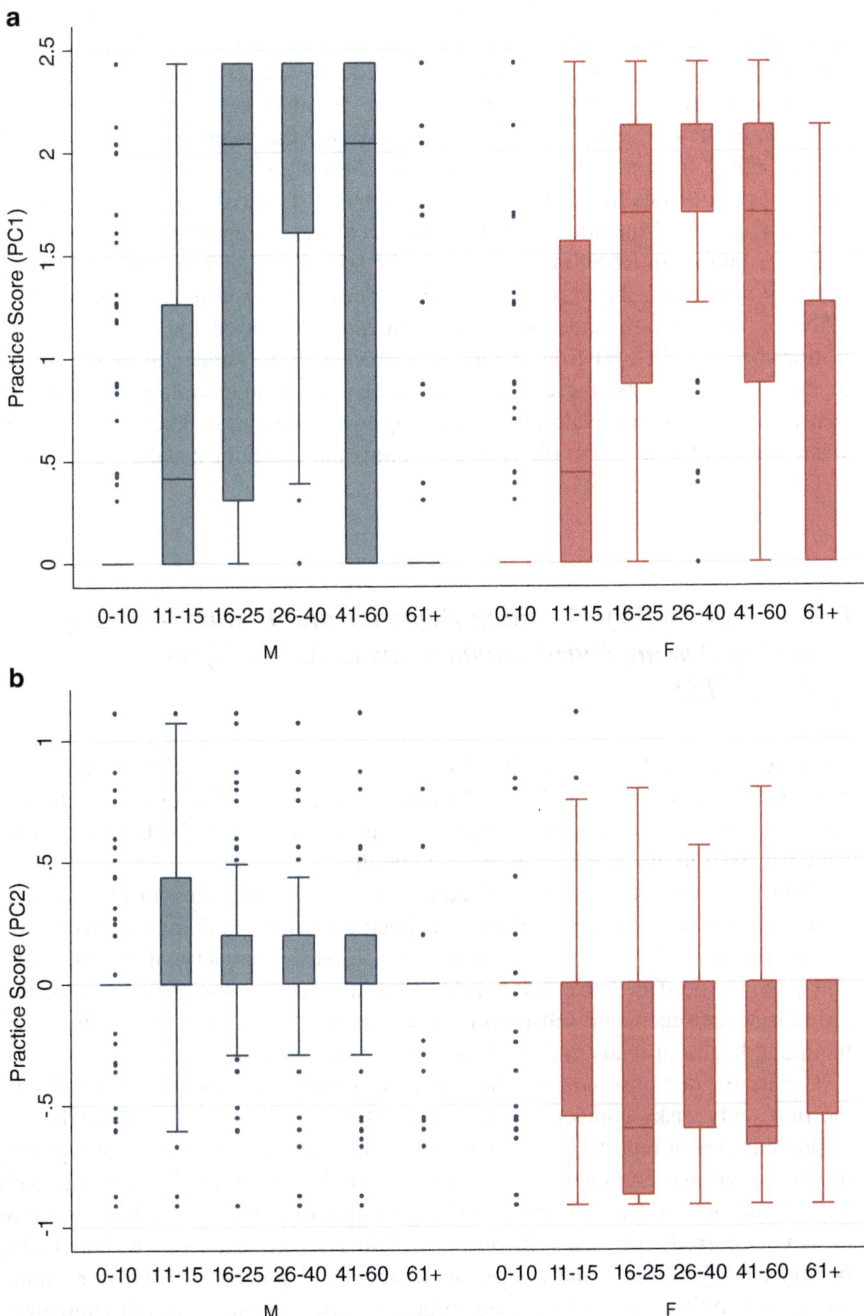

Fig. 13.2 (**a**) Practice 1-general food preparation-by age and gender. (**b**) Practice 2-slaughtering and removing internal organs-by age and gender. *Note*: The median value is indicated by the *horizontal bar* inside the box, the *upper* and *lower edges* of the box are the 75th and 25th percentiles, respectively; the *asterisks* are outliers (>1.5 times the IQR); the *upper* and *lower edges* of the whiskers (*lines*) are the largest and smallest nonoutlier values

Our results demonstrate that most of the population in rural Cambodia is in frequent contact with domestic poultry, with an estimated 58% of the population carrying out on a regular basis at least one of the practices that we considered of high risk of effective transmission if the bird is infected. There was substantial variation in the frequency of different practices and the potential risk of transmission of H5N1 from poultry to humans is not uniform across age and gender even amongst populations living in proximity to poultry. This risk assessment lacks the power of a formal quantitative risk assessment because epidemiologic data gaps and uncertainties of H5N1 pathogenesis in the host species currently exist. Data are urgently needed on the prevalence of H5N1 in poultry species in regions where H5N1 is recurrent or endemic in domestic poultry flocks; the potential routes of transmission of H5N1 from poultry to humans and prevalence of such practices in human populations; the contribution of genetic or immunological factors on transmission; virus survival in poultry during food preparation practices, in poultry waste, soil, and water under different environmental conditions; and the persistence of H5N1 in poultry tissues.

Changes in Poultry Handling Behavior and Poultry Mortality Reporting Among Rural Cambodians in Areas Affected by HPAI/H5N1

Since 2004, 25 HPAI H5N1 outbreaks in domestic poultry and eight human cases have been confirmed in Cambodia. As a result, a large number of avian influenza education campaigns have been ongoing in provinces in which H5N1 outbreaks have occurred in humans and/or domestic poultry.

Data were collected from 1,252 adults >15 years old living in two southern provinces in Cambodia where H5N1 has been confirmed in domestic poultry and human populations using two cross-sectional surveys conducted in January 2006 and in November/December 2007 (Table 13.5). Poultry handling behaviors, poultry mortality occurrence, and self-reported notification of suspect H5N1 poultry cases to animal health officials in these two surveys were evaluated.

Our results demonstrate that although some at risk practices have declined since the first study, risky contact with poultry is still frequent (Tables 13.6 and 13.7). Improved rates of reporting poultry mortality were observed overall, but reporting to trained village animal health workers (VAHW) decreased by approximately 50%. This research has shown that although some improvements in human behavior have occurred, there are still areas – particularly with respect to the handling of poultry among children and the proper treatment of poultry and the surrounding household environment – that need to be addressed in public health campaigns. Though there were some differences in the sampling methods of the 2006 and 2007 surveys, our results illustrate the potential to induce considerable, potentially very relevant, behavioral changes over a short period of time.

Table 13.5 Demographic characteristics of subjects >15 years old included in the 2006 and 2007 studies, Kampong Cham and Prey Veng, Cambodia

Characteristics	2006 ($n = 452$)	2007 ($n = 800$)	p-Value[a]
Gender (% male)	178 (39.4)	401 (50.1)	<0.001
Age (median, IQR)	38 (27–48)	36 (24–49)	
Occupation (% farmer)	400 (88.5)	557 (70.2)[b]	<0.001
Education (highest level reached) n (%)		[c]	
None	74 (16.4)	170 (21.3)	
Primary	258 (57.1)	413 (51.7)	
Secondary	95 (21.0)	164 (20.5)	
High School	21 (4.7)	40 (5.0)	
Beyond high school	4 (0.9)	4 (0.5)	
Pagoda	NA	8 (1.0)	0.06[d]
Asset ownership	[e]	[c]	
TV	173 (64.6)	133 (66.5)	0.66
Radio	132 (49.2)	96 (48.0)	0.79
Car	1 (0.4)	5 (2.5)	0.09
Bicycle	224 (83.6)	164 (82.0)	0.65
Poultry ownership n (%)	[g]	[e]	
Chickens	260 (97.0)	176 (88.0)	<0.001
Ducks	97 (36.2)	82 (41.0)	0.29
Any poultry	89 (33.2)	77 (38.5)	0.24

NA not assessed
[a]χ^2 or Fishers exact test, as appropriate, p-value comparing 2006 vs. 2007
[b]$n = 794$
[c]$n = 799$
[d]χ^2 test for trend
[e]Assessed only at household level $n = 200$
[g]Assessed only at household level $n = 268$

Table 13.6 Changes in source of avian influenza information in Kampong Cham and Prey Veng Provinces, Cambodia from January 2006 to December 2007

	Kampong Cham n (%)			Prey Veng n (%)		
Source of AI information	2006 ($n = 210$)	2007 ($n = 387$)	p-Value[a]	2006 ($n = 232$)	2007 ($n = 394$)	p-Value[a]
Television	166 (79.1)	294 (76.0)	0.06	188 (82.8)	359 (91.1)	0.002
Radio	155 (73.8)	311 (80.4)	0.006	187 (82.4)	296 (75.1)	<0.001
Village chief	2 (0.9)	21 (5.4)	0.01	3 (1.3)	27 (6.9)	0.001
Village veterinary staff	20 (9.5)	12 (3.1)	0.004	27 (11.9)	16 (4.1)	0.001
Health staff/health center	7 (3.3)	18 (4.7)	0.74	21 (9.3)	12 (3.1)	0.002
Newspaper	5 (2.4)	7 (1.8)	0.11	6 (2.6)	4 (1.0)	0.07
Public poster	21 (10.0)	39 (10.1)	0.77	28 (12.3)	51 (12.9)	0.53

[a]2006 vs. 2007 by province χ^2 or Fishers exact test p-value adjusted for gender

Table 13.7 Changes in poultry contact in Kampong Cham and Prey Veng provinces, Cambodia from January 2006 to December 2007

Reported practice	All subjects *n* (%) 2006 ($n = 450$)	2007 ($n = 800$)	*p*-Value[a]
Contact with domestic poultry			
Touch sick or dead poultry with bare hands	339 (75.3)	337 (42.1)	<0.001
Allow children in the household play (touch and catch) with poultry	92 (20.4)	205 (25.6)	0.06
Use dead domestic poultry from yard for household consumption	203 (45.1)	108 (13.5)	<0.001
Care or help care for poultry	319 (70.6)	588 (73.5)	0.03
Slaughter poultry	173 (38.3)	286 (35.8)	<0.001
Contact with poultry at live bird markets			
Ever bought poultry from the market for food during the study period	43 (9.4)	62 (7.8)	0.48
Contact with wild birds			
Eat wild birds	149 (33.1)	277 (34.7)	<0.001
Collect dead wild birds from the field for household consumption	37 (8.2)	36 (4.5)	0.002
Ever prepared wild birds for food	114 (31.2)	217 (27.1)	<0.001
Potential environmental contamination			
Prepare poultry near a pond, river, or water well	84 (23.0)	220 (27.5)	<0.001
Wash poultry products directly in the water source (pond/river)	6 (1.6)	99 (12.7)	<0.001
Use poultry feces for manure	347 (76.8)	494 (61.8)	<0.001

[a]χ^2 or Fishers exact test *p*-value adjusted for gender

Redefining FAOs Poultry Sectors for Countries with Large Sector 4 Holdings

Millions of people in South East Asia and around the world live in proximity to domestic poultry, which is predominantly reared for household consumption. Direct contact with infected birds is assumed to be the main source of infection to humans; however, the nature and patterns of contact with domestic poultry based on raising and holding practices likely vary from country to country. The Food and Agriculture Organization's (FAO) definitions for poultry sector holdings are used throughout the work to describe poultry production (FAO 2006a). These definitions are most valuable for comparing poultry production in countries with medium and/or large-scale commercial poultry production; however, they offer little distinction within countries primarily involved in village or backyard production. The following commentary offers newly defined subcategories within FAO's current definition for sector 4 poultry production for countries with large village/backyard poultry holdings (Table 13.8).

Using data of 600 randomly selected households from 115 randomly selected villages from six Provinces in Cambodia, Fig. 13.3 categorizes poultry raising using

Table 13.8 Poultry production sectors in Cambodia using expanded FAO definitions

Criteria		Newly defined categories: FAO sector 3	FAO sector 4, sub-category A	FAO sector 4, sub-category B	FAO sector 4, sub-category C	FAO sector 4, sub-category D
FAO defined	FAO defined poultry sector	3	4	4	4	4
	Poultry production	Commercial/village/backyard	Village/backyard	Village/backyard	Village/backyard	Village/backyard
	Biosecurity used	Low	Some-minimal	None	Some-minimal	None
Biosecurity measures	Contact with other animals					
	Domestic poultry	None	Some	Yes	Some	Yes
	Wild birds	None	Some	Some	Some	Some
	Other domestic animals	None	Some	Yes	Some	Yes
	Birds kept	Indoors closed system – animals kept in small buildings	Indoors closed system or kept enclosed part of the day	Outdoors open – free-ranging	Indoors closed system or kept enclosed part of the day	Outdoors open – free-ranging
Flock raising measures	Predominant species owned	Duck or chicken	Duck	Duck	Chicken	Chicken
	Flock size (median, range)[a]	Medium-Large	Medium	Small	Medium	Small
	Quantity of birds	>50	>50	1–50	>50	1–50
	Selling characteristics					
	Sell eggs	Yes	Some	No or rarely	Some	No or rarely
	Sell birds[b]	Yes	Some	No or rarely	Some	No or rarely

[a] Flock size: small = 1–50 animals, medium = 500–1,000, large 1,001–10,000
[b] Birds entering live poultry markets

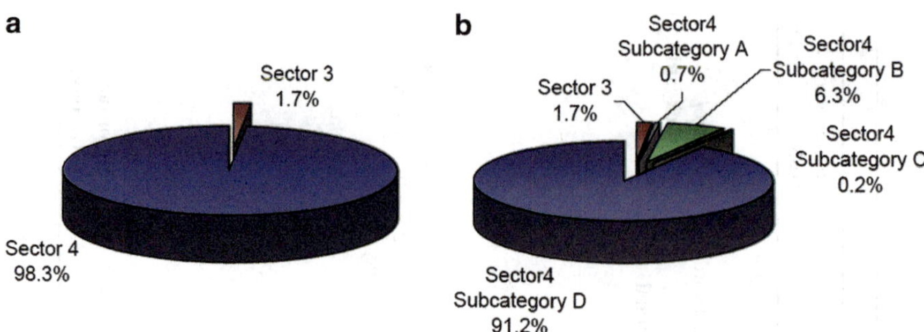

Fig. 13.3 Poultry production using FAO definitions in rural Cambodia (**a**) and newly defined subcategories described in Table 13.8 (**b**)

definitions provided in Table 13.8 compared to FAO's definitions. Using newly defined subcategories to describe sector 4 holdings further, poultry production in the rural Cambodia as 7% Sector 3, 0.4% Sector 4 Subcategory A, 6.5% Sector 4 Subcategory B, 0.2% Sector 4 Subcategory C, and 93.0% Sector 4 Subcategory D indicating that the majority of backyard holdings in rural Cambodia is very small chicken flocks.

The most important factors in describing sector 4 poultry holdings are direct measures of the raising habits including species raised, the mixing of domestic poultry species, mixing with other domestic animals (e.g., pigs, cats, or other mammals), mixing with wild bird species, and the use of biosecurity, i.e., the extent of free-ranging in the flock (e.g., free ranging within a specified area, free ranging on rice fields, in lakes, ponds) and use of containment of the flocks (fences, sheds, or other). Contacts with other domestic poultry (chickens and ducks) are likely the most important indicators characterizing poultry raising habits and use of biosecurity measures. Also important, but not as relevant in the case of Cambodia is the use of farm equipment on farms and sharing of such equipment. In Cambodia, this was rarely seen in rural areas; however, it is possible that if farm equipment became available in these areas, equipment would likely be shared within villages and perhaps within neighboring villages as well increasing chances of farm-to-farm transmission if the virus is present.

Study 2: Cross-Sectional Survey of 715 Rural Villagers, 123 Rural, Urban and Peri-Urban Market Sellers, and 139 Middlemen from Six Provinces and Phnom Penh

Since late 2003, more than 6,500 HPAI/H5N1 outbreaks have been reported in 61 countries and human cases have been identified in 15 countries (World Organization for Animal Health 2009; World Health Organization 2009). Although the mechanisms enabling persistence in poultry populations remain unclear, the movement of poultry through live-bird markets (LBM), which are common in Asian countries because of a cultural preference to consume freshly slaughtered meat, has been shown to be an important factor in the circulation of HPAI/H5N1 in Vietnam and Hong Kong (Kung et al. 2003a, 2007; Nguyen et al. 2005; Sims et al. 2003). The dense concentration and high turn-over rate of live birds in LBM provide ample conditions for virus amplification (Webster 2004) and therefore LBM may be an important reservoir for HPAI or "hub" of circulation (Senne et al. 1992). HPAI surveillance programs in several countries including Vietnam, Thailand, Cambodia, China, and Hong Kong have demonstrated that HPAI/H5N1 is circulating in LBM (Wang et al. 2006; Kung et al. 2003b, 2007; Nguyen et al. 2005; Amonsin et al. 2008; Guan et al. 2009; MAFF unpublished data).

The degree of connectedness of animal networks, that is the frequency with which links between premises and LBMs are made via people, animal movement, and/or sharing of equipment can determine the potential for widespread epidemics of disease (Kao et al. 2007). Thus an understanding of animal movement practices and their contact structures is important in designing targeted surveillance, disease prevention, and control activities (Dent et al. 2008; Green et al. 2008; Kiss et al. 2008; Truscott et al. 2007; Colizza et al. 2007; Hollingsworth et al. 2007; Hufnagel et al. 2004). This is particularly important in resource-limited settings where such activities may be limited (Stark et al. 2006). However, little is understood about poultry market chains in countries where HPAI/H5N1 is endemic or recurrent. We therefore conducted a comprehensive study to describe the current movements of live poultry throughout Southern Cambodia to understand how these movements could influence the potential spread of HPAI at local, regional and national levels.

Poultry Movement Networks in Cambodia: Implications for Surveillance and Control of Highly Pathogenic Avian Influenza (HPAI/H5N1) (Van Kerkhove et al. 2009a)

We conducted a cross-sectional survey of 715 rural villagers, 123 rural, peri-urban and urban market sellers and 139 middlemen from six provinces and Phnom Penh, to evaluate live poultry movement and trading practices. Direct trade links with Thailand and Vietnam were identified via middlemen and market sellers. Most poultry movement occurs via middlemen into Phnom Penh making live bird wet markets in Phnom Penh a potential hub for the spread of H5N1 and ideal for surveillance and control.

Our results have demonstrated that live poultry movement in Southern Cambodia is unidirectional, highly connected, and highly centralized (Figs. 13.4 and 13.5). We found the following:

- Approximately 83,000 live chickens and 35,000 live ducks are traded across the networks each week.

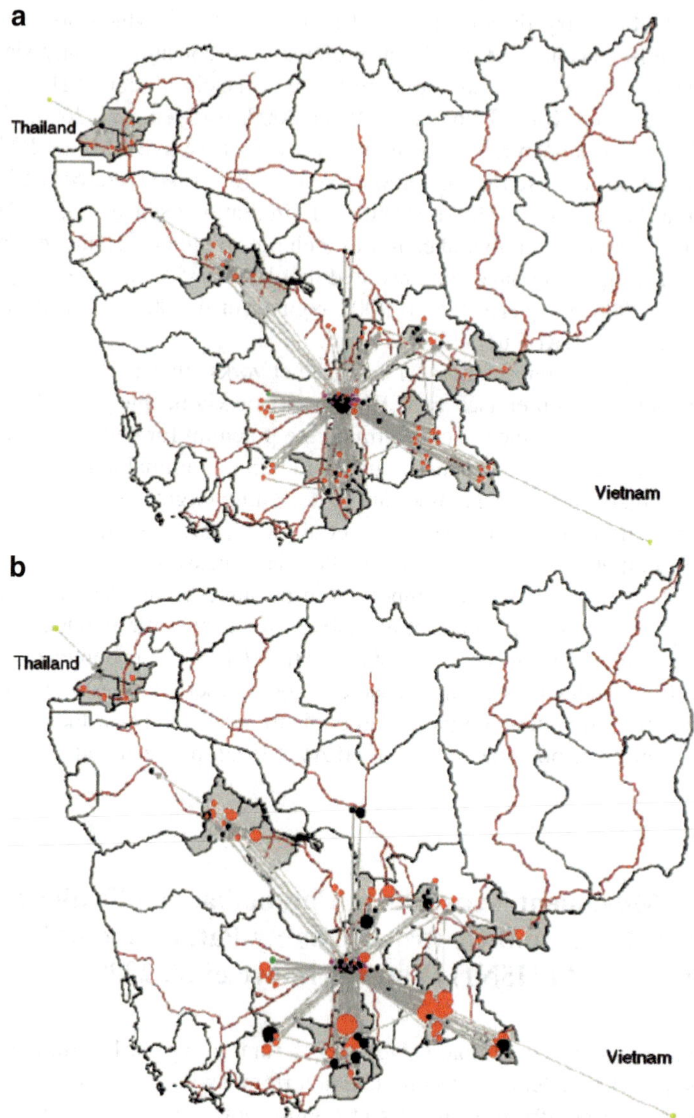

Fig. 13.4 Chicken trading networks in Cambodia with nodes weighted by (**a**) in-degree and (**b**) out-degree. The figures illustrate node sizes weighted by in-degree (**a**) and the same network weights nodes by OUT-degree (**b**). Node color indicates location type (*black* = market, *purple* = stock house, *red* = rural farm or household, *light green* = commercial farm, *gray* = semi-commercial farm, *yellow* = foreign source), ties show direction as indicated by the *arrow* and tie strength is indicated by the thickness of the *arrow*

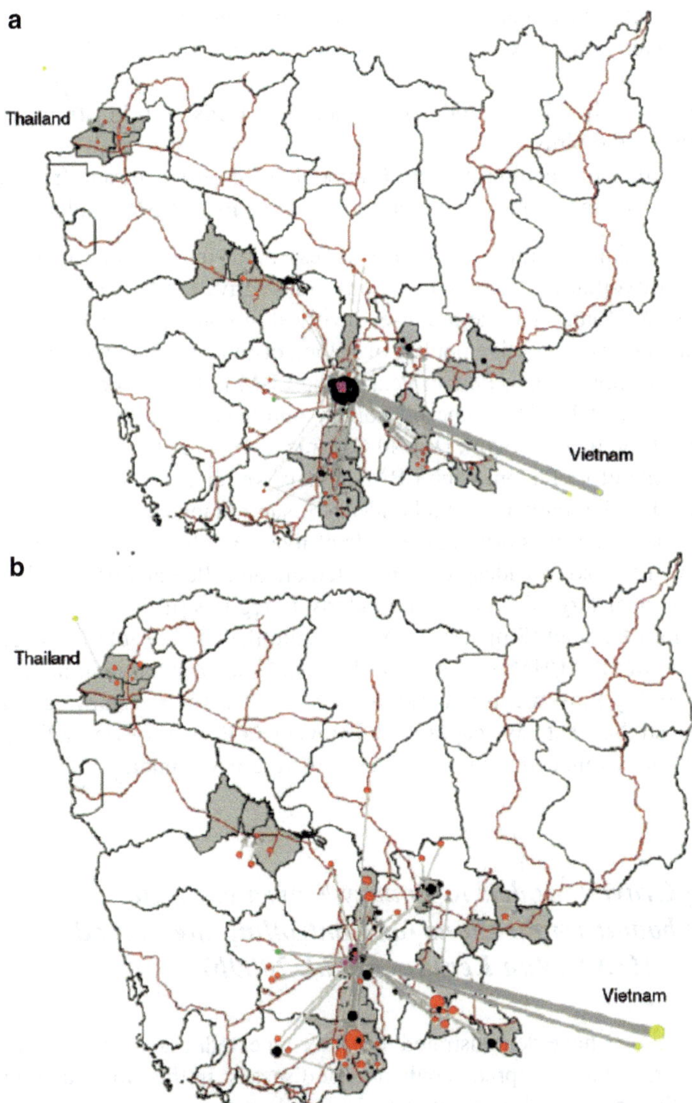

Fig. 13.5 Duck trading networks in Cambodia with nodes weighted by (**a**) in-degree and (**b**) out-degree. The figures illustrate node sizes weighted by in-degree (**a**) and the same network weights nodes by OUT-degree (**b**). Node color indicates location type (*black* = market, *purple* = stock house, *red* = rural farm or household, *light green* = commercial farm, *gray* = semi-commercial farm, *yellow* = foreign source), ties show direction as indicated by the *arrow* and tie strength is indicated by the thickness of the *arrow*

- Most poultry movement occurs via middlemen and market sellers on trucks and motorbikes into markets, semi-commercial farms, and stock houses located in Phnom Penh.
- Approximately 85% of middlemen trade live birds >10 km from where they purchased the birds.
- Live poultry originating in 11 of Cambodia's 24 provinces and from regions in Southern Vietnam are sold directly to the three main LBM in Phnom Penh.

Understanding poultry movement is essential to develop appropriate and targeted surveillance recommendations for active HPAI/H5N1 surveillance programs. We found that the premises involved in poultry trade are closely linked via middlemen carrying live poultry over long distances and that the unidirectional movement of poultry into Phnom Penh make LBM in Phnom Penh a potential hub for the spread of H5N1. Domestic poultry outbreaks of H5N1 have occurred in the areas of the main network (World Organization for Animal Health 2009; FAO 2008) and therefore Phnom Penh LBM, namely wet markets where live poultry are slaughtered at the market, would be ideal for surveillance and control.

The economic forces driving the trade of live animals and animal products have been shown to lead to widespread and often uncontrolled and/or illegal movement of animals over large distances, particularly in regions of the world where movement is not regulated (Sims 2007). Yet, despite their likely role in the circulation and spread of HPAI/H5N1 in Asia, little is understood about within and between country poultry movements via trading in the region. We are unaware of any other published studies that have captured the movement of poultry via trading, although one study has evaluated poultry trading in Northern Vietnam.

Fitting Gravity Models to Poultry Movement Data in Cambodia: Implications for Controlling the Spread of HPAI/H5N1 (Van Kerkhove et al. 2009b)

Previous studies have demonstrated that the connectedness of animal networks can result in large and widespread epidemics of diseases. In this study, a gravity model is fit to live poultry movement data in Cambodia using population data as an indicator of potential trade between the source where poultry are reared and destination of where poultry are sold to attempt to understand the potential driving forces behind the poultry movement patterns observed.

The main advantage of using a gravity model within transmission models rather than relying on the underlying movement data is that it can be applied outside the study areas. Thus, by fitting a model to the poultry movement data from Cambodia, it may be possible to predict trade flows in areas not covered by the study as well as in the wider Mekong Delta Region, which would be informative for HPAI control programs. Furthermore, as gravity models use information on the underlying populations, in theory they should be able to predict changes in movement patterns

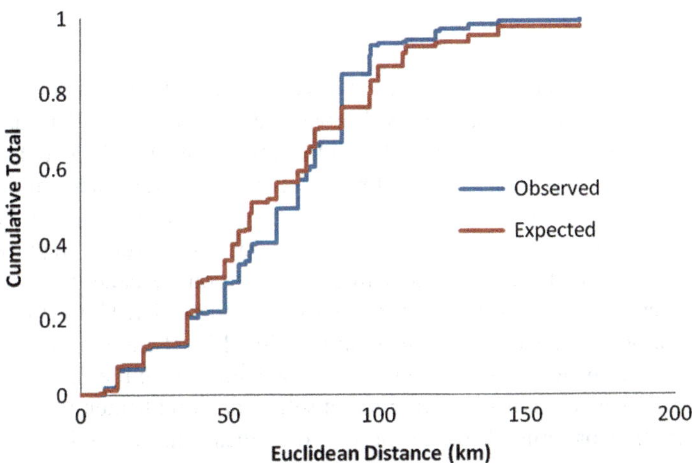

Fig. 13.6 Cumulative distribution of poultry movement by Euclidian distance. The plot shows the model $f(d, \mathrm{P1}, \mathrm{H2}) = GP_i^{\varepsilon}H_j^{b}k^*(d_{ij})$ fitted to all data where distance $\neq 0$; where P_i is an attribute of source location i (here the poultry population), H_j is an attribute of destination location j (here the human population), $k^*(d_{ij})$ is the normalized spatial kernel, ε and β are parameters which scale the influence of the source and destination populations, respectively, and G is a scaling parameter

following underlying changes in the population, although such predictions have yet to be validated within an infectious disease context.

Our model assumes that the movement of poultry is determined by a function of the distance between the two locations (the spatial kernel) and the populations at the source i and destination j. Thus the flow of poultry at distance d_{ij}, $f(d_{ij})$ is given by the equation:

$$f(d_{ij}) = GN_i^{\varepsilon}N_j^{\beta}k^*(d_{ij}),$$

where N_i is an attribute of source location i (either human or poultry population), N_j is an attribute of destination location j, $k^*(d_{ij})$ is the normalized spatial kernel, ε and β are parameters which scale the influence of the source and destination populations, respectively, and G is a scaling parameter. The model fit was assessed by comparing the deviance statistic under a Poisson likelihood for each model and also for individual data points to identify outliers.

Our results illustrate that poultry movement is best described using poultry populations at the source (representing the supply of poultry), human population at the destination (representing the demand for poultry), and an exponential kernel (mean distance 120 km) (Fig. 13.6). Population exponents ($\varepsilon = 0.5$, $\beta = 0.95$) indicate that poultry are bred primarily for local consumption with excess being traded, while the demand for poultry is proportional to population of the destination. This is a useful tool to predict poultry movement in wider areas and once validated, could be used to gain insight on how these movements could be controlled to prevent the spread of HPAI.

Conclusions

The research described in this chapter involved the collection of novel and original data from rural Cambodians, rural, peri-urban and urban market sellers, and poultry middlemen from six provinces and Phnom Penh between April 2006 and December 2007. In the first large-scale survey of randomly selected adults and children living in rural areas of Cambodia, we evaluated poultry ownership and husbandry practices, poultry mortality experienced, and poultry mortality reporting. This study also evaluated the extent and frequency of poultry handling behaviors of each subject and how these practices differ by age and gender. This is the first study to evaluate poultry contact patterns at an individual level and the first to evaluate poultry contact patterns of children. The second study identified and interviewed rural, peri-urban, and urban poultry market sellers and middlemen to evaluate their poultry trading patterns. Through these two studies, this research has evaluated poultry movement and the extent of interaction between humans and poultry in Cambodia to better understand the risks of sustained transmission of H5N1 in poultry and onward potential transmission to humans.

HPAI and Poultry Populations

The results from the first study demonstrated that most rural Cambodians own small quantities of chickens and/or ducks. Although a large majority of backyard flocks generally consist of less than a dozen chickens, most domestic poultry often mix with other domestic animals, including pigs, water buffalo, and dogs. Thus there is the potential for genetic mixing of an avian and nonavian influenza virus in these species (Alexander and Brown 2000; Horimoto and Kawaoka 2001).

The use of bio-security on farms with backyard poultry was found to be minimal or nonexistent and could be readily improved to a minimum level by keeping poultry species separated from other domestic species, separate from human areas, and restricted to forage for food within a specified and controlled area. One way to reduce the proportion of free-ranging animals and thus the potential for HPAI transmission between flocks is to improve the sealant properties of fences and netting that are used by approximately 25% of households in rural areas of the country. However, feasible options for providing feed to enclosed flocks would be necessary as flocks would no longer be allowed to freely forage for food. This presents a major challenge to HPAI control in developing countries.

Our research has highlighted the heterogeneity in poultry raising practices that would normally be classified as Sector 4 Poultry Production according to FAO definitions (FAO 2006b). To have more transparent and more informative descriptions of poultry raising in countries with predominant Sector 4 holdings and to design effective control measures for HPAI, further information is required, such as the size of poultry flocks, the species raised, and the output into the poultry market chain.

In Cambodia, more than 90% of backyard flocks are composed of small numbers of free-ranging chickens (Sector 4, Subcategory A). However, among households that raise poultry in Cambodia, approximately 1% of households included in our study rear ducks ($n > 50$) using a minimal level of bio-security (birds are kept indoors or allowed to be free-ranging for part of the day and kept indoors or fenced in at night; Sector 4, Subcategory B), whereas approximately 6% of households raise small numbers of free ranging ducks (Sector 4, Subcategory B). These two subcategories represent the greatest potential for sustaining H5N1 circulation in rural Cambodia since infected ducks can silently spread the virus (Chen et al. 2004; Hulse-Post et al. 2005). This is supported by research from Thailand, which has shown that free-grazing ducks were strongly associated with HPAI outbreaks in villages and thus were crucial for the persistence and spread of HPAI in Thailand in 2004 (Gilbert et al. 2006, 2007). Thus, it is especially important for HPAI surveillance to monitor the HPAI status of Sector 4, Subcategories A and B holdings since they represent the greatest HPAI risk. Control efforts in Thailand that focused on regulating movement and monitoring H5N1 infection via serologic and virologic testing of duck flocks may have led to the reduction of H5N1 outbreaks in 2006 (Webster et al. 2007).

Our research has also revealed that it is not uncommon for households to experience poultry mortality, with more than half of the households experiencing some degree of poultry mortality within the previous 8-month period. A key finding is that respondents reported an average of 50% within-flock mortality among both chicken and duck flocks over the study's recall period. Of cause for concern is that this level of mortality was not normally treated as suspect by villagers, even when within-flock mortality exceeded 50, 60, or 75%. It is probable that there has been considerable underreporting of HPAI in poultry populations in rural areas of the country and therefore there remains a strong need to improve the passive HPAI surveillance system in Cambodia. The current system relies on VAHW, who have been trained by FAO and NaVRI, to identify suspect poultry mortality. Given that our study population reported difficulties in distinguishing symptoms of AI from other poultry diseases including Newcastle disease, VAHW would provide a necessary link between rural subjects and the Ministry of Agriculture by providing information to the subject on AI and determining whether it is necessary to report the mortality event onwards to officials at the Ministry of Agriculture for follow-up. Without the VAHW, it would be increasingly difficult for subjects to know where or how to report poultry mortality.

It would be prudent to focus passive HPAI surveillance on high mortality events above a specified within-flock mortality threshold. For example, setting a threshold of within-flock mortality at >60% in chicken flocks or at >30% in duck flocks would help officials differentiate common poultry mortality from suspect mortality events. Using data from this study, if a threshold of 60% within-flock mortality were assumed to be the basis for which VAHW called upon ministry officials to investigate, 38 households should have been visited within the 8-month study period to investigate 30 occurrences of >60% mortality within a chicken flock, and 8 occurrences of >30% mortality within a duck flock events. Given the current limited personnel and financial resources of the Ministries of Agriculture in developing countries, investigating this many occurrences of suspect mortality that occurred in two provinces within an 8-month period would be difficult to achieve, even more unattainable would be to investigate suspect mortality events at the national scale.

Poultry Movement and HPAI

The second study evaluated poultry movement via middlemen and market sellers and illustrated that networks of poultry movement via trading in Cambodia are highly centralized, connected, and unidirectional. Most poultry movement occurs into Phnom Penh making the markets in Phnom Penh a potential hub for the spread of H5N1 and ideal for surveillance and control. Research has shown that live bird markets are an important reservoir for HPAI and an ideal environment for reassortment and transmission of HPAI from poultry-to-humans (Wang et al. 2006; Kung et al. 2003b, 2007; Nguyen et al. 2005; Amonsin et al. 2008; Woo et al. 2006). Thus, the movement of poultry through these markets is potentially important in the circulation and spread of HPAI (Kung et al. 2007; Sims et al. 2003). Domestic poultry outbreaks of H5N1 have occurred in areas of the main network and therefore Phnom Penh markets, namely wet markets, would be ideal for routine surveillance activities and control interventions.

Illegal cross-border trading of live poultry was also identified in this study, namely from Vietnam and to a lesser extent from Thailand. Live day-old ducks from Vietnam are directly traded with influential locations in Phnom Penh, including markets and semi-commercial farms, and make up approximately half of the total number of ducks traded weekly in Cambodia. Live and prepared poultry (birds that have been slaughtered, boiled, and defeathered) from Thailand are traded to markets in Banteay Meanchey province, but these movements are separated from the main network and I did not identify any direct links between Thailand and Phnom Penh.

Based on the results of this study, targeted surveillance recommendations were developed for NaVRI to improve their active HPAI surveillance activities. Two tiers of recommendations were developed to (1) monitor the HPAI status of poultry populations in rural areas (Tier 1 recommendations), and (2) identify locations that would allow for the early detection of HPAI incursion in markets, which have a high potential for spread throughout the network (Tier 2 recommendations). The purpose of the Tier 1 recommended locations for surveillance is to directly identify locations where HPAI can be rapidly detected if the virus is in the market system and would indirectly allow NaVRI to monitor the HPAI status of poultry populations in rural areas. These recommended locations include those with the highest in-degree values, and include markets, semi-commercial farms, and stock houses in Phnom Penh. If HPAI is identified in any of the Tier 1 recommended locations, an investigation to trace-back poultry back to its origin should be undertaken.

The purpose of the Tier 2 recommended locations is to identify locations that are most likely to interrupt poultry-to-poultry transmission of HPAI. The active surveillance of the locations that fall within Tier 2 would allow for the early detection of HPAI incursion in markets that have a high potential for spread. The Tier 2 recommended locations are those with the greatest number of connections and the largest out-degree scores. If HPAI is identified in any of the Tier 2 recommended locations, an investigation to trace-back and trace-forward poultry should be undertaken. This would require minimal effort to trace poultry forward as most are sold to markets in Phnom Penh and a moderate effort to trace poultry back to its origin as most are sold from nearby villages and markets.

An important finding of our second study, particularly when considering transmission dynamics of HPAI across the networks, is that poultry rarely spend more than 8 h in the market chain before they are slaughtered. This reduces the likelihood that active HPAI surveillance activities will capture an infected or diseased bird in the markets since poultry traders did not report selling visibly sick poultry. The wet markets in Cambodia provide an ample environment for HPAI viruses to persist (Wang et al. 2006; Kung et al. 2003b, 2007; Nguyen et al. 2005; Amonsin et al. 2008; Woo et al. 2006) and therefore the market environment should be included in the routine HPAI monitoring activities. This is supported by an investigation of live bird markets in China where investigators found no positive H5N1 samples from cloacal swabs of several bird species from several markets, but found one positive environmental sample from a goose cage (Wang et al. 2006). Therefore, further it would be prudent for Ministries to consider including monthly disinfection of wet market selling areas (e.g., stalls, selling platforms, locations where poultry are slaughtered) at their main live poultry selling markets in their HPAI control strategies.

A gravity model was fit to poultry movement data using population data as an indicator of potential trade between the source where poultry are reared and destination of where poultry are sold to attempt to understand the potential driving forces behind the poultry movement patterns observed. These results illustrated that poultry movement is best described using poultry populations at the source (representing the supply of poultry) and human population at the destination (representing the demand for poultry). The models also suggest that there may be different cost factors for poultry movement at different distances. A possible explanation for this trend may be that there are different economic drivers (e.g., mode of transportation, cost of diesel, profits from sale of poultry, journey time) for poultry traded over greater distances (poultry traded into Phnom Penh) when compared with shorter distances (localized poultry trading). This is a useful tool to predict poultry movement in wider areas and once validated, could be used to gain insight on how these movements could be controlled to prevent the spread of HPAI.

Given the rapid global spread of HPAI/H5N1 in recent years, surveillance of poultry populations will remain a high priority, particularly in the Mekong Delta Region where a considerable number of human deaths have occurred. These studies have been able to identify critical points for active HPAI surveillance and have informed Cambodia's HPAI surveillance activities. However, this does not replace the need for passive surveillance, which should be strengthened in rural areas of Cambodia by encouraging poultry owners to report poultry mortality to authorities. Since active surveillance in LBM is likely to remain a component of the surveillance and control efforts for HPAI in Cambodia and elsewhere, our results can be used to inform the selection of markets that best suits particular objectives of the surveillance system, in particular whether the objective is monitoring of the HPAI status of poultry populations in rural areas or early detection of incursion in markets with high potential for spread. Collection of similar data in other countries could prevent outbreaks or incursions of HPAI within their borders.

HPAI Transmission Risks at the Human/Animal Interface

The results from our large-scale cross-sectional survey demonstrated that most of the population in rural Cambodia is in frequent contact with domestic poultry. About half of the rural population sampled carried out, on a regular basis, at least one of the practices considered to be high risk for effective transmission if the bird is infected (e.g., slaughtering poultry, sharing water with fighting cocks, blowing into the beak of fighting cocks).

There was substantial variation in the frequency of different practices and thus the potential risk of transmission of H5N1 from poultry-to-humans is not uniform across age and gender even amongst populations living in proximity to poultry. In conducting a semi-quantitative risk assessment of the transmission potential of H5N1 from poultry-to-humans among rural Cambodians, we determined that males between the ages of 26 and 40, followed closely by males between the ages of 16 and 25 reported practices of contact with poultry that give rise to the greatest H5N1 transmission risk potential. Of the 3,600 subjects included in this assessment, approximately 16.2% ($n = 583$) had exposure risk scores above the 90th percentile and were largely male (72.3%) and had a median age of 29 (IQR: 21–42; range 6–69). These rural subjects have the greatest potential nonoccupational risk for poultry-to-human transmission of HPAI. This population group differs from the age and sex distribution of the total number of confirmed H5N1 human cases that have occurred to date, in which an excess of cases was observed in children and no differences observed between genders (Writing Committee of the Second World Health Organization Consultation on Clinical Aspects of Human Infection with Avian Influenza AV 2008). However, the group with highest exposure in this study is more similar to the age/sex distribution of the confirmed Thai cases ($n = 25$). The mean age of cases was 22 years and 64% of cases were male (WHO 2008) indicating that similar poultry contact patterns may exist in Thailand.

Such socio-demographic differences in human cases of H5N1 may be because contact patterns with poultry differ between countries. However, it is also suggestive that the variation in H5N1 incidence by age may not be due to exposure alone and that there may be differences by age in intrinsic immunologic susceptibility to infection, preexisting immunity against human influenza A virus and/or clinical presentation of disease. As my study is the first study to evaluate individual-level behavior of a large and randomly selected population, it would be useful to evaluate the poultry contact patterns of individuals living in rural areas of Vietnam, Thailand, Laos, China, and other Asian and African countries to evaluate whether behavior differs among these populations living in similarly close contact with poultry. No comparable studies have been conducted from which comparisons of regional contact patterns can be made, although contact with poultry is likely to be widespread.

This research found that poultry traders (i.e., poultry market sellers and middlemen) are in highly-concentrated contact with poultry and therefore have a greater potential poultry-to-human transmission risk when compared with rural Cambodians. All of the poultry traders included in my study reported practices that are considered to be high risk for effective transmission of H5N1 if a bird is infected with H5N1 since their daily activities regularly include contact with blood and bodily fluids through the practices of slaughtering, bleeding, and handling internal organs of poultry without the use of personal protective equipment.

Acknowledgements I would like to thank my coauthors; my interviewers, all of the study subjects and village chiefs; and the many district and provincial vetrinarians for their for their participation and assistance in these studies. I would also like to thank UNICEF and Insitute Pasteur du Cambodge for the funding to conduct these studies.

References

Alexander D, Brown I. Recent zoonoses caused by influenza A viruses. Rev Sci Tech 2000;19:197–225.

Amonsin A, Choatrakol C, Lapkuntod J, Tantilertcharoen R, Thanawongnuwech R, Suradhat S, et al. Influenza Virus (H5N1) in Live Bird Markets and Food Markets, Thailand. Emerg Infect Dis 2008;14(11):1739–42.

Apisarnthanarak A, Erb S, Stephenson I, Katz JM, Chittaganpitch M, Sangkitporn S, et al. Seroprevalence of Anti-H5 Antibody among Thai Health Care Workers after Exposure to Avian Influenza (H5N1) in a Tertiary Care Center. 2005. p. e16–e8.

Areechokchai D, Jiraphongsa C, Laosiritaworn Y, Hanshaoworakul W, O'Reilly M, (CDC) CfDCaP. Investigation of avian influenza (H5N1) outbreak in humans–Thailand, 2004. MMWR 2006;55(Suppl 1):3–6.

Biswas P, Christensen J, Ahmed S, Barua H, Das A, Rahman M, et al. Avian influenza outbreaks in chickens, Bangladesh. Emerg Infect Dis 2008;14(12):1909–12.

Bridges C, Katz J, Seto W, Chan P, Tsang D, Ho W, et al. Risk of influenza A (H5N1) infection among health care workers exposed to patients with influenza A (H5N1), Hong Kong. J Infect Dis 2000;181:344–8.

Bridges C, Lim W, Hu-Primmer J, Sims L, Fukuda K, Mak K, et al. Risk of influenza A (H5N1) infection among poultry workers, Hong Kong, 1997–1998. J Infect Dis 2002;185:1005–10.

Cai W, Schweiger B, Buchholz U, Buda S, Littmann M, Heusler J, et al. Protective measures and H5N1-seroprevalence among personnel tasked with bird collection during an outbreak of avian influenza A/H5N1 in wild birds, Ruegen, Germany, 2006. BMC Infectious Diseases 2009;9(1):170.

Cavailler P, Chu S, Ly S, Garcia J, Ha D, Bergeri I, et al. Seroprevalence of anti-H5 antibody in rural Cambodia, 2007. J Clin Virol. 2010 Jun;48(2):123–6.

Chen H, Deng G, Li Z, Tian G, Li Y, Jiao P, et al. The evolution of H5N1 influenza viruses in ducks in southern China. Proc Natl Acad Sci USA 2004;101(28):10452–7.

Colizza V, Barrat A, Barthelemy M, Valleron AJ, A Vespignani. Modeling the Worldwide Spread of Pandemic Influenza: Baseline Case and Containment Interventions. PLoS Medicine 2007; 4(1):95–110.

Dejpichai R, Laosiritaworn Y, Phuthavathana P, Uyeki T, O'Reilly M, Yampikulsakul N, et al. Seroprevalence of antibodies to avian influenza virus A (H5N1) among residents of villages with human cases, Thailand. Emerg Infect Dis2009; Available from http://www.cdc.gov/EID/content/15/5/756.htm.

Dent J, Kao R, Kiss I, Hyder K, Arnold M. Contact structures in the poultry industry in Great Britain: Exploring transmission routes for a potential avian influenza virus epidemic. BMC Vet Res 2008;4:27.

Dinh P, Long H, Tien N, Hien N, Mai L, Phong L, et al. and the World Health Organization/Global Outbreak Alert and Response Network Avian Influenza Investigation Team in Vietnam. Risk factors for human infection with avian influenza A H5N1, Vietnam, 2004. Emerg Infect Dis 2006;12(12):1841–7.

FAO. A Strategic Framework for HPAI Prevention and Control in Southeast Asia, Emergency Centre for Transboundary Animal Diseases (ECTAD), FAO, Bangkok. Bangkok2006a [15 August 2007]; Available from: http://www.fao.org/docs/eims/upload//224897/factsheet_productionsectors_en.pdf.

FAO. A Strategic Framework for HPAI Prevention and Control in Southeast Asia, Emergency Centre for Transboundary Animal Diseases (ECTAD), FAO, Bangkok. BangkokMay 2006b [15 August 2007]; Available from: http://www.fao.org/docs/eims/upload//224897/factsheet_productionsectors_en.pdf.

FAO. Outbreaks of Avian Influenza (subtype H5N1) in poultry. From the end of 2003 to 30 December 2008. Available at: http://www.oie.int/downld/AVIAN%20INFLUENZA/Graph%20HPAI/graphs%20HPAI%2031_12_2008.pdf. 2008 [cited 2009].

Fasina F, Ifende V, Ajibade A. Avian influenza A(H5N1) in humans: lessons from Egypt. Euro Surveill 2010;15(4):19473.

Gilbert M, Chaitaweesub P, Parakamawongsa T, Premashthira S, Tiensin T, Kalpravidh W, et al. Free-grazing ducks and highly pathogenic avian influenza, Thailand. Emerg Infect Dis 2006; 12(2):227–34.

Gilbert M, Xiao X, Chaitaweesub P, Kalpravidh W, Premashthira S, Boles S, et al. Avian influenza, domestic ducks and rice agriculture in Thailand. Agric Ecosyst Environ 2007; 119:409–15.

Green D, Kiss I, Mitchell A, Kao R. Estimates for local and movement-based transmission of bovine tuberculosis in British cattle. Proc R Soc B 2008;275:1001–5.

Guan Y, Peiris J, Lipatov A, Ellis T, Dyrting K, Krauss S, et al. Emergence of multiple genotypes of H5N1 avian influenza viruses in Hong Kong SAR. Proc Natl Acad Sci USA 2002; 99(13):8950–5.

Hinjoy S, Puthavathana P, Laosiritaworn Y, Limpakarnjanarat K, Pooruk P, Chuxnum P, et al. Low Frequency of Infection With Avian Influenza Virus (H5N1) Among Poultry Farmers, Thailand, 2004. Emerg Infect Dis 2008;14(3):499–50.

Hollingsworth T, Ferguson NM, Anderson R. Frequent Travelers and Rate of Spread of Epidemics. Emerg Infect Dis 2007;13(9):1288–94.

Horimoto T, Kawaoka Y. Pandemic Threat Posed by Avian Influenza A Viruses. Clin Microbiol Rev 2001 January 1, 2001;14(1):129–49.

Hufnagel L, Brockmann D, Geisel T. Forecast and control of epidemics in a globalized world. 2004. p. 15124–9.

Hulse-Post DJ, Sturm-Ramirez KM, Humberd J, Seiler P, Govorkova EA, Krauss S, et al. Role of domestic ducks in the propagation and biological evolution of highly pathogenic H5N1 influenza viruses in Asia. Proceedings of the National Academy of Sciences 2005 July 26, 2005;102(30):10682–7.

Kao RR, Green DM, Johnson J, Kiss IZ. Disease dynamics over very different time-scales: foot-and-mouth disease and scrapie on the network of livestock movements in the UK. J R Soc Interface 2007 October 22, 2007;4(16):907–16.

Katz JM, Wilina Lim, Bridges CB, Rowe T, Hu-Primmer J, Lu X, et al. Antibody Response in Individuals Infected with Avian Influenza A (H5N1) Viruses and Detection of Anti-H5 Antibody among Household and Social Contacts. The Journal of Infectious Diseases 1999;180:1763–70.

Kiss I, Green D, Kao R. The effect of network mixing patterns on epidemic dynamics and the efficacy of disease contact tracing. J R Soc Interface 2008;5:791–9.

Kung NY, Y. Guan, N. R. Perkins, L. Bissett, T. Ellis, L. Sims, et al. The impact of a monthly rest day on avian influenza virus isolation rates in retail live poultry markets in Hong Kong. Avian Dis 2003a;47:1037.

Kung N, Guan Y, Perkins N, Bissett L, Ellis T, Sims L, et al. The impact of a monthly rest day on avian influenza virus isolation rates in retail live poultry markets in Hong Kong. Avian Dis 2003b;47(3 Suppl):1037–41.

Kung NY, R. S. Morris, N. R. Perkins, L. D. Sims, T. M. Ellis, L. Bissett, et al. Risk for infection with highly pathogenic influenza A virus (H5N1) in chickens, Hong Kong, 2002. Emerg Infect Dis 2007;13:412–418.

Lu C, Lu J, Chen W, Jiang L, Tan B, Ling W, et al. Potential infections of H5N1 and H9N2 avian influenza do exist in Guangdong populations of China. Chin Med J (Engl) 2008;121(20): 2050–3.

Ministry of Agriculture, Fisheries and Forestry, Cambodia. Unpublished Data.

Mounts A, Kwong H, Izurieta H, Ho Y, Au T, Lee M, et al. Case-control study of risk factors for avian influenza A (H5N1) disease, Hong Kong, 1997. J Infect Dis 1999;180:505–8.

Nguyen DC, Uyeki TM, Jadhao S, Maines T, Shaw M, Matsuoka Y, et al. Isolation and Characterization of Avian Influenza Viruses, Including Highly Pathogenic H5N1, from Poultry in Live Bird Markets in Hanoi, Vietnam, in 2001. J Virol 2005;79(7):4201–12.

Normile D. Are Wild Birds to Blame? Science2005 October 2005;310(5747):426–8.

Ortiz J, Katz M, Mahmoud M, Ahmed S, Bawa S, Farnon E, et al. Lack of Evidence of Animan-to-Human Transmission of Avian Influenza A (H5N1) Virus among Poultry Workers, Kano, Nigeria, 2006. J Infect Dis 2007;196:1685–91.

Otte M, Pfeiffer D, Roland-Holst D, Inui K, Tung N, Zilberman D. Implications of Global and Regional Patterns of HPAI H5N1 Virus Clades for Risk Management. 2010 *in press*.

Santhia K, Ramy A, Jayaningsih P, Samaan G, Putra A, Dibia N, et al. Avian influenza A H5N1 infections in Bali province, Indonesia: a behavioral, virological andseroepidemiological study. Influenza and Other Respiratory Diseases 2009;3(3):81–9.

Schultsz C, Vo C, Nguyen V, Nguyen T, Lim W, Tran T, et al. Avian Influenza H5N1 and Healthcare Workers. Emerg Infect Dis 2005;11(7):1158–9.

Schultsz C, Van Dung N, Hai LT, Quang Ha D, Peiris JSM, Lim W, et al. Prevalence of Antibodies against Avian Influenza A (H5N1) Virus among Cullers and Poultry Workers in Ho Chi Minh City, 2005. PLoS One 2009;4(11):e7948.

Senne D, Pearson J, Panigrahy B. Live poultry markets: a missing link in the epidemiology of avian influenza. In: Beard C, Easterday B, editors. Proceeding of Third International Symposium on Avian Influenza. Richmond, VA: United States Animal Health Association; 1992. p. 50–8.

Sims L. Lessons learned from Asian H5N1 outbreak control. Avian Dis 2007;51(1 Suppl):174–81.

Sims LD, Ellis TM, Liu KK, Dyrting K, Wong H, Peiris M, et al. Avian influenza in Hong Kong 1997–2002. Avian Dis 2003;47:832–838.

Stark K, Regula G, Hernandez J, Knopf L, Fuchs K, Morris R, et al. Concepts for risk-based surveillance in the field of veterinary medicine and veterinary public health: Review of current approaches. BMC Health Services Research 2006;6(1):20–7.

Thanh N, World Health Organization International Avian Influenza Investigation Team V, Lim W. Lack of H5N1 avian influenza transmission to hospital employees, Hanoi, 2004. Emerg Infect Dis 2005;11(2):210–5.

Truscott J, Garske T, Chis-Ster I, Guitian J, Pfeiffer D, Snow L, et al. Control of a highly pathogenic H5N1 avian influenza outbreak in the GB poultry flock. Proc R Soc B 2007; 274:2287–95.

Van Kerkhove MD. H5N1/Highly Pathogenic Avian Influenza in Cambodia: Evaluating poultry movement and the extent of interaction between poultry and humans. London: London School of Hygiene and Tropical Medicine; 2009.

Van Kerkhove MD, Ly S, Holl D, Gutian J, Mangtani P, Ghani A, et al. An estimate of human exposure risk at the animal-human interface: Extent of poultry contact and potential risk of HPAI/H5N1 transmission in Cambodia Bangkok International Conference on Avian Influenza 2008: Integration from Knowledge to Control; 23–25 January: Bangkok, Thailand; 2008a.

Van Kerkhove M, Ly S, Holl D, Guitian J, Mangtani P, Ghani A, et al. Frequency and patterns of contact with domestic poultry and potential risk of H5N1 transmission to humans living in rural Cambodia. Influenza and Other Respiratory Viruses 2008b;2(5):155–63.

Van Kerkhove MD, Vong S, Guitian J, Holl D, Mangtani P, San S, et al. Poultry movement networks in Cambodia: Implications for surveillance and control of highly pathogenic avian influenza (HPAI/H5N1). Vaccine 2009a;27(45):6345–52.

Van Kerkhove MD, Truscott J, Garske T, Tatem A, Chhim V, San S, et al., editors. Fitting Gravity Models to Poultry Movement Data in Cambodia: Implications for controlling the spread of HPAI/H5N1. Epidemics; 2009b; Athens, Greece.

Van Kerkhove MD, Mumford E, Mounts AW, Bresee J, Ly S, Bridges C, Otte J, et al. Highly Pathogenic Avian Influenza (H5N1): Pathways of Exposure at the Animal-Human Interface, a systematic review. PLoS One 2011;6(1):e14582.

Vong S, Goghlan B, Mardy S, Holl D, Seng H, Ly S, et al. Low Frequency of Avian-to-Human Transmission of H5N1 in Southern Cambodia, 2005. Emerg Infect Dis 2006;Oct;12(10): 1542–7.

Vong S, Ly S, Van Kerkhove MD, Achenbach J, Holl D, Buchy P, et al. Risk Factors Associated with Subclinical Human Infection with Avian Influenza A (H5N1) Virus- Cambodia, 2006. Journal of Infectious Diseases 2009;199(12):1744–52.

Wang M, Di B, Zhou D, Zheng B, Jing H, Lin Y, et al. Food Markets with Live Brids as Source of Avian Influenza. Emerg Infect Dis 2006 November 2006;12(11).

Wang M, Fu C, Zheng B. Antibodies against H5 and H9 avian influenza among poultry workers in China. N Engl J Med 2009a;360(24):2583–4.

Wang Y, Liu Y, Jiang L, Liu Y, Yang Z, Hao A, et al. Risk assessment of H5N1 human infection after an outbreak of avian influenza in water fowl. Zhonghua Yu Fang Yi Xue Za Zhi 2009b 43(1):41–4.

Webster RG. Wet markets: a continuing source of severe acute respiratory syndrome and influenza? Lancet 2004;363:234–36.

Webster RG, Hulse-Post DJ, Sturm-Ramirez KM, Guan Y, Peiris M, Smith G, et al. Changing Epidemiology and Ecology of Highly Pathogenic Avian H5N1 Influenza Viruses. Avian Diseases 2007;51(s1):269–72.

WER. Human cases of avian infl uenza A(H5N1) in North-West Frontier Province, Pakistan, October–November 2007. Weekly Epidemiological Record (WER)2008 09 Dec 2008; 3 October 2008, 83(40):357–64.

WHO. World Health Organization: Avian Influenza. Website: http://www.who.int/csr/disease/avian_influenza/en/ Last accessed 27 October 2008. 2006–2009.

Woo P, Lau S, Yuen K. Infectious diseases emerging from Chinese wet-markets: zoonotic origins of severe respiratory viral infections. Curr Opin Infect Dis 2006 Oct;19(5):401–7.

World Health Organization. Cumulative Number of Confirmed Human Cases of Avian Influenza A/(H5N1) Reported to WHO as of 5 February 2009 [database on the Internet]. World Health Organization; Available at: http://www.who.int/csr/disease/avian_influenza/country/cases_table_2009_02_05/en/index.html. 2009 [cited 6 January 2009].

World Health Organization. Human avian influenza in Azerbaijan, February–March 2006. Weekly Epidemiological Record (WER) 2006; 5 May 2006, 81(18):183–8.

World Organization for Animal Health. Avian Influenza. Website Available at: http://www.oie.int/eng/info_ev/en_AI_avianinfluenza.htm [database on the Internet] 2009 [cited 30 June 2008]. Available from: http://www.oie.int/eng/info_ev/en_AI_avianinfluenza.htm.

Writing Committee of the Second World Health Organization Consultation on Clinical Aspects of Human Infection with Avian Influenza AV. Update on Avian Influenza A (H5N1) Virus Infection in Humans. N Engl J Med 2008 January 17, 2008;358(3):261–73.

Zhou L, Liao Q, Dong L, Huai Y, Bai T, Xiang N, et al. Risk Factors for Human Illness with Avian Influenza A (H5N1) Virus Infection in China. The Journal of Infectious Diseases 2009; 199(12):1726–34.

Part IV
Institutional Approaches to Controlling Animal Diseases

Chapter 14
AI Insurance/Indemnification: The Canadian Experience and Its Application to Developing Regions

Robert Burden

The Current Approach

The elements typically considered in foreign animal disease program management involve the development of policies and protocols for:

- Planning and prevention
  - Surveillance
  - Biosecurity
- Response
  - Pre-emptive culling
  - Emergency management/response (including stamping out)
- Recovery
  - Compensation

Although not an exhaustive list, there is little argument that each of these areas requires significant consideration in avian influenza (AI) policy development. Each area also requires different expertise in the design and implementation of the relevant policies and protocols. This division of expertise typically results in a separation in how and when the program for each area is delivered.

The ultimate impact is that the focus on individual components as separate entities effectively delinks them. As a result the resources allocated for compensation become associated with recovery rather than being incorporated into a strategic framework that motivates producers to increase biosecurity and surveillance activities. This has the unfortunate result of reducing the effectiveness of the overall

R. Burden (✉)
Serecon Management Consulting Inc., Edmonton, AB, Canada
e-mail: bburden@serecon.ca

D. Zilberman et al. (eds.), *Health and Animal Agriculture in Developing Countries*, Natural Resource Management and Policy 36, DOI 10.1007/978-1-4419-7077-0_14,

disease prevention, response, and recovery system as effective implementation of protocols is directly related to the commitment of the primary producer and their supply chain. Without this commitment, it is hard to administer an effective disease program. Ultimately the question becomes how to motivate appropriate actions on the part of the producer when these actions create costs for them and benefits that spread beyond them.

Main Strategic Considerations

The basic premise of a structured risk management approach (SRMA) is to motivate desired behavior by stakeholders in the areas of surveillance compliance, biosecurity, and emergency response by being more strategic in the design of risk transfer solutions. If done effectively the SRMA will reduce the potential of a disease outbreak, the cost of responding to it, and the need for compensation after it has been contained.

This can be accomplished by focusing on three main goals:

- Fully engage all stakeholders in the identification of the most effective prevention, control, and recovery policies and protocols that are realistic for a specific geographic region given the operating reality faced by the industry.
- Clearly outline how the benefits and resulting costs need to be, and in fact can be divided between public agencies (domestic), private companies/individuals, and the international community at large.
- Ensure that the program has both implicit and explicit forms of cross compliance between indemnification and desired behavior.

Engaging producers involves clearly understanding the operating reality they function in. It does not matter if the farmer is producing poultry for personal consumption, sale in a wet market, or as part of a larger collective. They are making market decisions and disease activities that disrupt this system come at a cost.

A second critical success factor in implementing a SRMA in poultry is to recognize that solutions will vary significantly across different production areas even within a single country. This has tended to be an issue for public agencies that typically like to develop a program and implement it on a broad scale which works for the "average producer." Unfortunately, the average producer rarely exists.

To address this, emphasis should be placed on the process used to develop and implement the solution rather than the solution itself. This approach is consistent with the concept of a separation between business and financing decisions. More specifically:

- Business decision
  - Let's figure out what we need/want based on the needs and abilities of the target production system as well as a clear understanding of its operating reality.

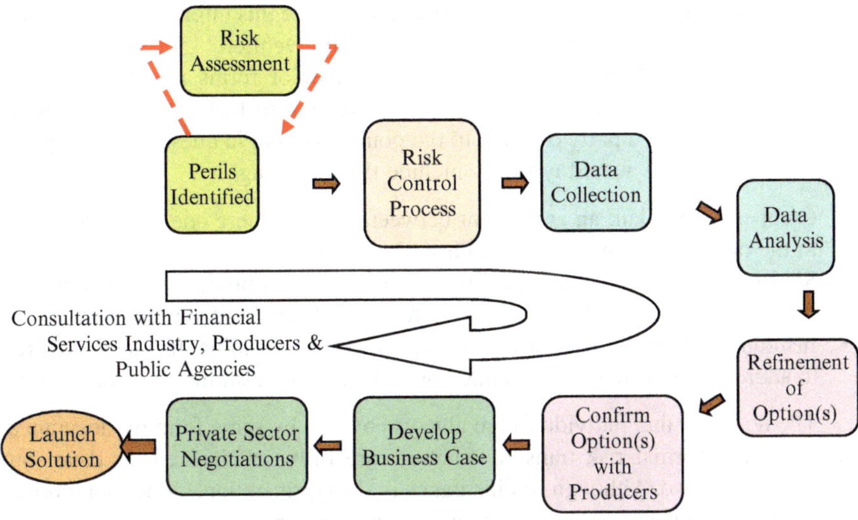

Fig. 14.1 Steps in insurance policy development. *Source*: Agriculture and Agri-Food Canada, Private Sector Risk Management Partnerships (PSRMP) program client booklet

- Financing decision
  - Figure out who should/would/can pay (involves consideration of public vs. private good).

If approached in this fashion, risk management and transfer markets can be used to address the concerns of the specific system in question. More importantly, the distinction in the roles of public and private stakeholders can also be dealt with in a systematic fashion.

Design Process: Coverage Requirements, Cross Compliance, and Evolving Roles

One of the first things to be accomplished in implementing a SRMA is to identify the needs of, and gaps in current risk management coverage. Of course this makes the assumption that there is some type of coverage at all. If not, there is still a need to identify what is required since this in effect becomes the gap analysis.

One effective approach is to use aspects of a typical insurance policy development process (Fig. 14.1). The main focus of the exercise is to identify the specific perils and outline actions that could mitigate their impacts. Once these are clearly understood in the context of the specific geographic area in question, there is a need to quantify the risk that is being addressed and then use this information as the basis for the development of the business case for an intervention.

The ultimate solution itself can take a number of forms ranging from pure compensation plans operated by a dedicated agency to that of a pure insurance option delivered by a party external to the poultry system in question. A simplified spectrum of options would typically include the following:

- Compensation plan: an agreement between parties where one provides support to the other under specifically defined situations.
- Reciprocal: A contractual agreement through which producers (subscribers to the policy) formally share risks for defined perils among themselves.
- Insurance: A more restrictive formal structure providing financial coverage to address specified losses sustained by stakeholders resulting from stated perils.

The reality is that individuals are always covered by some form of "insurance." They have a formal risk transfer structure, are fully self insured, or have some mixture of the two. Although self insurance is always an option, it does not motivate the development of formal risk management practices as would be required by insurance policy development.

The reality is that many of the perils facing developing nations as a result of AI are not likely to be transferable to a private insurance market due to the lack of formal, verifiable, and enforced risk mitigation practices in the industry. The process of developing these practices is of significant value to the industry. The fact is that a formal approach to risk management is always of value if there is a serious desire to reduce the impact of the disease.

A SRMA involves using elements of an insurance approach to initiate the risk identification and transfer process in a formal way while still recognizing the operating reality that the industry faces. This typically involves significant interaction of the stakeholders and numerous iterations in developing and implementing effective solutions.

The key factors involve establishing a formal risk mitigation team that is used to objectively link actions to risks taken by the sector stakeholders and then use this information as the basis for a formal risk assessment. It is important that this risk mitigation team include representatives from both the public sector as well as other stakeholders that are able to objectively assess the actual risks in the system. Producers, while obviously involved in the implementation of the standards may participate on this team, but it is essential that the team be allowed to objectively assess risk and not be biased through excess producer input.

Once the risk mitigation team completes the risk assessment, this information would form the basis of standards. They would also provide input on how these standards could be improved over time and how they should be monitored.

This then becomes the basis for the development of an indemnification process as it is at this point that there is a potential to engage the reinsurance industry to investigate the potential for their involvement in the system. While this is likely initially linked to public compensation programs, there is the potential to lead to a formal insurance policy. If not, the industry stakeholders are at least fully aware of

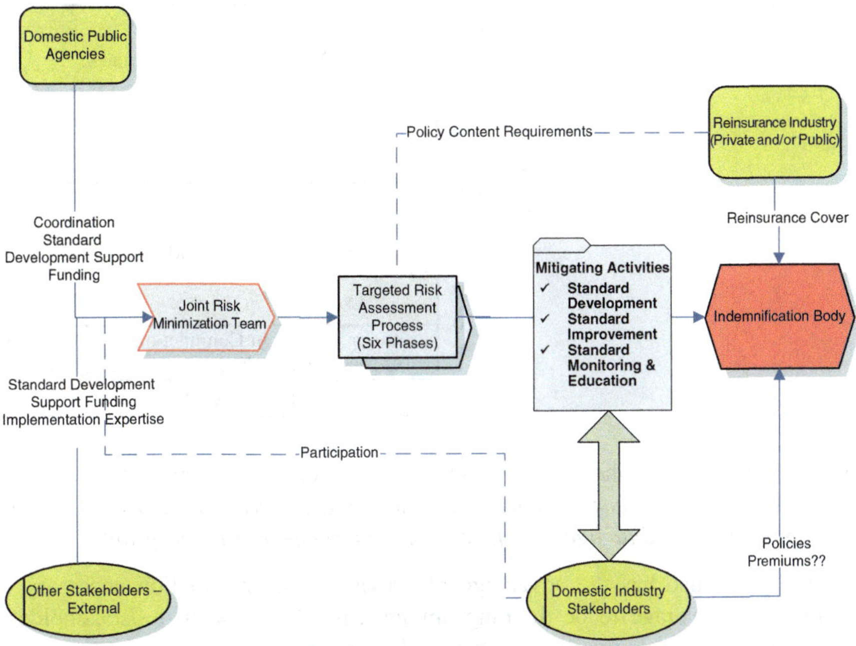

Fig. 14.2 Formal structure of the SRMA

how best to limit possibility of and the impacts from and AI outbreak, which effectively limits their exposure to the perils associated with AI, reducing the costs of self insurance.

An overview of the process can be observed in Fig. 14.2.

Design Considerations: Financing Options and Evolving Roles

The implementation of a SRMA as outlined in Fig. 14.2, results in a number of important benefits:

- It forces the stakeholders to clearly state the perils they face and clarify the specific elements of this that they want covered – *a recognition and quantification of the contingent liability they face.*
- It results in an assessment of the probable maximum loss resulting from an outbreak[1] – *helping to allocate costs to the main beneficiaries.*

[1] It is important to note that while the actual loss in developing countries may not be directly describable in a monetary fashion (genetic stock, protein source, etc.), the mitigation of the loss certainly can be.

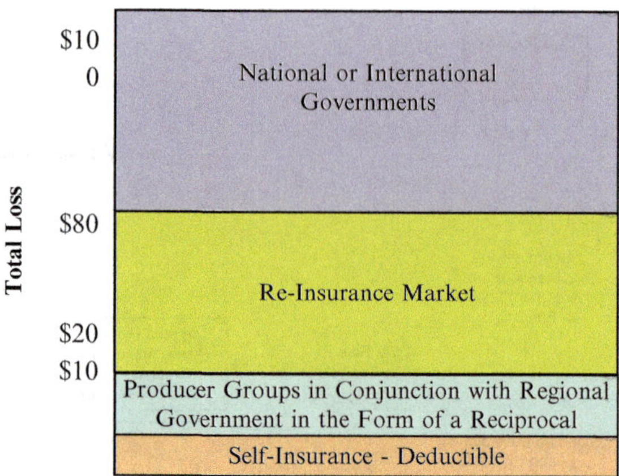

Fig. 14.3 Layered approach

- It helps to provide the incentive to motivate the control measures that would be implemented – *adjusting market structure in a desired way and reducing the exposure that public agencies face with their compensation programs.*

A SRMA also has the advantage of providing the information necessary to enable loss coverage to be dividing among a number of agencies/stakeholders. It also allows the role to evolve as the market matures.

Flexibility can be seen in the way that:

- Different agencies can select specific perils and provide full coverage.
- Costs can be shared within a specific peril.
- Exhaustion points (where one level of coverage runs out and another must kick in) can be identified for each level of coverage and/or each peril.

Ultimately, a mix of options can be designed for each layer of coverage by individual peril. These typically involve both involvement of individual producers, producer organizations, regional public agencies, national agencies, and the reinsurance market. It can also be extended to involve participation by international public agencies.

As a specific example, if the risk assessment has identified the need for a $100 pool of funds for a given pool category, responsibility can be allocated in a number of ways (Fig. 14.3).

In this case, the involvement of national or international governments would only occur if the outbreak cost exceeded $80. The first $10 would be self-insured by the producers. The second $10 loss would be covered by a reciprocal developed between the regional/local level of government and the individual producers themselves. Coverage under this reciprocal could be made contingent on meeting specific biosecurity and/or good production practices as specified during the SRMA.

This process is not rigid and can be adjusted to the specific needs of the stakeholders. As an example, the layers of coverage can be further broken into different segments of responsibility as outlined in Fig. 14.4.

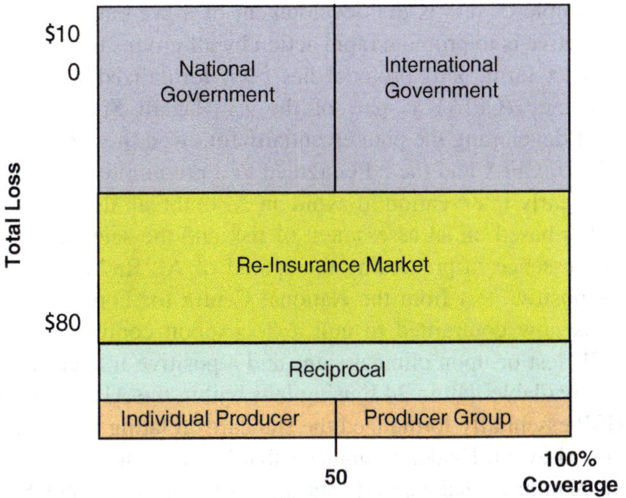

Fig. 14.4 Segmented approach

In this case, the individual would cover the first 50% of the first $10 loss, while the remainder would be covered by the producer group/corporation. This would be in the form of a deductible or program payment.

The second $10 in coverage would be managed by a formal reciprocal. This $20 in coverage would then be used to access an additional $60 from the reinsurance market. Any losses above this would be considered catastrophic and would be covered by the national government 50% and the international community.

The roles of the stakeholders evolve as the reciprocal fund is built. The initial years would require public agencies to finance the fund with producers contributing marginal dollars per unit sold. Producer dollars eventually replace the up-front public involvement to a preagreed level. If the fund has to be used, specific arrangements for refinancing the fund can be developed.

Canadian Experience

Experience in Canada has made the industry acutely aware of the relevance of taking a SRMA. Members of the National Poultry Group (NPG)[2] have worked hard to approach the issue in a more structured and strategic manner since 2004.

[2] This informal coalition of private sector poultry stakeholders includes: the Chicken Farmers of Canada (CFC), the Canadian Hatching Egg Producers (CHEP), the Egg Farmers of Canada (EFC), the Turkey Farmers of Canada (TFC), and the Canadian Poultry and Egg Processors Council (CPEPC).

A specific example of this is the development of a pre-emptive culling protocol where the objective is to promote rapid action by all groups to limit an AI outbreak to a single index farm. This protocol has been formalized with Canadian Food Inspection Agency (CFIA) as part of the AI Hazard Specific Plan (AIHSP). The process of developing the plan essentially followed the SRMA approach.

In early 2005, CFIA and the NPG agreed to a pre-emptive cull protocol as the cornerstone of early intervention to avoid an AI outbreak that gets out of control. This protocol is based on an assessment of risk and the scientific knowledge that time is of the essence in preventing the spread of AI. Rather than waiting for a confirmatory positive test from the National Centre for Foreign Animal Disease, the industry is now committed to cull a flock upon confirmation of H5 or H7 through a PCR test or upon clinical signs and a positive Influenza A test, both of which can be available within 24 h using labs within the AI laboratory network.

The AIHSP essentially formalized the standards relating to the actions taken in the during an AI event. Evidence suggests that having standards and a formal risk management plan have had a significant impact on disease exposure in Canada if the history of AI events is assessed.

In February 2004, the CFIA identified the presence of a low pathogenic H7 AI in the Fraser Valley in British Columbia. Subsequent tests in March revealed that the virus was also present in a highly pathogenic form. The CFIA depopulated all infected premises (42 commercial and 11 backyard premises) on which highly pathogenic AI was found and also destroyed all birds in the surrounding 3-km areas.

Direct government compensation under the Health of Animals Act (HofA) totaled $65 million. Even this large number, however, pales in comparison to the overall costs borne by industry, government, and local businesses. The total costs to the industry and the local economy of the 2004 outbreak in the Fraser Valley are estimated at more than $300 million ($170 million in direct costs).[3] This does not include the response and recovery costs borne by municipal, provincial, and federal governments.

Subsequent analysis conducted by the NPG suggested that the outbreak would have cost approximately $1.9 million[4] if it had been limited to the index farm and the four farms in proximity. Although difficult to determine exactly what would have happened, this estimate has been validated by the more recent experience in Canada with AI:

- In November 2005, the CFIA identified the presence of a suspected H5N2 virus on two duck farms in Yarrow, BC. The two duck farms were depopulated and the 78 commercial farms within a 5-km radius were kept under observation

[3] Serecon Management Consulting Ltd. "Economic Impacts on the British Columbia Poultry Industry Due to the Avian Influenza Outbreak." Prepared for the BC Poultry Industry Economic Impact Committee. August 2004.

[4] Internal documents prepared by the NPG in 2006 based on the knowledge of the size and nature of the index premise.

for a 3-week period. The surveillance period ended on December 10, 2005. By applying the pre-emptive cull protocol and using the risk management and mitigation process developed jointly by the CIFA and the poultry sector, the spread of AI was contained to the index and sister farm. Health of Animals Act compensation was estimated to be approximately $168,000 (maximum $35/duck × 4,800 ducks).

- This process was tested again in September 2007, when initial tests by the CFIA identified the presence of AI on a broiler breeder farm in southern Saskatchewan. The birds were destroyed and restrictions were applied to farms within 3 and 10 km of the infection site. Surveillance of all farms within 10 km was introduced and no further infections were discovered. The infected farm was cleaned and disinfected and was able to go back into production within 16 weeks. The final cost of this outbreak has not yet been determined; however, there is no question that it is substantially lower then the initial event in 2004.
- Finally, in January 2009, the CFIA identified the presence of a low pathogenic strain of H5 AI on a turkey operation in the Fraser Valley BC. The farm was depopulated and restrictions were placed on farms within the 3 km range. No additional infections were found and the farm went back into production for the following cycle. Again, the economic impacts of this outbreak were much smaller than in 2004.

Although it is difficult to validate a causal relationship between the development and implementation of a formal AIHSP after the outbreak in 2004 and the significant reduction in AI response and recovery costs, formalizing the process is obviously part of the solution. This is not to say that the problem is solved in Canada. In fact, the poultry industry and the CFIA continue to struggle with what is considered to be appropriate compensation to ensure the effective application of the protocol.

However, both sides do agree that the lynch pin of the pre-emptive cull protocol is on-going surveillance and early reporting of suspected flocks by farmers. In this regard, it is critical that farmers not hesitate in reporting, if they do the benefit of the pre-emptive cull protocol is completely negated. As a result, the poultry industry is committed to ensuring that compensation be considered as part of a SRMA where it is used as a tool in planning and prevention and not just response and recovery.

Canadian poultry producers have identified a number of additional coverage gaps in programs and services offered by both public and private sectors, and continue to raise this issue with their politicians and with Agriculture and Agri-food Canada (AAFC). Various levels of government and industry stakeholders now understand these gaps and have recognized that the financial services market plays an important role in developing and delivering risk management products and services. The market is now adjusting to account for this fact.

In response to issues like this, AAFC developed the Private Sector Risk Management Partnerships Program (PSRMP) as part of the Business Risk Management (BRM) suite of programs to work in partnership with the industry. The program supports (financial and technical assistance) producers and producer groups in the design and implementation of market based solutions. The ultimate goal is to have a fully private sector insurance solution, recognizing the reality that the larger the exposure the higher the likelihood of public involvement.

One of the most successful examples is the *Salmonella enteritidis* (S.e.) coverage program developed by the Ontario Broiler Hatching Egg and Chick Commission (OBHECC). Using the PSRMP process,[5] this group of primary producers designed and implemented financial protection for broiler breeder producers who were in full compliance with biosecurity requirements (as designed and monitored by the producers themselves). This involved the development of an administration structure, the Poultry Insurance Exchange (PIE), and funding of a reciprocal. The development of the reciprocal with its associated self discipline resulted in the attraction of the reinsurance market. Poultry producers control the Board of the PIE and can thus adjust the policy based on their needs.

This PSRMP process is now being replicated by the commercial laying sector commercial laying sector, the pre-lay (pullet) sector, as well as the leghorn breeder sector for S.e. across Canada, and provides evidence of how the links in a SRMA work:

1. The best way to control S.e. within a flock is to prevent infection from ever entering the premises. Application of HACCP based food safety and biosecurity programs are essential control measures.
2. Once a flock or its environment has been confirmed positive for S.e., early slaughter is often the best control option.
3. Understandably, farmers are reluctant to allow the testing of their facilities in the absence of a program to compensate for the loss of income due to early flock disposal.
4. However, to obtain adequate reinsurance cover for the reciprocal insurance plan (compensation), it will be necessary to demonstrate that the industry has an effective on farm food safety/biosecurity program.
5. The pullet grower, through their HACCP based On Farm Food Safety program, will likely demand that the chicks provided by the hatchery to produce pullets, are from a source certified as S.e. free. Eventually this demand would work its way back to the primary breeder who would be required to supply the breeder grower with chicks certified to be S.e. free.
6. Because of this likely chain reaction, a coordinated system of best management practices and insurance, from the breeder grower, hatching egg producer through to the hatchery and pullet grower, would be a worthwhile goal to place at the heart of a multiyear plan to reduce the potential for S.e. in the leghorn industry.
7. Any program to formulate best management practices and insurance coverage across the target industry must have the full cooperation and assistance of the hatcheries, EFC, and the provincial/territorial egg producer boards with

[5] The development of this risk management program in Ontario actually predated the introduction of the PSRMP program; however, it was the nature of the approach used for the industry/ government discussions that formed the basis for the PSRMP program.

regard to the gathering of statistics, formulation of best management practices, design of any insurance or compensation package and last but not least, the implementation and verification of the resulting systems.
8. Because of all of these links, a successful project will require good communications, consultation, and confirmation of agreement from industry leaders at each step in its development.

The Canadian poultry sector is committed to the SRMA in the development of risk management options. PSRMP program options have recently been used to develop risk transfer solutions for AI in both Ontario and British Columbia. It is important to emphasize that it is the process of developing a risk transfer product vs. the product itself that is the key to limiting the exposure of the industry to the impacts of AI.

Potential Application in the Developing World

This approach is applicable in the developing world and needs to be considered in developing disease management policies. This belief is based on two principles that are relevant to any farm production system:

- Principle 1: Public and private capacity limitations
- Principle 2: Farm decision processes and market involvement

Public and Private Capacity Limitations

There are not enough veterinarians and labs to have full flock cycle surveillance in any country in the World. This is not to deny the importance of a statistically valid surveillance sampling system, and in fact a strong system of veterinary health professionals is a critical piece in AI prevention.

On the contrary, there are simply not enough resources to provide the necessary public infrastructure to cover the world's poultry production systems. While this is true in developed situations, it is especially true for developed nations as well. This reality places increased importance on the need to motivate desired behavior, since the ability to enforce it externally is limited.

The issue of building capacity in veterinary science must not be ignored. However, given the limitations outlined above, it is obvious that it must be done in a strategic way and ensures the mobilization of the best resource that the public agencies have at their disposal – the cooperation of poultry producer and other direct stakeholders.

If this can be accomplished, you effectively increase surveillance exponentially by ensuring that birds are closely observed daily and any potential problems identified immediately.

Farm Decision Processes and Market Involvement

Farm production decisions assess the cost of inputs against the production output created by them. Farmers in the developed world typically value output based on the prices defined by open markets. Valuation in the developing world tends to rely on less formal markets for both inputs and outputs. However, the negotiation and decision process itself is similar.

One of the main differences is that farmers in less developed countries tend to have to accept a higher risk level in the production process. Limited resources, both public and private, reduce the ability for the development of an infrastructure to mitigate risk. As a result, the SRMA will obviously have to involve increased levels of public involvement to be successful.

Despite the need for an increase in public involvement, it does not mean that these contributions cannot be made in a fashion that provides motivation and some of the necessary infrastructure for public/private sector linkages. If we consider the situation outlined in Figs. 14.2 and 14.3, the necessary level of total coverage will decline, and there may be a need to significantly increase the public involvement in the development of the reciprocal.

The key is that regardless of the initial level of producer input, the process reenforces the need for industry supported actions that prevent disease and mitigates the cost if one occurs. Self-discipline replaces the need for regulation, and a significant portion of the self-discipline is motivated by market forces.

Summary and Conclusions

Poultry markets continue to evolve and AI prevention, response, and recovery activities must be seen as part of this evolution. Effective motivation for change is directly related to the need for producers to protect their assets. On the contrary, there is also a significant potential public benefit associated with effective disease response, which creates difficulty in determining how to deal with the issue and who should pay for it.

It is strongly recommended that a formal SRMA be considered, extending the traditional role of compensation from its historical role in recovery to the areas of disease prevention, response, and recovery process. This allows for a more natural evolution of market solutions and a more systematic way of identifying roles and responsibilities. It also allows markets to evolve in a way that increases stakeholder responsibility, while still ensuring that public benefits are paid for by the public.

The first step in designing any coverage option, regardless of who is paying for the ultimate program, is work closely with stakeholders to determine what the perils are and how they can be mitigated. Once this is done, there is a significant diversity of options available to address the problem.

A SRMA provides public and private stakeholders a way of leveraging market tools to enable other forms of producer/stakeholder indemnification including but not limited to:

- Farmer levy funding
- Bank guarantee funding
- Retrospective payment
- Reciprocal insurance
- Reinsurance options

The extent of public involvement in designing and implementing a SRMA in developing countries would have to be higher than that provided for producers in the developed world. Fortunately, this in no way limits the use of available insurance and reinsurance markets once the policy and protocols are designed. Cost sharing can be arranged based on need and social objectives. This does not limit the ability for the process to initiate market evolution to a desired end.

In addition to helping ensure the survival of affected producers, the process has the added benefit of developing a means of cross compliance to desired actions that help to reduce the potential of an outbreak in the first place.

Finally, while the options selected/used and the funding necessary may vary, the SRMA provides a consistent approach of developing the options and the resulting business cases. If the business case development can become more consistent, it will ultimately facilitate funding decisions, which will have to be made on them.

Chapter 15
Understanding Knowledge, Attitude, Perceptions, and Practices for HPAI Risks and Management Options Among Kenyan Poultry Producers

Marites Tiongco, Clare Narrod, Rosemarie Scott, Mimako Kobayashi, and John Omiti

Introduction

Since its emergence, HPAI H5N1 strain has attracted considerable public and media attention because the virus has been shown to be capable of causing fatal disease in humans. Although there is fear that the virus may mutate into a strain capable of sustained human-to-human transmission, the greatest impact to date has been on the highly diverse poultry industries in affected countries. In response to this, policies against HPAI have so far focused on implementing prevention, control, and eradication measures in poultry populations, with more than 175 million birds culled in Southeast Asia alone.

Until recently, significantly less emphasis has been placed on understanding producer behavioral factors that may alter smallholder's willingness to adopt good practices and disease prevention and control measures. Understanding the factors affecting behavior is important because in disease control setting the conditions required to achieve the efficient outcome are often absent due to information problems

M. Tiongco (✉)
Research Fellow, Markets, Trade, and Institutions Division, International Food Policy Research Institute, Washington, DC, USA
e-mail: m.tiongco @cgiar.org

C. Narrod
Senior Research Fellow, Markets, Trade, and Institutions Division, International Food Policy Research Institute, Washington, DC, USA

R. Scott
Research Assistant, Markets, Trade, and Institutions Division, International Food Policy Research Institute, Washington, DC, USA

M. Kobayashi
Research Assistant Professor, Department of Economics, Reno, NV, USA

J. Omiti
Kenya Institute for Public Policy Research and Analysis, Nairobi, Kenya

D. Zilberman et al. (eds.), *Health and Animal Agriculture in Developing Countries*, Natural Resource Management and Policy 36, DOI 10.1007/978-1-4419-7077-0_15,

resulting in market failures and/or coordination failures. Information problems arise due to (a) incomplete information, where the strategies and payoffs of the agents are not known and without information being complete the outcomes of agents' actions may not be efficient; (b) imperfect information, where the actions of all players are not known, which is likely given the structure of the poultry industry with a large number of actors along the value chain; and (c) asymmetric information, where one party (e.g., private producers) has more information than the other party (e.g., regulators). Moreover, one's actions against contagious disease can cause spill-over effects to the others (called externalities). Under these conditions, market failure is known to arise, and government regulators may choose to intervene to correct the market failure and achieve more efficient levels of disease control efforts.

Because of stochastic forces and often complex interactions among players in the poultry value chains, it is not always clear to regulatory decision makers how to intervene optimally, particularly to ensure that poor producers participate in efforts to reduce the risk of a disease. This chapter discusses the initial findings of producer's knowledge, attitude, perception, and practices (KAP) of HPAI in Kenya, a country that has not had the disease, but a scare in bordering Sudan and a media alert have had negative economic impacts. Information revealed by KAP analysis regarding HPAI is important for policy makers in assessing poor people's willingness to adopt prevention and control measures that are cost-effective in reducing risk. The analysis presented here is based on primary data collected through a household survey in 2009. This work feeds into cost-benefit and cost-effectiveness analyses to evaluate the effects of alternative and hypothetical HPAI control strategies in an environment where a number of potential control measures are currently not in place and there is uncertainty over their effectiveness. The full analysis can be found in the report titled Evaluating Risk Management Options to Reduce the Risk of HPAI for Kenya Poor's Producers (Narrod et al. 2010).

Poultry Production in Kenya and HPAI Disease Situation

Understanding the structure of poultry industry is particularly important in Kenya where, of the estimated 28 million domestic birds (Table 15.1), indigenous chicken constitutes about 70% and exotic birds 28% (Ministry of Livestock Annual Report 2006). Other poultry species such as ducks, turkeys, geese, quails, and ostrich make up the remaining 2% of the population. Of the 28% exotic birds, broilers constitute 20% while hybrid layers make up 8% of the total population (Department of Livestock Production Annual Report 2003). The Rift Valley and Nyanza are estimated to have over six million poultry of which over five million are indigenous chicken while in Eastern and Central there are close to four million birds in each region. Because of the rather harsh semi-arid agro-climatic conditions, Northeastern Kenya has the lowest bird population.

Smallholder poultry farms mainly raise indigenous chicken and are concentrated in the Rift Valley, Nyanza, and Eastern regions of Kenya. They supply 50% of the

Table 15.1 Estimated geographical distribution of poultry populations (thousand birds) in Kenya

Region	Indigenous	Broilers	Layers	Others	Total
Nyanza	5,944.8	203.6	48.2	46.8	6,243.28
Rift Valley	5,776.4	1,137.1	283.4	167.8	7,364.57
Eastern	3,628.8	163.9	112.6	21.3	3,926.68
Western	2,517.6	116.5	23.6	159.7	2,817.40
Coast	2,153.5	248.0	79.4	133.6	2,614.5
Central	1,787.0	1,079.2	440.9	35.6	3,342.70
Nairobi	141.4	188.1	957.8	10.0	1,297.25
Northeastern	165.0	0.2	0.3	0.0	165.46
Total	22,114.3	1,946.2	3,136.5	574.9	27,771.8

Source: Ministry of Livestock Annual Report (2006)

eggs and 60% of the poultry meat in the country (MoLFD Annual Report 2004). Indigenous chicken are kept under low input systems, hence offering high returns relative to inputs and investments. Over 90% of all rural households keep poultry, either for subsistence or commercial purposes (Nyange 1995; Ndegwa and Kimani 1996). Flock size ranges between 10 and 30 chicken on average, and exceeds 30 birds for commercial units. Commercial poultry production is concentrated in urban areas where the markets for poultry products are. Day-old chicks are sourced from commercial hatcheries mainly located in urban and peri-urban areas of Nairobi, Kisumu, Nakuru, and Mombasa.

The Kenyan poultry sector in reference to HPAI has been classified according to the level of integration, production technology used, and the level of biosecurity going from continuum of sector 1 for farms with high levels of biosecurity to sector 4 for farms with low levels of biosecurity. Larger, more commercialized actors involved in formal arrangements tend to have much stronger biosecurity practices than smaller-scale farmers. In Kenya, sector 1 tend to be large integrated industrial producers, sector 2 hatcheries, sector 3 smallholder semi-commercial farmers, and sector 4 village or "backyard" (traditional) poultry producers (Omiti and Okuthe 2008; Okello et al. 2010).

Though Kenya to date has not experienced an HPAI case, there were media reports in September 2005 showing dead birds in Nakuru (Rift Valley) and Kasarani (Nairobi) resulting in a HPAI scare in which fear and panic spread throughout the country from January to March 2006. This scare was accentuated due to an outbreak of HPAI in Southern Sudan. Kimani et al. (2006) estimated that the cost of this scare in Kenya was around a Kshs of 2.3 billion. In addition the country is considered at high risk of an HPAI occurrence due to its location along the migratory route of wild birds and because of the confirmed case in Sudan, Kenya's northern neighbor. Kenya further suffers from porous borders where illegal trade in poultry and poultry products continues unabated, a weak surveillance and regulatory system, and a large backyard poultry population, which nevertheless is an important component of people's livelihood particularly in the rural areas.

The Government of Kenya listed HPAI as a notifiable disease in 1998 and since then has put in contingency plans in 2005 and implemented measures for early detection, prevention and control of HPAI. The measures they implemented include enhancing surveillance and epidemiologic investigation, culling and compensation, enhancing biosecurity, implementing targeted vaccination programs, quarantine and movement control, and developing appropriate regulations and mechanisms for their implementation. Okuthe (2008), however, notes the Department of Veterinary Services suffers from inadequate funding, personnel, and equipment. Thus the question is, are these proposed and implemented measures the most cost-effective approach, given a large number of households in Kenya keep chickens in their backyards? Or would some other mechanism such as education improve farmer's adoption of low-cost risk mitigation measures? To investigate what behavioral factors influence their practices we examine knowledge, attitude, perceptions, and practices (KAP) of HPAI for producers with different production scales.

Hypotheses of This Study

The following ten hypotheses are tested in this study:

1. Larger producers have better knowledge of HPAI symptoms, more correct beliefs about safe practices in handling poultry and poultry products, and more correct perception about disease transmission risks than small and backyard producers.
2. Poultry producers, whether small or large, who have experienced poultry diseases in the past have better knowledge of how to prevent or reduce the risk of HPAI than those producers who have not experienced any poultry diseases in the past.
3. Producers' knowledge regarding HPAI is influenced by their level of education.
4. Perception about disease transmission risks is influenced by past disease experiences.
5. Producer's beliefs about good practices and perceptions about disease transmission risks are associated with their knowledge about HPAI symptoms.
6. Producers with experience of disease in the past are more concerned about disease spread risks in their village.
7. KAP indices explain actual adoption of biosecurity actions.
8. Smaller and poorer producers adopt fewer biosecurity actions.
9. Producers that have layers, which have longer production cycle and often are raised in cages, have more biosecurity measures.
10. Producers' education and household income level influence the number of biosecurity measures implemented.

Literature Review and Conceptual Framework

Literature Review

Knowledge, attitude, perception, and practices analyses (KAP) are widely used to evaluate the effectiveness of information campaigns and to assist policymakers in customizing educational programs to fit the public's needs. They are especially instrumental in assessing a country's vulnerability to animal diseases because in order to prevent rapid disease spread, both agricultural producers and consumers must understand the risks involved and necessary precautions must be taken by all actors along the value chain. Information generated by KAP studies in these situations greatly augments policymakers' ability to implement educational campaigns that efficiently fill the gaps in the public's understanding of the problem and minimize the risk of disease outbreak. KAP studies target the following information to capture the public's relationship with the given issue in three complimentary ways:

1. Knowledge: the degree of factual understanding of the topic and associated issues.
2. Attitude and perceptions: feelings toward the subject, including judgment of its importance and influence on people's lives.
3. Practices: current actions taken as a result of the knowledge, attitude, and perception toward the issues.

There have been numerous attempts to investigate KAP levels for Highly Pathogenic Avian Influenza (HPAI) on the general population (Fielding et al. 2005; Olsen et al. 2005; UNICEF-Georgia 2007; Suphunnakul and Maton 2009; Di Giuseppe et al. 2008; Leslie et al. 2008) and on target groups (UNICEF-Myanmar 2006; Leggat et al. 2007). These authors' conclusions are summarized below and a comparison of their methodologies follows.

Fielding et al. (2005) surveyed households in Hong Kong over the phone and estimated exposure and risk perception of avian influenza from live chicken sales. Likert scales were used across the questionnaire to identify attitudinal and knowledge predictors of risk perceptions. Questions on risk perception were related to respondents' views of catching AI from buying live chicken and likelihood of self and family members getting sick from buying live chicken. Questions with categorical response formats were asked such as whether respondents purchased live chicken (yes/no), whether respondents was touching chicken during purchase (yes/no), and how likely will respondent get sick from buying live chickens (Likert scale). Results indicated that one-third of respondents perceived some risks from purchasing live chicken but the magnitude of exposure to hazards associated with AI seldom exceeded 60%. Respondents who expressed anxieties about AI were also worried about getting sick due to buying live chickens, which implies that they perceive live bird markets as health threats. Older respondents viewed purchasing live birds during the recent AI outbreak as low risk possibly because of past experience of disease outbreaks, which implies that hazard familiarity and experience could reduce associated risk perceptions. The authors concluded that in the long run public awareness campaigns have little effect in reducing high-risk behavior, while in the short run the effect was large but temporary.

Similarly Olsen et al. (2005) surveyed residents of rural Thailand regarding HPAI knowledge and attitude, and practices (KAP) about preventive measures to avoid getting sick due to poultry disease such as AI, 6 months before and 6 months after attending an educational campaign. KAP questions asked include whether it is safe to touch sick or dead birds with bare hands, whether it is safe to eat dead birds from farm, and whether it is safe to eat pinkish chicken meat or runny egg yolk. The percentage of respondents who believed that it was safe to touch sick or dead birds with bare hands decreased significantly after hearing about AI. However, their practices and actions around poultry and poultry products changed less significantly, which implies that certain actions were already appropriately put into practice or the changes in producer perception and beliefs were not met by changes in behavior. Practices that did not change included preparation of dead poultry for household consumption despite widespread education campaign about AI.

A similar study done by UNICEF in Georgia and Myanmar used structured questionnaires to collect data which were analyzed using paired *t*-test. Findings showed that although the majority of respondents were familiar with HPAI, few could list the symptoms of HPAI in poultry and transmission modes of the disease. Similar findings, using similar methods, were reported by Suphunnakul and Maton (2009) in their HPAI KAP study in Thailand, where the authors recommended that, in addition to a general a public awareness campaign, a detailed and targeted campaign would be beneficial in reducing the risk of HPAI spread.

Di Giuseppe et al. (2008) conducted a KAP study on HPAI of Italian adults that were randomly drawn from a list of parents from 40 randomly-selected schools in Naples, Italy, in terms of knowledge about modes of transmission, practices such as hand-washing with soap before and after touching raw poultry meat and glove usage, and the perception about the correct definition of AI and perception of risk of contracting HPAI (whether it could be transmitted by eating and touching raw eggs and poultry products). Similar to Fielding et al. (2005), the results suggested that there were moderate levels of HPAI knowledge, a limited understanding of details of symptoms, transmission and prevention, and a high perception of being at risk of contracting HPAI. Higher mean total score of respondents' level of perceived risk of contracting AI indicates higher risk perception, i.e., if respondents feel very much at risk, they would report a score of 10. Results from multivariable logistic k regression model showed that perceived risk of being infected is significantly higher if respondents were from lower socioeconomic classes, had lower educational level, if they did not know the definition of avian influenza, if they knew that avian influenza could be transmitted by eating and touching raw eggs and poultry foods, and if they believed that they did not need additional information about the disease. Compliance with precautionary behavior was more likely in those who perceived themselves to be at higher risk, who knew the hygienic practices, who knew the modes of transmission for HPAI, and who received information from health professionals and scientific journals. Leslie et al. (2008) found that though overall knowledge of HPAI in five provinces of Afghanistan was low, individuals from higher socioeconomic classes had high knowledge levels.

An examination of the methodologies adopted to come to these conclusions is helpful in evaluating the strengths and weaknesses of various statistical tools that accommodate different types of research questions authors may pursue. Most of the studies described above used a Likert scale in surveys and to create KAP indices, for a group of questions, the answer to each question was rated and variables that capture the total points earned were created. The studies differed mostly in terms of the statistical methods used to analyze these scores, the number of questions used in the construction of KAP indices, and whether they were identifying current KAP levels or evaluating an educational program's impact on KAP.

Some studies only utilized *t*-tests to identify significant differences in KAP scores between interest groups (Mahmoodabad et al. 2008; Ly et al. 2007). This method is most basic because it establishes a simple mean difference without controlling for other factors that may be influencing the KAP variations. It is even inappropriate when there are large differences between the groups besides those variables in the *t*-test that may affect respondents' KAP scores. The conclusions made were much more valuable to policymakers when authors used multivariate regression analyses to measure the impact of multiple factors on respondents' KAP scores (Imai et al. 2005; Di Giuseppe et al. 2008; Xiang et al. 2010; Leslie et al. 2008; Liebenehm et al. 2009; Fielding et al. 2005).

Some studies created binary KAP variables by categorizing KAP levels into groups (often negative and positive groups) (Kumar and Popat 2010; Leggat et al. 2007; Lau et al. 2007; Fielding et al. 2005); these studies restricted the scope of their analyses because regression coefficients could not capture the full variation in KAP levels or in factors that influence those scores. Instead, other authors used indicators that captured the percentage of total possible KAP points when several different questions were used to construct KAP indices (Leslie et al. 2008; Xiang et al. 2010; Suphunnakul and Maton 2009). Some other authors limited the KAP indices to two or three points and hence did not capture as much variation in dependent variables and may not have fully measured the respondents' KAP (Imai et al. 2005; Mahmoodabad et al. 2008; Fielding et al. 2005). Leslie et al. (2008) improved the precision of their indices by weighting the responses to questions used in each index based on each question's importance in determining superior knowledge, attitudes, and practices on HPAI.

Another major difference in the methodology used in each KAP study was whether the authors were trying to evaluate an educational program or the current KAP levels to assist in the creation of future educational programs. Suphunnakul and Maton (2009) and Mahmoodabad et al. (2008) conducted KAP surveys before and after 6- and 3-month programs, respectively, and used *t*-tests to determine whether the differences between KAP scores were statistically significant. This method is limiting for two reasons. First, the use of *t*-tests, as previously mentioned, does not allow for the controlling of other influencing factors. Second, these authors assume that experiencing the educational program was the only event that occurred in the time between the two surveys that could have affected respondents' KAP levels. This assumption is usually unreasonable, especially when the length of time in between the surveys and the length of the program are as long as they were in the two previously mentioned studies. Liebenehm et al. (2009) avoid these assumptions by using cross-sectional data and propensity score matching to construct a counterfactual group to conduct an ex-post evaluation of a technology adoption program on participants' KAP scores.

An examination of past KAP studies shows that the most effective methodologies use categorical KAP indices, conduct multivariate regressions to identify and control for multiple influencing factors, and use advanced project evaluation methods when attempting to determine and measure the effectiveness of educational programs. Additionally, the results of previous KAP studies on animal diseases suggest that it is important to control for socioeconomic classes, rural and urban settings, education levels, age, and previous experience with animal diseases. We consider these variables in our KAP analysis in addition to information about beliefs and practices surrounding the management of sick or dead birds.

Description Data

Conceptual Framework

The approach taken in this study is adapted from the theoretical frameworks developed by Huang (1993) and Jolly et al. (2009). Their models for economic analysis and decision making take into consideration the psychological, social, and other noneconomic factors that guide decision makers' behavior. Huang's (1993) approach assumed that individuals' perceptions were formulated from available information, knowledge, experiences as well as personal, social, and cultural backgrounds. Jolly et al. (2009) extends Huang's approach and assumes that individuals' perception about the problem affects knowledge and awareness, and in turn develops an attitude that will promote action to minimize risks. In this study, we assume that individual's perception about disease spread in the village is influenced by socioeconomic and demographic factors as well as his knowledge and beliefs about HPAI, before any action is taken to minimize risks.

Distribution of Households Surveyed in Kenya

A total of 453 households were sampled in eastern, western, and central regions of Kenya. Of the households sampled there were 266 households with less than 50 birds and were free ranging (these are classified as "free range producers"), 93 households with 51–500 birds that are classified as "small-scale commercial producers," and 38 households with 501–1,000 birds that are classified as "medium scale commercial producers." There were also 56 households with 1,001–5,000 birds that are classified as "large scale producers." Table 15.2 summarizes the

Table 15.2 Distribution of broilers, layers, and indigenous chickens by flock size

	Free range (≤50 birds)	Small scale (51–500 birds)	Medium scale (501–1,000 birds)	Large scale (1,001–5,000 birds)	All house holds
Native chicken owners	257	38	7	1	303
Layer owners	6	50	28	32	116
Cock owners	27	13	2	0	42
Broiler owners	8	22	14	29	73
All households	266	93	38	56	453

Source: ILRI/ILRI/KIPPRA survey

Table 15.3 Average number of birds by production type in Kenya

All households	Free range (≤50 birds) ($n = 266$)	Small scale (51–500 birds) ($n = 93$)	Medium scale (501–1,000 birds) ($n = 38$)	Large scale (1,001–5,000 birds) ($n = 56$)	All households ($n = 453$)
Owns native chicken	97% (0.18)	41% (0.49)	18% (0.39)	2% (0.13)	67% (0.47)
Owns layers	2% (0.15)	54% (0.50)	74% (0.45)	57% (0.50)	26% (0.44)
Owns broilers	3% (0.17)	24% (0.43)	37% (0.49)	52% (0.50)	16% (0.37)
Owns cocks	10% (0.30)	14% (0.35)	5% (0.23)	0% (0.00)	9% (0.29)
Owns other birds	6% (0.24)	11% (0.31)	3% (0.16)	0% (0.00)	6% (0.24)
Total birds (number of birds)	16.69 (12.70)	237.70 (139.78)	758.29 (155.92)	4,091.88 (5,181.95)	628.05 (2,238.92)

Source: ILRI/ILRI/KIPPRA survey
Numbers in parenthesis in this row are standatd deviations.

distribution of the sampled households in terms of poultry production scale and poultry types.

Table 15.3 presents the variation in the poultry types in percentage terms. Nearly all (97%) of the free range producers raised local breeds, for which the average flock size was 14 birds (not shown in the table). Small scale and medium scale producers reported a larger number of layers than native birds and broilers. Large scale producer households were nearly evenly split between broilers and layers, though the average number of total head of broilers was larger.

Constructing Indices for Knowledge, Attitudes, and Perceptions (KAP) About HPAI

In the household survey a total of 40 questions on knowledge, attitudes, perception, and practices (KAP) were asked. These questions were grouped into five categories: knowledge, beliefs, actions, reporting, and perception. These questions were framed as dichotomous questions (yes/no) or multiple choice questions that allowed multiple answers. For example, questions on practices or actions taken in preventing or controlling disease outbreaks were structured as dichotomous choice so as to capture differences or common practices of households within the study area. A Likert scale was used to elicit risk perceptions. For each category of KAP questions, responses were scored by awarding 1 point for each acceptable or correct answer and 0 for each wrong answer, and then scores were summed by category and by household to come up with an index as described in Table 15.4.

We construct the following KAP indices:

1. Knowledge KAP: this index captures the producers' knowledge about typical HPAI symptoms.
2. Beliefs KAP: this index measures the producers' beliefs about good practices in handling poultry and poultry products.
3. Action KAPs: three action KAPs were created; one of handling sick birds, one on disposal of dead birds, and one on other practices.
4. Reporting sick birds: this is constructed as index of the bad practices HH denies doing.
5. Perception KAP: this index captures how concerned the producers are about potential disease spread when there is an infection within a village.

Estimating KAP Indexes

In conducting multivariate regression analyses of KAP indices, we focus on three of the seven KAP indices (listed in Table 15.4), and the analyses are conducted sequentially. First, we estimate what influences knowledge KAP; in particular we are interested in what types of producers have more knowledge about HPAI symptoms, which in turn reveal at least partially why some producers are more motivated to acquire such knowledge than others. Second, the determinants of beliefs KAP are estimated. In doing so, we maintain the hypothesis that beliefs about good practices are in part influenced by the knowledge about animal diseases. Thus, we include knowledge KAP index as an explanatory variable in the regression model. Third, the determinants of perception KAP are estimated. We hypothesize that perception KAP is influenced by knowledge and beliefs KAPs, and the latter are included in the regression model. Other explanatory variables included in estimation are discussed for each regression model below.

One of the questions we ask in the series of KAP analysis is whether and how the past experience with poultry disease affects the KAP index levels. However, there is a possibility that the disease experience and KAP levels are *endogenously* determined. In other words, past disease experience may affect KAP levels of a producer, but KAP levels may also have affected whether the producer's poultry had disease in the past. Because the presence of endogeneity can affect the statistical nature of the results, for each of the three regression models, we test for the

Table 15.4 Construction of KAP indices

Knowledge KAP	Given 1 point for each of the following symptoms identified with AI
	Sudden or unexpected death of healthy birds
	Ruffled feathers
	Minimal food intake
	Swollen or bluish comb
	Bloody diarrhea
	Difficulty breathing
	Reduction of ceasing of egg production
Beliefs on safe practices handling poultry and products	Given 1 point for each of the following bad practices HH denies doing
	Adults touch sick or dead poultry with bare hands
	Children in household touch sick or dead poultry
	Prepare raw poultry and other foods with the same cutting boards and utensils
	Use sick or dead poultry for meal preparation
	Eat poultry that is pink in the middle
	Eat eggs with runny yolk
Action KAP 1: Handling sick birds	Given 1 point for providing the response in parentheses to each of the following practices
	Slaughter sick fowl for food or gift (No)
	Burn or destroy sick fowl (Yes)
	Sell sick fowl (No)
	Give antibiotics to sick fowl (Yes)
	Apply vaccine to sick fowl (Yes)
Action KAP 2: Disposal of dead birds	Given 1 point for providing the response in parentheses to each of the following practices
	Consume or gift dead fowl (No)
	Sell dead fowl (No)
	Sell the drippings of dead fowl (No)
	Burn carcasses (Yes)
	Bury carcasses (Yes)
	Leave carcasses in open (No)
	Dispose carcasses in river/pond (No)
Action KAP 3: other practices	Given 1 point for providing the response in parentheses to each of the following risk mitigation practices HH reports doing
	Touch sick or dead poultry with bare hands (No)
	Wash hands with soap and water immediately after touching sick/dead poultry (Yes)
	Children in household touch sick or dead poultry with bare hands or otherwise play with them (No)
	Take sick or dead poultry and prepare it for a meal (No)
	Prepare raw poultry and other foods with the same cutting boards and utensils without washing (No)
	Wash hands with soap and water immediately after preparing poultry for cooking (Yes)
	Wash the cage/pen (Yes)
	Spray disinfectant in cage/pen (Yes)

(continued)

Table 15.4 (continued)

Perception on disease transmission risks	Given the equivalence of the following answers corresponding to the question: "If your neighbor told you that there are sick poultry in the village or nearby (while your birds are still healthy) how likely do you think it would be for your birds to get sick?" Completely or very likely = 4 (high risk), somewhat likely = 3 (moderate risk), somewhat unlikely = 2 (low risk), completely or very unlikely = 1 (no risk)
Reporting KAP	Given 1 point for each of the following answers Friend or neighbor Village head/village staff Local veterinary doctor District veterinary office/agricultural agency Agent/distributor If contact/report (immediately/minutes/hours less than 24 h)

Source: ILRI/ILRI/KIPPRA (2009)

endogeneity between disease experience and KAP index levels. Table 15.5 lists the definition and summary statistics of all variables used in the Kenyan empirical analyses by production scale.

Estimating the Determinants of Actual Biosecurity Actions; Results and Discussions of KAP Analysis

Knowledge KAP

We begin our empirical analysis with the estimation of the determinants of knowledge KAP index. Although the theoretical value of this index is between 0 and 7, the actual levels of the index for Kenya producers in the sample range between 0 and 3. Using the knowledge KAP index as the dependent variable, we run a count regression model (Poisson regression). We consider the dependent variable level as an outcome of three related but separate forces: (1) access to information, (2) ability to obtain and process information, and (3) eagerness to obtain information. Accordingly, we specify the regression model to include the following independent or explanatory variables:

Number of years of poultry production to capture the ability of producers to obtain and process information.

Household size to capture exposure of a larger household to a potentially greater number of information sources.

Gender of household head to capture potential difference between male-headed and female-headed households in the level of HPAI knowledge.

Education level of household head to capture the ability to obtain and process information.

Table 15.5 Definition and descriptive statistics of all variables used in the Kenyan empirical analyses

Variables		Free range (≤50 birds) ($n = 266$)	Small scale (51–500 birds) ($n = 93$)	Medium scale (501–1,000 birds) ($n = 38$)	Large scale (1,001–5,000 birds) ($n = 56$)	All households ($n = 453$)
Dependent variables						
Perception_kap	Index on perceptions about the level of concern about disease spread within a village	3.506 (0.697)	3.370 (0.822)	3.053 (0.899)	3.339 (0.721)	3.419 (0.754)
Beliefs_kap	Index on beliefs about good practices	6.733 (1.420)	7.140 (0.962)	7.079 (1.194)	7.411 (0.987)	6.929 (1.291)
Knowledge_kap	Index on knowledge on AI symptoms	0.414 (0.712)	0.462 (0.731)	0.395 (0.755)	0.679 (0.741)	0.455 (0.726)
Explanatory variables						
b1	number of years in poultry raising	12.578 (11.546)	10.482 (12.884)	10.014 (10.756)	11.907 (9.327)	11.793 (11.508)
Hhsize	Number of people in the household	5.981 (2.455)	5.086 (2.357)	4.605 (1.925)	5.036 (2.288)	5.565 (2.422)
Hhhgen	1 if head is female; 0 otherwise	0.774 (0.419)	0.828 (0.379)	0.895 (0.311)	0.911 (0.288)	0.812 (0.391)
Hhhedu	Head's education (years)	8.213 (4.297)	11.516 (3.966)	10.447 (4.979)	12.571 (3.103)	9.655 (4.501)
Hhhedu2	Head's education (years), squared	85.835 (68.959)	148.183 (93.366)	133.290 (87.459)	167.500 (77.556)	113.442 (83.983)
Child	1 if has child younger than 12; 0 otherwise	0.335 (0.473)	0.237 (0.427)	0.184 (0.393)	0.250 (0.437)	0.291 (0.455)
ln_totinc	Log of total household income	11.243 (1.662)	12.418 (1.637)	12.898 (1.850)	13.249 (1.769)	11.871 (1.856)
ln_poultry_total	Log of total poultry flock size	2.503 (0.858)	5.271 (0.667)	6.609 (0.220)	7.921 (0.794)	4.089 (2.174)
Dis_farm	Distance to the nearest poultry farms	0.342 (0.427)	0.420 (0.651)	0.676 (1.278)	0.649 (1.560)	0.418 (0.770)
Dis_ahealth	Distance to animal health shop	6.701 (7.912)	6.616 (9.472)	7.563 (7.750)	5.235 (5.783)	6.580 (8.024)
Layer_tot	1 if HH raised layer breed; 0 otherwise	0.023 (0.149)	0.538 (0.501)	0.737 (0.446)	0.571 (0.499)	0.256 (0.437)
Ifdisease	1 if already experienced poultry disease; 0 otherwise	0.885 (0.320)	0.791 (0.409)	0.658 (0.481)	0.782 (0.417)	0.831 (0.376)
Bioctc_rs		3.985 (2.115)	6.247 (2.677)	7.895 (2.275)	8.036 (2.149)	5.278 (2.786)

(continued)

Table 15.5 (continued)

Variables		Free range (≤50 birds) (n = 266)	Small scale (51–500 birds) (n = 93)	Medium scale (501–1,000 birds) (n = 38)	Large scale (1,001–5,000 birds) (n = 56)	All households (n = 453)
	number of actual biosecurity measures adopted at the time of interview					
Western	1 if western region; 0 otherwise	0.447 (0.498)	0.280 (0.451)	0.105 (0.311)	0.018 (0.134)	0.331 (0.471)
Eastern	1 if eastern region; 0 otherwise	0.523 (0.500)	0.108 (0.311)	0.026 (0.162)	0.000 (0.000)	0.331 (0.471)
Div_flock	Average flock size by division	91.789 (313.080)	968.796 (991.659)	1,584.395 (1,012.803)	1,956.839 (729.896)	627.603 (946.146)

Source: ILRI/ILRI/KIPPRA Survey (2009)
Note: Numbers in parentheses are standard deviations

Total household income to capture the variation in the means or resources to obtain available information.

Total poultry flock size to capture the importance of poultry production for the household, hence potential eagerness to obtain available information.

Distance to the nearest poultry farm and distance to animal health shop to capture the availability of and access to information regarding HPAI.

A dummy variable that indicates that the household raised layer breed in the past year to test whether layer owners have higher motivation for obtaining information as layer breeds have longer life cycle and thus these birds are a longer-term asset.

Past experience with poultry disease in the household's own flock to test whether it influences information acquisition.

Two dummy variables that indicate the western and eastern regions to capture variations in knowledge KAP index because these regions had producers of smaller size relative to those in the central region. We applied count model estimation (negative binomial and Poisson regressions). Since the statistical test rejects the null hypothesis of over-dispersion, the Poisson model is found to fit well with the data.

Subsequently, we tested for endogeneity of the past poultry disease experience and knowledge KAP by applying an endogenous switching model described in Miranda and Rabe-Hesketh (2006), where we hypothesized that those producers with previous disease experience had different response regarding knowledge KAP (i.e., previous disease experience is the switching variable). The result of the endogenous switching model indicates that the switching is indeed exogenous. The null hypothesis of over-dispersion was also rejected. Thus, we return to the original Poisson regression.

The column (1) of Table 15.6 lists the results of Poisson regression for knowledge KAP. The coefficients on the explanatory variables are interpreted as the contributions of the explanatory variables to a higher knowledge KAP score. A positive coefficient implies that a subject with a larger value of the independent (explanatory) variable has a higher knowledge KAP score. A negative coefficient implies that a subject with a larger value of the independent variable has a lower knowledge KAP score.

Although the overall predictive power of the estimation is relatively low (Pseudo-$R^2 = 0.0686$), there are some important findings. First, education level of household head exhibits nonlinear effect on knowledge KAP index: the number of years of education has positive impacts on the knowledge KAP score at an increasing rate. This implies that producers with higher levels of education are likely to be better informed and thus may be more aware of the symptoms of HPAI than those less educated. Second, we find that knowledge about HPAI symptoms is higher for households with larger poultry flocks, likely reflecting that owners of larger flocks are more motivated to acquire information about poultry diseases since more is at stake for these producers in poultry health management. Third, the regression results indicate that knowledge KAP is lower for those producers that have had poultry

Table 15.6 Determinants of Kenyan knowledge, belief, and perceptions

	(1)	(2)	(3)
	Poisson	Ordered probit	Ordered probit
Variables	Knowledge KAP	Beliefs KAP	Perception KAP
Index on knowledge on AI symptoms		0.114 (0.0834)	0.375[a] (0.0954)
Index on beliefs about good practices			0.00586 (0.0504)
Number of years in poultry raising	−0.00547 (0.00744)	0.00478 (0.00551)	0.00565 (0.00593)
Household size	0.0465 (0.0365)		
1 if head is female	−0.0911 (0.203)	−0.00647 (0.156)	−0.149 (0.172)
Head's education (years)		0.0255 (0.0420)	−0.0420 (0.0487)
Head's education (years), squared	0.00208[b] (0.000854)	−0.000452 (0.00224)	0.00199 (0.00260)
1 if has child younger than 12; 0 otherwise		−0.0748 (0.134)	0.437[a] (0.153)
Log of total household income	−0.0342 (0.0551)	0.0265 (0.0449)	−0.0617 (0.0442)
Log of total poultry flock size	0.194[a] (0.0647)	0.0958[c] (0.0495)	−0.0701 (0.0533)
Distance to the nearest poultry farms	0.0655 (0.0959)		0.193 (0.122)
Distance to animal health shop	0.00621 (0.0100)		
1 if HH raised layer breed	0.275 (0.228)		−0.330[c] (0.177)
1 if poultry disease in past 5 years	−0.492[b] (0.204)	−0.222 (0.166)	0.370[b] (0.169)
1 if western region; 0 otherwise	0.272 (0.347)	0.239 (0.232)	0.0341 (0.307)
1 if eastern region; 0 otherwise	1.435[a]	0.0479 (0.248)	−0.106 (0.336)
Average flock size by division			0.000178 (0.000117)
Constant	−1.996[b] (0.796)		
Cut1		−1.581[b] (0.688)	−2.652[a] (0.783)
Cut2		−1.302[c] (0.674)	−1.825[b] (0.772)
Cut3		−0.690 (0.664)	−0.734 (0.766)
Cut4		−0.291 (0.661)	
Cut5		0.365 (0.659)	
Cut6		0.999 (0.660)	
Observations	328	353	352
Log likelihood	−288.1	−479.9	−333.1
Pseudo-R-squared	0.0686	0.0250	0.0669
χ^2	42.43	24.65	47.74
p-Value	2.82e − 05	0.0103	2.80e − 05

Source: ILRI/ILRI/KIPPRA survey
Standard errors in parentheses
[a] $p < 0.01$
[b] $p < 0.05$
[c] $p < 0.10$

disease in their flocks in the past. Given the exogeneity of this variable and knowledge KAP, this likely implies that these producers experienced poultry disease in their flock due to factors other than a lack of knowledge about disease symptoms. Fourth, we find that knowledge KAP is higher in the eastern region relative to the central region. No such difference was found for western region.

Beliefs KAP

Next we estimate the determinants of beliefs KAP index, which characterize the number of "good practices" that the producers believe in regarding handling of poultry and poultry products. Because many of the items in the list of practices pertain to those as consumers of poultry products, we also include relevant household characteristics as explanatory variables in the regression. The explanatory variables included in the estimation model are the following:

Knowledge KAP index to test whether the knowledge about HPAI symptoms influences beliefs about good practices.

Number of years of poultry production to capture the ability of producers to identify good practices.

Gender of household head to capture potential difference between male-headed and female-headed households in the beliefs about good practices.

Education of household head to capture the ability to obtain and assess information about good practices.

Dummy variable of whether the household has a child to capture potential impact of child presence on the household's beliefs about good practices.

Total household income to capture the variation in the means or resources to obtain available information.

Total poultry flock size to capture the importance of poultry production for the household, hence the eagerness to form correct beliefs about practices and implement good practices.

Past experience with poultry disease in the household's own flock to test whether and how it is associated with forming correct beliefs about good practices.

Two dummy variables that indicate the western and eastern regions with producers of smaller scale relative to those in the central region; it is expected that there may be some factors that are specific to these regions that cannot be captured by the other explanatory variables such as general and marketing infrastructure.

We started with applying count model estimation (negative binomial and Poisson regressions) and found that the models were statistically insignificant in that all coefficients were not collectively significantly different from zero. We then implemented ordered probit regression. The endogeneity of beliefs KAP and the past poultry disease experience was again tested and rejected. Thus, the result of standard ordered probit regression are presented here. The coefficients on the explanatory variables are interpreted as the contributions of the variables to the probability of falling in a "higher" bin or category (in this case the number of correct beliefs about good practices). A positive coefficient implies a larger probability that a subject with a larger value of the independent variable will be observed in a higher bin or category. A negative coefficient implies that a subject with a larger value of the independent variable is likely to be observed in a lower bin or category. The "cuts" in the table are interpreted as the cutoff points between the bins: cut1 is a cutoff point between beliefs KAP = 0 and beliefs KAP = 1, cut2 between beliefs KAP = 1 and beliefs KAP = 2, and so on. For example, the probability that beliefs KAP falls in the bin "beliefs KAP = 0" is denoted as Pr(beliefs KAP = 0) = Pr(Xb + u < cut1), where Xb represents the linear combination of explanatory variables and u is the error term.

The column (2) of Table 15.6 lists the results of ordered probit regression for beliefs KAP. Again the predictive power of the estimation is low (Pseudo-$R^2 = 0.0250$), but the model is significant in that the all the coefficients are collectively significantly different from zero (p-value = 0.0103). The only significant variable in this regression model is flock size; the result suggests that the poultry flock size and beliefs KAP are positively associated. This may reflect the hypothesis that poultry production is relatively more important for these producers and they tend to have higher incentives to be aware of good practices and to form correct beliefs. The knowledge KAP index is found to have no influence on the beliefs KAP index.

Perception KAP

Finally, we estimate the determinants of perception KAP index, which is a categorical variable that takes the value of 1 when the producer is least concerned about disease spread within a village when there is a disease case in the village and the value of 4 when the producer is most concerned. We consider that the level of concern about disease spread within a village is an outcome of how correctly and rationally the producers can assess the risk of disease spread as well as the circumstances in which the producers operate. We estimate an ordered probit model with the following explanatory variables:

Knowledge KAP index to test whether and how the producers' knowledge about HPAI symptoms influences the level of concern about disease spread.

Beliefs KAP index to test whether and how the producers' beliefs about good practices are associated with the level of concern about disease spread.

Number of years of poultry production to capture how the experience with poultry production affects disease risk perception.

Gender of household head to capture potential difference between male-headed and female-headed households in the perception about disease spread risks.

Education of household head to control for the ability to make rational judgment about the risks of disease spread.

Dummy variable of whether the household has a child to capture potential impact of child presence on the household's assessment of disease spread risks.

Total household income to control for the potential variation in risk perception formation due to varied financial ability to cope with disease and other risks.
Total poultry flock size to capture the importance of poultry production for the household, which likely affects the way in which the risk of disease spread is assessed.
Distance to the nearest poultry farm to test if geographically more isolated poultry producers have lower risk perception.
A dummy variable that indicates that the household raised layer breed in the past year to test whether layer owners form risk perception differently.
Past experience with poultry disease in the household's own flock to test whether and how it influences the way the disease spread risks are assessed.
Two dummy variables that indicate the western and eastern regions because these regions had producers of smaller size relative to those in the central region.
Average flock size at the village level to test whether producers in a village dominated by small producers perceive higher disease spread risks.

The column (3) of Table 15.6 lists the results of ordered probit regression for perception KAP. Again, endogeneity between perception KAP and past disease experience was rejected. Several important findings emerge from the regression results. First, knowledge KAP index is significantly and positively associated with perception KAP. This implies that those producers with higher knowledge about HPAI symptoms are found to have *higher* concerns about disease spread risks. If we were to posit that those producers with higher knowledge score are able to make more informed and rational assessment of true disease spread risks, the regression results indicate that less informed producers tend to *underestimate* the risks of disease spread should a disease occur in a village. Beliefs KAP index has no significant influence on risk perception among the Kenyan poultry producers. Second, producers with young children in the households or with experience with poultry diseases are more concerned with disease spread within a village. On the contrary, the concerns for disease spread are lower among producers with layer breeds.

Estimating the Determinants of Actual Biosecurity Actions

To conclude the KAP analysis, we estimate the determinants of actual biosecurity practices currently adopted by the producers. Using the responses to the questions regarding the biosecurity measures, we implement two regressions. First, we construct a count variable that represents the total number of biosecurity actions among the 11 that the producers currently implement. Using the count variable as the dependent variable, a count model is estimated. Second, by stacking the binary (yes/no) responses to all of the 11 biosecurity actions, we implement panel-data probit (random-effects probit) to estimate what influences the probability that the producers adopt the 11 biosecurity practices. In this regression, each producer has 11 observations.

The two estimation models include the same set of explanatory variables. First, all three KAP indices are included in the estimation model to analyze how knowledge, beliefs, and perception about disease and disease risks influence actual biosecurity decisions. We also include all the other variables used in the empirical KAP analysis in the estimation models. For the random-effects probit estimation, we also include dummy variables representing each of the biosecurity actions.

For the count model specification, a Poisson regression was implemented. Because of difficulties in convergence, test for over-dispersion or endogeneity of past disease experience could not be conducted. The test for endogeneity was not conducted for panel-data probit model. Thus caution is warranted in interpreting both the count model and the panel-data probit results.

The regression results are listed in Table 15.7. The qualitative results (i.e., signs of the estimated coefficients) are almost identical between Poisson regression and the random-effects probit. One important finding from these regressions is that KAP indices are found to be important in explaining the actual biosecurity decisions of the Kenyan producers. Beliefs KAP index is significantly and positively associated with the number and probabilities of biosecurity measures taken. This indicates that those who believe in good practices in handling poultry and poultry products are more motivated to take actual biosecurity measures. But then, the perception KAP index is significantly and negatively associated with biosecurity behavior; those who perceive higher risk of disease transmission actually implement fewer measures or with lower probabilities. This may indicate that when a producer "correctly" perceives a higher disease spread risk, s/he may also form a perception that biosecurity actions are ineffective in mitigating the risks, resulting in lower actual biosecurity behavior. The former result for beliefs KAP may imply that household education and outreach program may be effective in influencing household biosecurity behavior, while the latter result for perception KAP indicates that efforts to influence risk perception formation, as identified in regression (3) in Table 15.6, may not bring about desired impacts on biosecurity adoption in Kenya.

As was expected, the biosecurity adoption levels and probability are positively associated with poultry flock size. The distance to the nearest poultry farm is negatively associated with household biosecurity behavior in the count model, possibly indicating that geographical isolation lowers subjective probability of own flock getting infected, which is slightly different from perceived disease spread risk within a village captured in perception KAP. In the panel-data probit regression, the estimated coefficients on the average flock size by division is positive and significant, indicating that in villages that have, on average smaller producers biosecurity probabilities are also lower. The latter effect may arise because the perceived effectiveness of biosecurity measures is lower in those villages if smaller producers are considered to be riskier in terms of disease transmission. Finally, we find that producers in the western and eastern region implement fewer biosecurity actions or adopt biosecurity actions with lower probabilities compared to producers from the central region. This is likely because western and eastern regions are dominated with free-range and small backyard poultry compared with the central region where one finds commercial poultry producers.

Table 15.7 Estimation results of actual biosecurity actions amongst poultry producers in Kenya

		(4)	(5)
Method		Poisson	Random-effects probit
Variables	Labels	Bioctc	Bioaction
Knowledge_kap	Index on knowledge on AI symptoms	0.0396 (0.0355)	0.101 (0.0648)
Beliefs_kap	Index on beliefs about good practices	0.0407[a] (0.0216)	0.0670[a] (0.0363)
Perception_kap	Index on perception of spread	−0.0762[b] (0.0318)	−0.154[b] (0.0614)
B1	Number of years in poultry raising	−0.000539 (0.00226)	0.00354 (0.00421)
Hhsize	Household size	0.0194 (0.0121)	0.0291 (0.0217)
Hhhgen	1 if head is female	0.0153 (0.0671)	−0.00858 (0.117)
Hhhedu	Head's education (years)	0.00468 (0.0193)	0.0509 (0.0326)
Hhhedu2	Head's education (years), squared	−0.000272 (0.000977)	−0.00289[a] (0.00173)
Child	1 if has child younger than 12; 0 otherwise	−0.0165 (0.0589)	−0.0211 (0.106)
ln_totinc	Log of total household income	0.0184 (0.0161)	0.0693[b] (0.0337)
ln_total_poultry	Log of total poultry flock size	0.0329[a] (0.0191)	0.0729[a] (0.0386)
Dis_farm	Distance to the nearest poultry farms	−0.0599[a] (0.0321)	0.0833 (0.0796)
Dis_ahealth	Distance to animal health shop	0.00442 (0.00294)	0.00526 (0.00557)
Layer	1 if HH raised layer breed	−0.0147 (0.0617)	−0.0622 (0.138)
Ifdisease	1 if poultry disease in past 5 years	0.00107 (0.0611)	0.0166 (0.125)
Western	1 if western region; 0 otherwise	−0.203[a] (0.109)	−0.593[b] (0.232)
Eastern	1 if eastern region; 0 otherwise	−0.735[c] (0.125)	−1.483[c] (0.254)
Div_flock	Average flock size by division	1.09e − 05 (3.99e − 05)	0.000202[b] (9.27e − 05)
Constant		1.388[c] (0.324)	−3.096[c] (0.646)
Biosecurity action 1	Closed doors in poultry house all the time		2.152[c] (0.176)
Biosecurity action 2	Checked poultry house daily for dead or sick birds		3.994[c] (0.206)
Biosecurity action 3	Kept same poultry cage during the outbreak in village		2.887[c] (0.184)
Biosecurity action 4	Quarantined new purchased poultry		2.720[c] (0.185)
Biosecurity action 5			3.780[c] (0.198)

(continued)

Table 15.7 (continued)

Method		(4) Poisson	(5) Random-effects probit
Variables	Labels	Bioctc	Bioaction
	Checked the symptoms of diseases before purchase		
Biosecurity action 6	Conducted all in and all out method for each type of poultry		2.364[c] (0.181)
Biosecurity action 7	Monitored contact between your and neighbors' poultry		1.977[c] (0.183)
Biosecurity action 8	Monitored contact between your and wild poultry		1.034[c] (0.176)
Biosecurity action 9			2.965[c] (0.181)
Biosecurity action 11	Frequently cleaned floors and cages from feces		1.773[c] (0.177)
Observations		327	3,070
Log likelihood		−698.0	−1,240
R-squared		0.1642	
χ^2		274.2	713.3
p-Value		0	0

Standard errors in parentheses
[a]$p < 0.1$
[b]$p < 0.05$
[c]$p < 0.01$

Summary and Conclusions

Analyses of the three KAP indices on knowledge about HPAI symptoms (knowledge KAP), beliefs about safe practices handling poultry and products (beliefs KAP), and perception on disease risk transmission (perception KAP) reveal their important determinants. For knowledge KAP, important factors that contribute to higher level of knowledge about HPAI symptoms are education level of household head, total flock size, if the household had a poultry disease in the past, and if location of poultry farm is in the eastern region. For beliefs KAP, larger flock size significantly influences individual's correct belief about safe handling practices. For perception KAP, those producers with higher knowledge about HPAI symptoms and with an experience of poultry disease in the past are found to have higher concerns about disease transmission risks.

The role of knowledge, beliefs, and perception about disease and disease risks in actual decisions regarding biosecurity practices by the producers are also investigated. We find that the actual biosecurity actions are influenced by beliefs and perception KAP indices. We also find that smaller and poorer producers adopt fewer biosecurity actions, thus they are considered to be riskier in terms of infection and transmission risks. The biosecurity adoption levels and probability are positively associated with household income (but not significantly in the Poisson model).

The findings generated in this study are important for policy makers as they formulate effective strategies to prevent and control disease outbreaks. For example, information in this study will help decision makers in the proper allocation of funds in improving poor individuals' knowledge about the nature of animal diseases and how to prevent and control them, which eventually would result to significant changes in behavior such as adoption of biosecurity measures. The role of other socio-economic factors including marketing practices and infrastructure facilities need to be explored to obtain a better understanding of the determinants of perceptions about disease transmission risks.

References

Di Giuseppe, G., R. Abbate, L. Albano., P. Marinelli, and I. F. Angelillo. 2008. A Survey of Knowledge, Attitudes and Practices towards Avian Influenza in an Adult Population of Italy. *BMC Infectious Diseases* 8(36).

Fielding, R., W. W. T. Lam, E. Y. Y. Ho, T. H. Lam, A. J. Hedly, and G. M. Leung. 2005. Avian Influenza Risk Perception, Hong Kong. *Emerging Infectious Diseases* 11(5) 677–682.

Huang, C.L. 1993. Simultaneous Equation Model for Estimating Consumer Risk Perceptions, Attitudes, and Willingness to Pay for residue-free Produce. The Journal of Consumer Affairs. Vol. 27, No. 2, pp.377–396.

IFPRI/ILRI/KIPPRA Survey 2009. A Household survey conducted for HPAI DfID Project. IFPRI, Washington, D.C.

Imai, T., K. Takahashi, T. Hoshuyama, N. Hasegawa, M. Lim, D. Koh. 2005. SARS Risk Perceptions in Healthcare Workers, Japan. *Emerging Infectious Diseases* 11(3): 404–410.

Jolly, C. M., B. Bayarda, R. T. Awuahb, S. C. Fialor, J. T. Williams. 2009. Examining the structure of awareness and perceptions of groundnut aflatoxin among Ghanaian health and agricultural professionals and its influence on their actions. The Journal of Socio-Economics 38 (2009) 280–287.

Kimani, T., Obwayo, M., Muthui., L and (2006). Avian Flu threat: Socio-Economic Assessment of the impacts on Poultry- Related Livelihoods in Selected Districts in Kenya. Pan-African Programme for the Control of Epizootics (PACE), Nairobi, 106pp.

Kumar, G., Popat, M. (2010). Farmers' perceptions, knowledge and management of aflatoxins in groundnuts (Arachis hypogaea) in India. Crop Production 29: 1534–1541.

Lau, J. T. F., J. H. Kim, H. Tsui, and S. Griffins. 2007. Perceptions related to human avian influenza and their associations with anticipated psychological and behavioral responses at the onset of outbreak in the Hong Kong Chinese general population. *American Journal of Infection Control*. 35(1): 38–49

Leggat, P. A., D. Mills, and R. Speare. 2007. Hostellers' Knowledge of Transmission and Prevention of Avian Influenza When Traveling Abroad. *Travel Medicine and Infectious Disease* 5: 53–56.

Leslie, T., J. Billaud, J. Mofleh, L. Mustafa, and S. Yingst. 2008. Knowledge, Attitudes, and Practices regarding Avian Influenza (H5N1), Afghanistan. *Emerging Infectious Diseases* 14(9): 1459–1461.

Liebenehm, S., H. Affognon, and H. Waibel. 2009. Assessing the Impact of Agricultural Research on Farmers' Knowledge About African Animal Trypanosomosis: An Application of the Propensity Score Matching Approach. Paper delivered at conference GEWISOLA, Institute of Development and Agricultural Economics, Kiel, Germany, 30 November 2009.

Ly, S., M. D. Van Kerkhove, D. Holl, Y. Froehlich, and S. Vong. 2007. Interaction Between Humans and Poultry, Rural Cambodia. Emerging Infectious Diseases 13(1): 130–132.

Mahmoodabad, S. S. M., A. Barkhordari, M. Nabizadeh, and J. Ayatollahi. 2008. The Effect of Health Education on Knowledge, Attitude and Practice (KAP) of High School Students' Towards Brucellosis in Yazd. *World Applied Sciences Journal* 5(4): 522–524.

Ministry of Livestock and Fisheries Development (MOLFD) 2003. Department of Livestock Production Annual Report: 2003. Nairobi, Kenya: Republic of Kenya Ministry of Livestock and Fisheries Development.

Ministry of Livestock and Fisheries Development (MOLFD) 2004. Annual Report: 2004. Nairobi, Kenya: Republic of Kenya Ministry of Livestock and Fisheries Development.

Ministry of Livestock and Fisheries Development (MOLFD) 2006. Livestock Annual Report: 2006. Nairobi, Kenya: Republic of Kenya Ministry of Livestock and Fisheries Development.

Miranda, A. and S. Rabe-Hesketh. 2006. Maximum likelihood estimation of endogenous switching and sample selection models for binary, ordinal, and count variables. *Stata Journal* 6(3): 285–308.

Narrod, C.; M. Tiongco, M. Kobayashi, R.Scott, A. Saak, and J. Omiti 2010 Evaluating Risk Management Options to Reduce the Risk of HPAI for Kenya's Poor. Draft Report for HPAI DfID Project.

Ndegwa J.M. and Kimani C.W. 1996. Rural Poultry Production in Kenya: Research and Development Strategies. In: Fungoh, D.O. and G.C.O. Mbandi (Eds). Focus on Agriculture for Sustainable Development in a Changing Economic Environment. Proceedings of the 5th KARI Scientific Conference. 14–16 October 1996. KARI Headquarters, Nairobi, Kenya.

Nyange, R.K. 1995. Poultry development in Kenya. In: African network for rural poultry Development Workshop, 13–16th June, 1995. Addis Ababa, Ethiopia: 31–35.

Okello, J.J., Z. Gitonga, J. Mutune, R. Mutuli, M. Afande, and K.M. Rich. 2010. Value Chain Analysis of the Kenya Poultry Industry: The Case of Kiambu, Kilifi, Vihiga, and Nakuru Districts. Pro-Poor HPAI Risk Reduction Strategy Paper. Washington, D.C. and Nairobi, Kenya: International Food Policy Research Institute and International Livestock Research Institute.

Olsen, S.J., Y. Laosiritaworn, S. Pattanasin, P. Prapasiri, S. F. Dowell. 2005. Poultry-Handling Practices During Avian Influenza Outbreak, Thailand. *Emerging Infectious Diseases* 11(10) 1601–1603.

Omiti, J.M. and Okuthe, S.O. 2008. An Overview of the Poultry Sector and Status of Highly Pathogenic Avian Influenza (HPAI) in Kenya – Background Paper. Pro-Poor HPAI Risk Reduction Strategy Paper. Washington, D.C. and Nairobi, Kenya: International Food Policy Research Institute and International Livestock Research Institute.

Suphunnakul, P., and T. Maton. 2009. Community Participation as a key element in Prevention and Control of Avian Influenza in Song Phi Nong District, Suphan Buri Province. Journal of Public Health 39(1): 61–73.

UNICEF-Georgia. 2007. Study of Knowledge, Attitudes, Practices, and Behaviors to Inform the Avian Influenza Prevention and Containment Communication Strategy in Georgia. Georgia: United Nations International Children's Emergency Fund. pp. 201.

UNICEF-Myanmar. 2006. Knowledge-Attitudes-Practices (KAP) Study on Poultry Rearing and Other Practices Pertaining to Avian Influenza. Myanmar: United Nations International Children's Emergency Fund. pp. 65.

Xiang, N., Shi, Y., Wu, J., Zhang, S., Ye, M., Peng, Z., Zhou, L., Zhou, H., Liao, Q., Huai, Y., Li, L., Yu, Z., Cheng, X., Su, W., Wu, X., Ma, H., Lu, J., McFarland, J., Yu, H. 2010. Knowledge, attitudes, and practices (KAP) relating to avian influenza in urban and rural areas of China. *BMC Infectious Disease* 10(34): 1–7.

Chapter 16
Controlling Animal Disease in Africa

Karl M. Rich and Brian D. Perry

Introduction

Africa presents a number of unique challenges in the field of animal health, which distinguishes the continent from many other regions of the world. Africa is home to a diverse range of agro-ecological and production systems, with significant interactions between them that are mediated by several elements: the movements of wildlife and pastoralist cattle; the endemic presence of disease vectors such as ticks, flies, and mosquitoes; the variability in climate that can accentuate conditions favorable for disease spread; and the contrasting market relationships and interactions between smallholder and commercial systems alike. Moreover, there is significant heterogeneity in the capacity, resources, and incentives of actors within the different livestock value chains, including producers, traders, market agents, processors, retailers, and support services (including government), to mitigate disease which, given these ecological and market interactions, further complicates effective disease control efforts by the public and private sectors (Rich and Perry 2011a). Declining public budgets allocated to animal health and often erratic donor priorities toward specific diseases muddle the situation even more (Winter-Nelson and Rich 2008).

The scope of animal diseases in Africa is wide-ranging. In 2008, 105 diseases were reported to the Africa Union-Inter-African Bureau for Animal Resources

K.M. Rich (✉)
Department of International Economics, Norwegian Institute of International Affairs (NUPI), P.O. Box 8159 Dep., 0033 Oslo, Norway

International Livertock Research Institute, Nairobi, Kenya

B.D. Perry
Nuffield Department of Clinical Medicine, University of Oxford, Oxford, UK

College of Medicine and Veterinary Medicine, University of Edinburgh, Edinburgh, UK

Department of Veterinary Tropical Diseases, University of Pretoria, Pretoria, South Africa

D. Zilberman et al. (eds.), *Health and Animal Agriculture in Developing Countries*, Natural Resource Management and Policy 36, DOI 10.1007/978-1-4419-7077-0_16,

(AU-IBAR) from 44 of AU-IBAR's 53 member countries. Nine of the reported 105 diseases were zoonotic, including anthrax, avian influenza, brucellosis, cysticercosis, rabies, Rift Valley fever, and tuberculosis (AU-IBAR 2009). Another 12 diseases were transboundary in nature (the presence of which and means of control can often have severe implications on international market access), including foot-and-mouth disease (FMD), contagious bovine pleuropneumonia (CBPP), African swine fever (ASF), Newcastle disease (ND), and *Peste des Petits ruminants* (PPR). Indeed, the presence of endemic disease can limit the incentives for value chain actors within Africa to invest heavily in livestock, despite significant research that has shown that livestock can represent an important pathway out of poverty for smallholders (Perry et al. 2002a, 2003, 2005). More fundamentally, the constant control of, and losses from, animal diseases impose severe burdens on resource-constrained countries within Africa. For instance, Kristjanson et al. (1999) estimated that the direct costs to producers and consumers from trypanosomosis, a disease spread by tsetse flies, were over US$1.3 billion per year, and this did not include any indirect costs in related markets (e.g., by-products such as manure). Musisi et al. (2004) cited a figure of US$2 billion per year related to the direct and indirect costs of CBPP in Africa, and specifically notes that the CBPP outbreak in Botswana in 1995 imposed losses of over US$500 million due to the slaughter of 320,000 cattle and other ancillary costs within the value chain.

The ability to control animal diseases depends on the technological, infrastructural, and institutional tools available to veterinarians and producers. The control of rinderpest from a technical standpoint (e.g., vaccination), for example, is much simpler than the control of FMD, as there is only one serotype of rinderpest, it does not have a carrier state in nature, and the vaccine employed is both highly effective and heat thermostable, allowing it to be administered in a host of difficult conditions (Roeder and Rich 2009). The control of transboundary diseases often relies on the use of movement controls, which require not only good management of veterinary services but also well-maintained infrastructure in the form of fences and other barriers that impede animal contacts. Conversely, the use of fencing itself can be controversial in terms of its impact on ecological and wildlife habitats (Thomson 2008), not to mention the *de jure* separation of production systems behind fences that limits opportunities for those producers on the "wrong" side of the fence (Perry et al. 2003; Rich and Perry 2011b). Public veterinary services are a critical dimension in the control of animal disease as well, and further play animportant accreditation role (in partnership with the OIE, the World Organization for Animal Health) in international circles to report the presence of disease and to certify disease freedom. However, in the wake of structural adjustment policies (SAPs) implemented in the 1980s and 1990s in Africa, public budgets for veterinary services have come increasingly under pressure and declined over the past couple of decades, placing a larger burden on the private sector to fill the void (Leonard et al. 1999). At the same time, innovative programs to improve service delivery and surveillance such as the use of community animal health workers (CAHW) and participatory epidemiology (PE), respectively, provide potential means to bridge existing gaps, although their role vis-à-vis traditional mechanisms (and actors) in the veterinary public health sphere has not always been fully demarcated (Scoones and Wolmer 2006).

In this chapter, we analyze the different types of control strategies that have been implemented in Africa to manage various types of animal diseases. We start by reviewing the major animal diseases confronting African producers and some of the different challenges they impose. We next review specific types of strategies that have been employed in Africa. Three characteristics of strategies will be dissected: technical aspects, infrastructural elements, and institutional capacity, drawing on different case studies to illustrate the application of strategies in the field. From this analysis, some emerging issues and challenges will be elucidated in light of changing demands on veterinary services resulting from globalization and rising standards for livestock products.

Overview of Major Livestock Diseases in Africa

Africa is a home to a large number of diverse animal diseases. In Table 16.1, we summarize some of the more important diseases found in Africa, although we caution that this list is not at all exhaustive and omits a number of important diseases (e.g., rabies, brucellosis) that impact African producers and consumers alike. Table 16.1 also omits rinderpest, which has now been officially declared as eradicated (Roeder and Rich 2009). However, the discussion that follows will provide some insights from the eradication of rinderpest that may (or, in some instances, may not) have applicability in the context of controlling other animal diseases. In Table 16.2, we provide some information on the perception of certain diseases in different parts of Africa based on rankings made in collaboration with stakeholders.

The majority of diseases found in Table 16.1 are endemic throughout Africa, with outbreaks fuelled by a combination of animal movements, insect vectors, ecology, and climatic fluctuations. Not surprisingly, the highest numbers of cases are found in rural smallholder production of poultry, pigs, and ruminants. ND of poultry is particularly widespread throughout Africa, affecting 31 of the reporting 44 countries in the AU-IBAR (2009) database, with PPR (over 244,000 cases) affecting 19 countries, and ASF (with nearly 200,000 cases) in 18 countries (AU-IBAR 2009). Diseases such as FMD and LSD are similarly widespread in terms of their geographic scope (23 and 29 countries, respectively), but the depth of cases is less, owing to fewer number of cattle kept (the species most impacted by these diseases) vis-à-vis small-stock. Many of these diseases are characterized by high levels of mortality, which in some production systems can impose significant losses on smallholder producers in particular. AHS, ASF, East Coast fever (ECF), HPAI, ND, and rinderpest are particularly noteworthy in this regard. For instance, while AHS is regularly associated with disruptions in commercialized horse racing and showing in South Africa, it affects large numbers of rural horses and other equines in Ethiopia and Eritrea (which have a population of over eight million equines), so

Table 16.1 Subset of major animal diseases currently found in Africa

Disease	Species affected	Mode of transmission	Degree of prevalence in SSA	Main effects of disease	Control strategies employed	Zoonosis?	Implications on international trade
African horse sickness	Equines (horses, mules)	Midges; zebras act as host reservoir	Endemic in western, eastern and southern Africa; main cases in 2008 in Ethiopia (319 outbreaks) and S. Africa (401)	Mortality (50–95% in horses); 47% case fatality in 2008 in reporting countries	Vaccination, but complicated by nine disease serotypes	No	Bans on movements of horses from infected areas
African swine fever	Pigs	Viral: animal contacts, soft ticks, infected feed	High throughout SSA; nearly 200,000 cases reported in 2008	High mortality depending on virulence; 50% mortality rate in 2008 in reporting countries	Stamping out and disinfection of infected premises; no vaccine available	No	
Anthrax	Cattle, sheep, pigs	Bacterial	Widespread throughout SSA. 31,591 cases reported in 2008, of which 27,610 in Ethiopia	Very high case fatality	Vaccination	Yes	
CBPP	Cattle	Bacterial: animal contacts, particularly with carrier (recovered but still infected) animals	High, particularly in West and Central Africa. 47,405 cases reported in 2008	High morbidity in infected animals, disruption of lactation, 29% mortality rate in 2008 in reporting countries	Vaccination (short-duration immunity and some complications); stamping out, movement control, zonation. Misuse of antibiotic treatments occurs	No	Movement bans of animals from infected regions

East coast fever (Theilerosis)	Cattle, buffalo	Tick-borne	Mainly found in Southern and Eastern Africa. 37,584 cases reported in 2008	Mortality in unexposed animals, morbidity/ carrier state in surviving animals	Acarcides to control ticks; live vaccines in limited use	No	Movement controls on animals
FMD	Cattle, sheep, goats, pigs	Viral: Animal contacts, fomites, infected feed. Buffaloes act as asymptomatic host reservoir	Endemic in most of Africa except for cordoned off FMD-free zones in Namibia, Botswana, S. Africa, and Swaziland. 46,562 cases reported in 2008	High morbidity in infected animals, mastitis in dairy animals. Can cause mortality in young (<1 year) animals. Less than 2% mortality rate in 2008 in reporting countries	Movement control, vaccination, contact slaughter and stamping out mainly in commercial settings	No	Segmentation of international meat (beef) markets based on FMD status (free without vaccination, free with vaccination, endemic). Higher prices and greater market access to FMD-free countries
HPAI	Poultry, wild birds	Viral: bird contacts, ducks as possible reservoir	Sporadic in SSA. Only Togo reported cases in 2008 to AU-IBAR. Endemic in Egypt. Past cases found in Ghana, Nigeria	High mortality in infected birds	Mainly stamping out and vaccination	Yes, WHO has reported 34 human deaths in Egypt to date	Temporary trade bans imposed on affected countries in some instances
Lumpy skin disease (LSD)	Cattle, buffalo	Biting flies and ticks	Widespread throughout Africa, particularly Southern and	High morbidity in infected animals, including reduced milk	Mainly vaccination	No	Import restrictions, movement controls

(continued)

Table 16.1 (continued)

Disease	Species affected	Mode of transmission	Degree of prevalence in SSA	Main effects of disease	Control strategies employed	Zoonosis?	Implications on international trade
			Eastern Africa. 66,314 cases reported in 2008	production, damage to hides, and abortions			
Newcastle disease	Poultry	Viral: bird contacts	Widespread throughout Africa. Over 594,000 cases in 31 African countries in 2008	Periodic epidemics with high mortality. 51% mortality rate in 2008 in reporting countries. Morbidity in surviving chickens (reduced egg production)	Vaccination, stamping out, and quarantine	No	Import restrictions, movement controls
Peste des Petits Ruminants (PPR)	Goats, sheep	Viral: animal contacts	Endemic throughout Western, Central, and Eastern Africa in particular. 244,054 outbreaks in 2008	Morbidity and mortality, particularly in young animals	Mainly vaccination, occasionally stamping out	No	

Rift valley fever	Cattle, sheep, goats, buffaloes	Mosquitoes	Epidemics usually associated with abnormal climatic conditions (El Niño). 1,155 cases reported in 2008. Major outbreak occurred in Kenya/Tanzania in 2006/2007. Recent outbreak reported in S. Africa in 2010	High mortality in young animals, high levels of abortions in pregnant animals (up to 85%)	Movement controls, vaccination, stamping out	Yes, 148 deaths associated with 2006/2007 outbreak in Kenya; 87 deaths in 2010 S. Africa	Trade bans imposed by trading partners; can be lengthy as witnessed by Saudi bans of livestock products from Horn of Africa for 18 months, 1998–2000
Trypanosomiasis	Cattle, sheep, goats, pigs, horses, camels	Tsetse flies	Widespread throughout sub-Saharan Africa. Nearly 118,000 cases reported in 2008	Mainly morbidity in cattle, with mortality possible if untreated	Vector control of tsetse flies, use of preventive and therapeutic medications	Yes, as humans can acquire certain types from tsetse flies	

Sources: Umali et al. (1994); OIE Technical Disease Cards (http://www.oie.int/eng/maladies/en_technical_diseasecards.htm); DEFRA A–Z Index of Diseases (http://www.defra.gov.uk/foodfarm/farmanimal/diseases/atoz/index.htm); AU-IBAR (2009)

Table 16.2 Top twenty diseases/conditions ranked based on their impact on the poor in Africa and by production system in the developing world (alphabetically listed by ranking group)

	West Africa (poverty impacts)	Eastern/Central/Southern Africa (poverty impacts)	Pastoral systems	Mixed crop-livestock systems	Peri-urban systems
Top 10	Anthrax	ECF	CBPP	Anthrax	Coccidiosis
	Blackleg	Ectoparasites	Ectoparasites	Ectoparasites	Ectoparasites
	CBPP	GI parasitism	GI parasitism	FMD	FMD
	Dermatophilosis	Haemonchosis	Haemonchosis	GI parasitism	GI parasitism
	Ectoparasites	Infectious coryza	Neonatal mortality	Liver fluke	Haemonchosis
	Gastro-intestinal (GI) parasitism	ND	Nutritional/ micronutrient deficiencies	Neonatal mortality	Infected coryza
	Heartwater	Neonatal mortality	Respiratory complexes	Neonatal disease virus (NDV)	Neonatal mortality
	Liver fluke	Nutritional/micronutrient deficiencies	RVF	Nutritional/ micronutrient deficiencies	ND
	Respiratory complexes	Respiratory complexes	Trypanosamaevansi	Reproductive disorders	Nutritional/ micronutrient deficiencies
	Trypanosomosis	RVF	Trypanosomosis	Toxocaravitulorum	Respiratory complexes
Next 10	Anaplasmosis	Babesiosis	Anthrax	Brucella abortus	Anthrax
	Brucellosis	CBPP	CCPP	Coccidiosis	Fowl cholera
	Contagious caprinepleuro-pneumonia (CCPP)	Coccidiosis	Foot problems	Haemonchosis	Fowl pox
	FMD	Foot problems	Heartwater	Haemorrhagic septicaemia	Foot problems
	Foot problems	Fowl pox	Liver fluke	Mastitis	Heartwater
	Haemorrhagic septicaemia (HS)	Heartwater	Mange	PPR	Hog cholera
	ND	Liver fluke	ND	Respiratory complexes	PPR
	PPR	Reproductive disorders	PPR	Rinderpest	Reproductive disorders
	RVF	Tick infestation	Sheep and goat pox	*T. evansi*	RVF
	Sheep and goat pox	Trypanosomosis	Tick infestation	Trypanosomosis	Trypanosomosis

Source: Perry et al. (2002b)

impacting transport in isolated rural communities. Rinderpest pandemics in Africa have imposed significant losses, particularly on pastoral communities, with the 1979–1983 pandemic affecting some 100 million cattle (Roeder and Rich 2009). Costard et al. (2009) note that ASF disproportionately impacts smallholder pig producers in Africa that have neither the means nor commercial incentives to implement biosecurity measures necessary to control or prevent the disease; there are no vaccines available to protect against ASF, or any effective therapeutics. Moreover, the absence of compensation for pig producers forced to cull pigs during outbreaks often fuels its spread, as producers will either move pigs away from affected regions or slaughter the pigs and cook/sell the meat, which itself can fuel the spread of disease if such scraps are consumed (FAO 1998). ND likewise impacts the rural poor that rely on backyard poultry as a source of cheap protein and quick income (particularly for women) needed for school fees and other social obligations; conversely, the free-range production system employed in traditional settings also fuels its spread (Kitalyi 1998). Even those diseases that are less severe from their direct impacts on animal mortality have important effects on production caused by disease-related morbidity. CBPP, for example, in addition to causing mortality, limits the productivity of livestock and crops (through draught power) alike, while FMD and LSD reduce milk yields and increase maintenance costs for livestock.

No less important are the impacts that certain animal diseases have on international trade. FMD probably receives the most attention in this regard given its contagious nature and rapid spread. International markets for beef are segmented on the basis of a country's (or a region within a country) FMD status, as certified by the OIE. Countries that are FMD-free and do not vaccinate their animals have the highest level of market access, including potential access to Europe, Japan, and Korea. The distinction between FMD-free with and without vaccination is important, because it is difficult to discern the difference between an animal and carcass that has the FMD virus or has been vaccinated against FMD and generated an immune response (Rich and Winter-Nelson 2007). Countries such as Japan and Korea employ "zero risk" standards that mandate FMD-free without vaccination status for potential exporters.[1] Not surprisingly, exporting countries that are FMD-free without vaccination can obtain both higher prices and the ability to more flexibly market different cuts of meat based on demand and preferences. Major beef exporters like Brazil are (mostly) FMD-free with vaccination and can export to most, but not all countries (Japan, Korea, and the United States prohibit imports from Brazil, for example). Countries that are not FMD-free are much more limited in the countries to which they can export, and in receiving the potential price premiums in the developed world, most of which only accept meat from sources that FMD-free with (or without) vaccination.

[1] Ironically, this standard failed to prevent recent FMD outbreaks occurring in both countries in 2010!

In Africa, only parts of Namibia, Botswana, South Africa, and Swaziland are recognized as FMD-free without vaccination. FMD freedom has been maintained by an elaborate and expensive system of cordon fencing and surveillance to limit animal movements and contacts with endemic zones further north. Perry et al. (2003) note that in South Africa, the maintenance and personnel costs associated with its fencing system are between 12.5 and 14.5 million Rand (about US$1.6–1.9 million) per year; this does not include the initial capital investment for the fences themselves. Zimbabwe had a similar system that fell into rapid disrepair in the early 2000s, as budget constraints severely limited veterinary services, particularly limited foreign exchange to acquire vaccines and perform maintenance on vehicles (Perry et al. 2003). These countries maintain such an elaborate system to safeguard their duty-free, quota-free access to the European Union; a small duty-free quota is also allocated by Norway. Such a traceability system is costly: an estimate from ODI (2007) found that the start-up costs for the traceability program used in Botswana cost 166 million Botswanan Pula (roughly US$30 million) and another 15 million Pula to maintain each year. Rich and Perry (2011b) cite earlier estimates that compliance with EU standards adds US$5.50 per exported carcass. Such a system allows Namibia, for example, to cross-subsidize exports to South Africa that it sells at a loss (Rich and Perry 2011b).

Elsewhere in Africa, the presence of endemic disease limits the potential for international trade. In endemic areas north of the cordon fence (or "Red Line") in Namibia, home to nearly one-half of the country's cattle herd, market opportunities are limited only to local markets or the South African market after a period of quarantine (Rich and Perry 2011b). RVF outbreaks have severely limited the market access of livestock products from the Horn of Africa. Two trade bans were imposed in 1998 and 2000 after RVF outbreaks in the region; the last ban was not formally removed until 2003 (Nin Pratt et al. 2005). Trade from this region is predominately informal and pastoral-based, with supplies of small ruminants serving demand for the Hajj in the Arabian Peninsula. This trade, while not formally documented, is not insignificant, with Nin Pratt et al. (2005) citing estimates of live animal exports at 1.3–3 million, of which 60–80% originate from Ethiopia. Further estimates from Nin Pratt et al. (2005) on the Somali region in Ethiopia found that the 16-month trade ban imposed starting in 1998 after an occurrence of RVF reduced value-added by US $132 million, or 42% of the value-added created in the Somali region. CBPP also limits international trade and is a high priority for control in export-oriented areas of Southern Africa (Musisi et al. 2004).

The next section will discuss in detail some of the strategies that have been employed for disease control in Africa, and a summary from Table 16.1 provides a useful *précis*. There are vaccines available for many of the important diseases affecting Africa, but their effectiveness can vary considerably. The infrequent nature of certain diseases like RVF provides a further challenge in terms of the logistics required to maintain and distribute adequate quantities of strategic stocks of vaccines. Conversely, diseases that strike frequently and ideally require annual vaccination to control them, such as blackleg and anthrax, impose their own logistical challenges. Movement controls and quarantines for transboundary diseases are often complicated by the limited human resources necessary to effectively maintain such controls. Stamping out is often mandated for those diseases that have no specific treatment (e.g., ASF, HPAI) or where international trade considerations make it commercially viable (e.g., CBPP, FMD), but difficulties often arise in terms of the administration of compensation payments necessary to institutionalize these efforts as a part of public policy. Moreover, in most contexts in Africa, the justification for stamping out is limited, as international trade access is limited and the livelihood losses associated with it are too great. Other treatments have important environmental externalities (e.g., acarcides for tick-borne diseases) or can potentially accentuate the disease (e.g., antibiotics for CBPP).

Strategies and Challenges for Animal Disease Control in Africa

Role of Technology

For many diseases, preventive technological interventions such as vaccines are the first line of defense against animal diseases. The success of a vaccination campaign, however, relies not only on the quality of the science but also on the ability of veterinary institutions to adequately deliver and administer vaccines in the field to the target production system or age group, and at the appropriate interval. This section will focus primarily on the former, with delivery issues addressed later in the section.

The effectiveness of vaccination from the standpoint of the science behind different vaccines varies considerably by disease. Rinderpest is an example of an animal disease that has been successfully eradicated in Africa, aided considerably by an extremely effective vaccine, but also by a disease that has a relatively straightforward epidemiology. The technology behind the development of a rinderpest vaccine dates back to the 1880s with the discovery of the protective ability of the serum from recovered animals (reviewed by Roeder and Rich 2009). This was later adapted into an attenuated goat-adapted (as well as rabbit-adapted) vaccine that was successfully used to control and eradicate the disease in Asia. Further breakthroughs came in the 1960s with vaccines generated from tissue cultures, which avoided significant side effects or disease revision (reviewed by Roeder and Rich 2009). The main disadvantage was the need for a proper cold chain to adequately store the vaccine, which complicated its delivery in Africa. However, with the development of the heat-stable (up to 4 weeks) Thermovax vaccine in 1990, the ability to safely administer and deliver vaccines in a host of difficult environments significantly facilitated the final stages of the eradication campaign in East Africa (reviewed by Roeder and Rich 2009).

As noted earlier, an important advantage in the control of rinderpest is that there is only one serotype and that the vaccines that have been developed for it confer lifelong immunity. By contrast, other diseases are much less amenable to vaccine control by virtue of complications related to either the disease or its vaccine technology. Contrastingly, AHS has nine different serotypes, and while a polyvalent cell culture attenuated vaccine is available, it is a live virus vaccine, does not fully protect horses against the prevalent serotype, and carrier animals (particularly zebras) can maintain virus pools in an ecosystem. Control of AHS thus relies heavily on adequately identifying the causal serotype in circulation at the time of an outbreak. There are six of the seven different strains of FMD in Africa: O, A, C, and SAT, each necessitating different vaccines to confer immunity (Vosloo et al. 2002). Current FMD vaccines are further constrained by their relatively short duration of immunity (approximately 6 months) and the lack of heat-stable vaccines at present (Perry and Sones 2007). Two vaccines exist for the control of CBPP (T1Sr and T1/44), but neither are ideal for inducing long-term immunity. The latter vaccine in particular is often associated with severe side effects, and was reportedly not very effective during an outbreak in Botswana in the mid-1990s (FAO 2004). Vaccination for RVF poses a more problematic logistical challenge. Major RVF outbreaks typically occur once every decade or so in the Horn of Africa, generally coinciding with *El Niño* weather patterns. Although vaccines for RVF are available, the shelf life of such vaccines is typically about 4 years, making the holding of strategic stocks unviable, while the time required to create more vaccine is likely to be several months (ILRI/FAO 2009).

Vaccines are not the only form of technological mechanism for the control of disease. Tick-borne disease control involves the use of acaracides, or chemical treatments, applied to control tick vector populations. While generally effective on ticks, such control measures are expensive and also come at relatively high environmental cost that are often not incorporated in economic evaluations of such diseases (Perry and Rich 2008). Trypanosomiasis is more often controlled with drugs given at the farm level to animals rather than treatments aimed at tsetse fly control, with resistance to such drugs increasing in recent years (Affognon 2007).

A challenge in Africa is the need to motivate improved investments in better technologies needed to control animal diseases. A particular problem is that the relatively small scope of animal disease vis-à-vis other global health priorities reduces incentives for private funding of long-term vaccine initiatives. One mechanism to mitigate this has been through the creation of the Global Alliance for Livestock Vaccines (GALVmed), a public–private partnership underwritten by the Bill and Melinda Gates Foundation and the Department for International Development (DFID) of the United Kingdom aimed at developing improved vaccines and delivery systems to reach the poor in the developing world, specifically Africa. GALVmed recently contributed US$28 million to support the registration and commercialization of its activities, including the development of a vaccine for ECF, developed in collaboration by the International Livestock Research Institute (ILRI) and the Kenyan Agricultural Research Institute (KARI) (see http://www.galvmed.org/news-resources/content/east-coast-fever-vaccine-registered-tanzania). This vaccine employs an "infection-and-treatment" method whereby animals are both infected with the parasite and given antibiotic treatment to modulate the severity of the disease, so inducing a protective response. This technique has reportedly reduced mortality from ECF by over 95%, while lowering production costs associated with expenditures on acaracides (see http://www.galvmed.org/news-resources/content/east-coast-fever-vaccine-registered-tanzania). Recent research initiatives have further been launched to try to develop better vaccines for FMD (Perry and Sones 2007) and CBPP (see http://www.au-ibar.org/ach_animhealth/vacnadaLivDisCBPP.html). At the same time, Scoones and Wolmer (2006) argue that research funding within the animal health subsphere is often devoted more toward diseases of trade (e.g., FMD) instead of more pedestrian (but equally damaging) diseases such as ND, with prioritization of diseases often driven from researcher and donor mandates instead on stakeholder needs. Conversely, Perry and Rich (2008) note that entrenched pharmaceutical interests promoting acaracides for tick-borne diseases can actually work against vaccine development, despite the multiplicity of benefits on production and the environment such vaccines would have.

Role of Infrastructure

Concomitant with technology is the need for proper support infrastructure in the form of veterinary services necessary to diagnose, report, and contain disease incursions. In Africa, such infrastructure has often been woefully lacking, with a number of important exceptions. As noted previously, southern African countries have invested significant resources in the development a complex system of fencing, surveillance, and movement control, often at high cost to traditional production systems and to the ecology of wildlife habitats. The success of these measures largely depends on the ability of countries to maintain such controls over time. Events over the past decade in Zimbabwe have devastated its ability to adequately maintain disease-free zones from which to export. Perry et al. (2003) report that as early as 2002, high staff vacancy rates (30%), ill-maintained transport (78% of vehicles not in commission), damage to fences, and limited foreign exchange have made surveillance effort and movement controls difficult to maintain. The presence of such countries alongside those such as Botswana and Namibia has been a cause of concern and tensions in the past (Scoones and Wolmer 2006). Mangani (2004) notes that the spread of CBPP in Zambia stemmed in part from lax movement controls and illegal movements of cattle from control zones to previously free areas.

Infrastructure to support animal health mitigations also includes the need for laboratory support for diagnosis and vaccine production. During the recent global pandemic of avian influenza, substantial amounts of money were invested in building laboratory diagnostic capacity, both in terms of equipment and reagents but also in training, in many countries; for example in 2008, a US$7.14 million grant was given by the World Bank to the government of Egypt specifically for laboratory capacity strengthening.

Institutional Strategies

A critical component to the control of any animal disease is the existence of strong veterinary services. In Africa, veterinary services were initially established during colonial periods as a means of protecting new commercial livestock enterprises using imported livestock breeds, highly susceptible to many of the diseases endemic in African breeds, and of facilitating meat trade between Africa (particularly Southern Africa) and Europe through the eradication and control of disease (Scoones and Wolmer 2006). Indeed, much of the infrastructure discussed in the previous section had its origins in colonial veterinary policy that was continued after independence, often backstopped by foreign donor support (Scoones and Wolmer 2006). By the 1980s, however, financial crises beset many African countries and overall government budgets came under increased pressure. Leonard et al. (1999) notes that budget cuts in veterinary services typically targeted supplies and transport rather than staff, with some countries spending over 90% of their veterinary budget on salaries. SAPs were implemented throughout much of sub-Saharan Africa, which both radically decreased public budgets and privatized many of the functions previously administered by the public sector. With respect to the veterinary sector, a number of reforms were conceived including (1) movements toward cost recovery mechanisms for services rendered; (2) better balance in budgetary allocations within the public sector between salaries and operating costs; (3) privatization of certain veterinary services and greater clarity on the roles of public and private sectors in the veterinary field; (4) privatized trade in animal health inputs such as medicines and vaccines; and (5) modifications toward regulations to allow for greater private sector participation (Gauthier et al. 1999; Sidibe 2003).

One of the challenges from structural adjustment reforms has been in defining clarity in the relative roles between public and private sectors, particularly pertaining to the role of paraprofessionals such as CAHW. A number of studies from the late 1990s highlighted the strong resistance in the public veterinary community from the use of paraprofessionals and the private sector in administering certain types of veterinary activities and in the accreditation of CAHWs more generally (Ashley et al. 1996; FAO 1999; Gauthier et al. 1999). While some resistance remains today, and regulations to harmonize practices and guidelines remain an area for policy (Allport et al. 2005; PACE 2006; Perry and Sones 2009), the use of community-based systems for privatized service delivery has started to take hold, with OIE recommendations made in 2003 to strengthen linkages between CAHWs, private sector, and public veterinary services in areas of vaccination, surveillance, and disease control (Riviere-Cinnamond 2005), and some countries, such as Ethiopia, formally recognizing CAHWs.

An important impetus for this change in mindset was the use (and success) of CAHWs in the final stages of rinderpest eradication. From a technical standpoint, Scoones and Wolmer (2006) cite the greater efficiency achieved by CAHWs in administering vaccination coverage in Ethiopia during the 1990s compared to government veterinarians (84% for CAHWs vs. 72% for the government, despite limited infrastructure in terms of transport used by CAHWs). But a more important contribution of CAHWs was in their leveraging of local knowledge of pastoralists and other stakeholders to root out the last reservoirs of disease. During the rinderpest campaign, the use of a tool known as PE emerged that sought out, engaged, and empowered farmers and communities to assess their knowledge of disease and provide important insights on disease dynamics and risk factors in data-scarce environments (Catley 2000, 2006; Mariner and Paskin 2000; Jost et al. 2007). This became particularly useful as a surveillance tool and a means to confirm the absence of disease in a particular site (Jost et al. 2007). Although these tools are largely based on semistructured interviews in the same fashion that participatory rapid appraisal techniques are used in the rural development and livelihoods area (Catley 2009), such tools can also be used to calibrate quantitative models of disease spread (Mariner et al. 2006). Indeed, a major part of the surveillance and response strategy for avian influenza in Indonesia has been based on PE tools, while significant attention to PE has been given by a number of veterinary schools and research centers (especially ILRI) in Africa to mainstream this as a part of animal health curriculum and research (Jost et al. 2007; Catley 2009).

From a policy perspective, one of the challenges facing the livestock sector is the need for strong, evidence-based decision support mechanisms that both promote and support the sector and are sensitive to the multifaceted constraints inherent in the African context. Rich and Perry (2011a) note the dissonance that exists between first-world perspectives and interventions and developing world realities, with the institutional capacity necessary to plan and act upon shocks to the system often lacking. The case of the 2007 outbreak of RVF in Kenya is an interesting case in point. As documented in ILRI/FAO (2009), RVF posed a number of unique challenges when it reemerged in late 2006 following the heavy *El Niño* rains that affected the Horn of Africa. First, response to the outbreak was beset by a number of administrative and logistical delays that hampered response, despite various warnings and signals from stakeholders on the ground. Second, and related the first point, the Department of Veterinary Services lacked any emergency planning mechanism necessary to mobilize resources to respond to the outbreak, with decision-makers disinclined to spend scarce resources in the event of a "false alarm." Third, given that RVF strikes infrequently based on climatic patterns governing the *El Niño* rains, there was little in the way of "institutional memory" within front-line ministries that could have provided valuable insights into supporting decision making.

To deal with this glaring lacuna, a decision-support protocol was developed among stakeholders in a participatory fashion to provide a framework for how policy makers should respond based on the occurrence of a sequence of events that portend greater risk for the emergence of RVF. Table 16.3 summarizes these events and the dimensions of interventions that need to be employed at each event period, with specific interventions depending on the stage of an outbreak. For instance, prior to any disease occurring, interventions would primarily involve having appropriate protocols and tools ready in the event of an outbreak. In the area of capacity,

Table 16.3 Elements of a decision-support tool for response to RVF outbreaks in Kenya

Event sequence (in order of increasing, then decreasing risk)	Actions to be taken for each event
Normal situation between outbreaks	Capacity building and training
Early warning of RVF issued, based on GIS or other meteorological forecasts	Communication
Localized, prolonged heavy rains reported by eyewitnesses	Coordination
Localized flooding reported by eyewitnesses	Early warning systems
Localized mosquito swarms reported by eyewitnesses	Surveillance
First detection of suspected RVF in livestock by active searching and/or rumors from herders	Disease control
Laboratory confirmation of RVF cases in animals	Vector control
First rumor/report of human RVF cases	Trade and markets
Laboratory confirmation of human RVF cases	Funding
No new human cases for 6 months	Post outbreak recovery and reflection
No clinical livestock cases for 6 months	Institutions and policy
Postoutbreak recovery and reflection	Research, impact assessment, and risk assessment

Source: ILRI/FAO (2009)

this could include ensuring that appropriate risk assessment training has occurred and laboratory facilities are in place in the event of an RVF outbreak. Similarly, from the standpoint of communication, this could include ensuring that appropriate messages are in place, while coordination would aim at linking appropriate stakeholders to streamline response procedures. As early risk factors emerge (e.g., rains, increased mosquito populations), greater monitoring and response would occur, including coordinating supplies and surveillance strategies and having key partners and networks on standby, particularly in high-risk areas. An important part of this includes ensuring that funding resources are ready should they be required from national and international sources. Once the disease has struck, the framework would move into dedicated control efforts in local and adjoining regions, but leveraging the earlier response activities to better utilize resources and minimize delays. As the disease is controlled, stock-taking plays an important role to reevaluate the effectiveness of measures and messages, and to see where improvements can be made in the future.

Emerging Issues

A number of important issues face the livestock sector in Africa in the future. There is increased recognition of the role of livestock as a source of poverty alleviation (Randolph et al. 2007; Perry and Grace 2009), and with that emerge challenges in terms of how to address the logistics of disease control and the orientation of veterinary services themselves. Scoones and Wolmer (2006) note the potential tradeoffs that exist under the guise of "pro-poor" policy, in terms of balancing livelihoods considerations on the one hand and market access (particularly to developed country markets) on the other; these are further amplified by the diversity of production systems present in Africa. Perry and Rich (2007) note that the control of FMD, for example, can have significant pro-poor effects across different contexts, even taking into account often stark differences in the production setting. For instance, FMD control can have important poverty impacts in pastoral systems by reducing risk and vulnerability associated with low productivity of animals, while in more commercialized settings, increased market access can open up employment opportunities downstream throughout the supply chain that can have important pro-poor effects.

Changes in global demand and the potential for changes in global regulations could have an important impact on the livestock sector in Africa and the scope for such priority setting. Rich (2009) notes that while demand for meat products is rising (including strong demand from Africa), as a supplier, Africa remains a relatively small player in global meat markets. Historically, this has been due to the presence of endemic animal diseases, with market access only coming from commercial zones in Southern Africa, and even there in relatively small volumes. Much has been made about the concept of "commodity-based trade" in Africa as a way to ease some of the disease burdens associated with market access (Rich and Perry 2011b). Under commodity-based trade (CBT), the primary issue concerns the inherent safety of the product that is traded, not the origin of the animal from which that product is derived. In the case of FMD, meat that has been properly chilled and matured poses negligible risk for the spread of FMD. Advocates for CBT thus argue that a change in international trading regulations that accept a CBT approach could increase Africa's potential for market access into lucrative export markets (Thomson 2008). Recent work by Rich and Perry (2011b) demonstrates that CBT may not benefit Africa in the short-run by virtue of putting increased emphasis on supply-chain processes necessary to ensure food safety and freedom from disease that could render African products uncompetitive in export markets. Indeed, most meat exports from Southern Africa are only competitive in the EU because of preferential trade arrangements vis-à-vis competitors such as Brazil. Moreover, given that CBT would benefit all potential suppliers of meat products, the gains under a CBT world are likely to come from low-cost, high-volume suppliers in South America and South Asia, rather than from Africa (Rich and Perry 2011b). Having said that, the move toward a CBT world combined with increased investments in productivity-enhancing technologies and supply-chain improvements could be an important mechanism to value-add meat from Africa, whether serving domestic, regional, or high-value markets. In such a context, the control of disease is complementary toward risk management strategies, and will place an increased focus on synergies between the public and private sectors in certifying and facilitating greater trade.

A second emerging issue, particularly in the context of zoonotic diseases, is the move toward a "One Health" approach (WHO 2009). The idea of One Health is to conceive of more integrated ways of response and control of zoonotic diseases between the animal and public health realms. This would necessarily be a much more interdisciplinary approach that merges the efforts of veterinarians, doctors, policymakers, and researchers alike. A discussion of the One Health proposal hosted by the Steps Centre at the Institute of Development Studies, University of Sussex provided some perspectives on the approach, using the response to avian influenza in Asia as a backdrop. Particular attention was placed on reconciling bottom-up approaches, perspectives, and knowledge of stakeholders in their livelihoods and poverty context with the current constellation of international actors and organizations tasked with more top-down coordination of animal and public health issues. Context will play an important role in determining the success of a One Health approach, as will the governance mechanisms within and between various institutions necessary to put it into practice (Steps Centre 2009).

Conclusions

The control of any animal disease is a challenge regardless of the context; the diversity, resource constraints, and infrastructural challenges present in Africa compound such efforts, rendering market access more challenging for African livestock producers. Nonetheless, there are important lessons and innovations that have been developed in the African context, and indeed the eradication of rinderpest occurred despite many of the challenges inherent in Africa. An emerging, central theme will be the interface between strong credible veterinary services, with new demands from clients in the context of market access and more integrated, "One Health" approaches.

References

Affognon, H. 2007. Economic Analysis of Trypanocide Use in Villages under Risk of Drug Resistance in West Africa. Ph.D. Dissertation, University of Hannover.

Africa Union Interafrican Bureau for Animal Resources (AU-IBAR). 2009. Pan African Animal Health Yearbook 2008. AU-IBAR, Nairobi.

Allport, R., Mosha, R., Bahari, M., Swai, E., Catley, A. 2005.The use of community-based animal health workers to strengthen disease surveillance systems in Tanzania.Rev. sci. tech. Off. int. Epiz. 24(3), 921–932.

Ashley, S.D., Holden, S.J., Bazeley, P.B.S., 1996. The Changing Role of Veterinary Services: A Report of a Survey of Chief Veterinary Officers' Opinions. Report to the Office International des Epizooties, Livestock In Development, Somerset, UK.

Catley, A. 2000. The use of participatory appraisal by veterinarians in Africa.*Office international des epizooties revue scientifiqueet technique* 19(3), 702–714. http://www.oie.int/eng/publicat/rt/1903/A_R1932.htm.

Catley, A. 2006. The use of participatory epidemiology to compare the clinical and veterinary knowledge of pastoralists and veterinarians in East Africa. *Tropical Animal Health and Production* 38, 171–184.

Catley, A., 2009. From Marginal to Normative: Institutionalizing Participatory Epidemiology. Working Paper, Feinstein International Center, Tufts University. Available at http://www.future-agricultures.org/farmerfirst/files/T3c_Catley.pdf.

Costard S., Wieland B., de Glanville W., Jori F., Rowlands R., Vosloo W., Roger F., Pfeiffer D.U., Dixon L.K., 2009. African swine fever: how can global spread be prevented? *Phil. Trans. R. Soc. B* 364, 2683–2696.

Food and Agriculture Organization of the United Nations (FAO). 1998. African Swine Fever in Nigeria hits rural poor. Article found at Internet address: http://www.fao.org/english/newsroom/highlights/1998/981201-e.htm, retrieved 2 July 2010.

Food and Agriculture Organization of the United Nations (FAO). 1999. The effect of structural adjustment programmes on the delivery of veterinary services in Africa. Working paper, Animal Production and Health Division, Food and Agriculture Organization (FAO) of the United Nations, FAO, Rome. Available at ftp://ftp.fao.org/docrep/fao/010/ah933e/ah933e.pdf.

Food and Agriculture Organization of the United Nations (FAO), 2004. Proceedings of Towards sustainable CBPP control programmes for Africa, FAO-OIE-AU/IBAR-IAEA consultative group on contagious bovine pleuropneumonia, Third Meeting, Rome 12–14 Nov. 2003.

Gauthier, J., Simeon, M., de Haan, C., 1999. The effect of structural adjustment programmes on the delivery of veterinary services in Africa. *In* Comprehensive reports on Technical Items presented to the International Committee or to Regional Commissions. OIE, Paris, 133–156.

International Livestock Research Institute and Food and Agriculture Organization of the United Nations (ILRI/FAO), 2009. Decision support tool for prevention and control of Rift Valley fever epizootics in the Greater Horn of Africa. Version 1.ILRI Manuals and Guides No. 7, ILRI, Nairobi and FAO, Rome.

Jost, C.C., Mariner, J.C., Roeder, P.L., Sawitiri, E., Macgregor-Skinner, G.J., 2007. Participatory epidemiology in disease surveillance and research. Rev. sci. tech. Off. int. Epiz. 26(3), 537–549.

Kitalyi, A.J., 1998. Village chicken production systems in rural Africa: Household Food Security and Gender Issues. FAO Animal Production and Health Paper 142 (Rome: Food and Agriculture Organization of the United Nations), found at http://www.fao.org/DOCREP/003/W8989E/W8989E00.htm#TOC, retrieved 4 October 2006.

Kristjanson, P.M., Swallow, B.M,Rowlands, G.J.,Kruska, R.L., de Leeuw, P.N., 1999. Measuring the costs of African animal trypanosomosis, the potential benefits of control and returns to research. Agricultural Systems 59, 79–98.

Leonard, D.K., Koma, L.M.P.K, Ly, C., Woods, C.S.A., 1999. The new institutional economics of privatising veterinary services in Africa. Rev. sci. tech. Off. int. Epiz. 18(2), 544–561.

Mangani, M.P.C., 2004. Contagious bovine pleuropneumonia in Zambia. In FAO (2004) Proceedings of Towards sustainable CBPP control programmes for Africa, FAO-OIE-AU/IBAR-IAEA consultative group on contagious bovine pleuropneumonia, Third Meeting, Rome 12–14 Nov. 2003.

Mariner, J.C., McDermott, J., Heesterbeek, J.A., Thomson, G., Martin, S.W., 2006. A model of contagious bovine pleuropneumonia dynamics in East Africa. Preventive Veterinary Medicine 73(1), 55–74.

Mariner, J.C., Paskin, R., 2000. Manual on participatory epidemiology: Methods for the collection of action-oriented epidemiological intelligence. Food and Agriculture Organization of the United Nations (FAO) Animal Health Manual No. 10, FAO, Rome.

Musisi, F.L., Dungu, B., Thwala, R., Mogajane, M.E., Mtei, B.J., 2004. The threat of contagious bovine pleuropneumonia and challenges for its control in the SADC region. In FAO (2004) Proceedings of Towards sustainable CBPP control programmes for Africa, FAO-OIE-AU/IBAR-IAEA consultative group on contagious bovine pleuropneumonia, Third Meeting, Rome 12–14 Nov. 2003.

Nin Pratt, A., Bonnet, P., Jabbar, M., Ehui, S., de Haan, C. 2005. Benefits and costs of compliance of sanitary regulations in livestock markets: The case of Rift Valley fever in the Somali Region of Ethiopia. International Livestock Research Institute, Nairobi, Kenya.

Overseas Development Institute (ODI), 2007. Analysis of the Economic and Social Effects of Botswana's Loss of Preferential Market Access for Beef Exports to the European Union. Final Study, August 2007, Overseas Development Institute, London.

Pan African Control of Epizootics (PACE), 2006. Booklet on PACE Success Stories. Available at: http://www.fao.org/ag/againfo/programmes/documents/grep/PACE_booklet.pdf.

Perry, B.D., Gleeson, L.J., Khounsey, S., Bounema, P., Blacksell, S., 2002a. The dynamics and impact of foot and mouth disease in smallholder farming systems in South East Asia: a case study in the Lao Peoples Democratic Republic. Rev. sci. tech. Off. int. Epiz. 21, 663–673.

Perry, B.D., Randolph, T.F., McDermott, J.J., Sones, K.R. and Thornton, P.K. (2002b). Investing in Animal Health Research to Alleviate Poverty. International Livestock Research Institute (ILRI), Nairobi, Kenya, 140pp plus CD-ROM.

Perry, B.D., Randolph, T.F., Ashley, S., Chimedza, R., Forman, T., Morrison, J., Poulton, C., Sibanda, L., Stevens, C., Tebele, N., Yngstrom, I, 2003. The impact and poverty reduction implications of foot and mouth disease control in southern Africa, with special reference to Zimbabwe. International Livestock Research Institute (ILRI), Nairobi, Kenya.

Perry, B.D., Nin Pratt, A., Sones, K., Stevens, C., 2005. An appropriate level of risk: balancing the need for safe livestock products with fair market access for the poor. Pro-Poor Livestock Policy Initiative Working Paper No. 23, Food and Agriculture Organisation (FAO) on the United Nations, Rome.

Perry, B.D., Rich, K.M. 2007. The poverty impacts of foot and mouth disease and the poverty reduction implications of its control, *Veterinary Record* 160, 238–241.

Perry, B.D., Rich, K.M. 2008. "Economic impacts of tick-borne diseases in Africa," Presentation for the Onderstepoort Centenary: Pan-African Veterinary Conference 2008, University of Pretoria, South Africa, 8 October 2008.

Perry B.D., Sones, K.R., eds., 2007. Global Roadmap for improving the tools to control foot-and-mouth disease in endemic settings. Report of a workshop held at Agra, India, 29 November-1 December 2006. International Livestock Research Institute, Nairobi, Kenya.

Perry, B.D., Sones, K. 2009. Strengthening Demand-Led Animal Health Services in Pastoral Areas of the IGAD Region. IGAD LPI Working Paper No. 09–08, Food and Agriculture Organisation (FAO) on the United Nations, Rome.

Perry, B.D. and Grace, D. 2009. The impacts of livestock diseases and their control on growth and development processes that are pro-poor. *Philosophical Transactions of the Royal Society, B*, 364, 2643–2655.

Randolph, T.F., Schelling, E., Grace, D., Nicholson, C.F., Leroy, J.L., Cole, D.C., Demment, M.W., Omore, A., Zinsstag, J., Ruel, M., 2007. Role of livestock in human nutrition and health for poverty reduction in developing countries. J. Anim. Sci. 85, 2788–2800.

Rich, K.M. 2009. What can Africa contribute to global meat demand: Opportunities and constraints, *Outlook on Agriculture* 38 (3), 223–233.

Rich, K.M., Perry, B.D., 2011a. The economic and poverty impacts of animal diseases in developing countries: new roles, new demands for economics and epidemiology, *Preventive Veterinary Medicine*, 101(3–4), 133–147.

Rich, K.M., Perry, B.D., 2011b. Whither Commodity-Based Trade. Development Policy Review 29(3), 331–356.

Rich, K.M., Winter-Nelson, A. 2007. An Integrated Epidemiological-Economic Analysis of Foot and Mouth Disease: Applications to the Southern Cone of South America. *American Journal of Agricultural Economics* 89 (3): 682–697.

Riviere-Cinnamond, A. 2005. Animal Health Policy and Practice: Scaling-up Community-based Animal Health Systems, Lessons from Human Health. PPLPI Pro-Poor Livestock Policy Initiative Working Paper No. 22, Food and Agriculture Organisation (FAO) on the United Nations, Rome.

Roeder, P., Rich, K.M. 2009. The global effort to eradicate rinderpest. Discussion Paper 923, International Food Policy Research Institute, Washington, DC.

Scoones, I., Wolmer, W. 2006. Livestock, Diseases, Trade, and Markets: Policy Choices for the Livestock Sector in Africa. Working Paper 269, Institute of Development Studies at the University of Sussex, Brighton, UK.

Sidibe, A.S., 2003. Organisationactuelleet future des Services veterinaires en Afrique. Rev. sci. tech. Off. int. Epiz. 22(2), 473–484.

Steps Centre 2009. One World, One Health – From principles to action. Perspectives from the expert meeting held in Brighton, UK, 26–27 February 2009. Institute of Development Studies at the University of Sussex, Brighton, UK. Available at http://www.steps-centre.org/PDFs/One%20World%20-%20principles%20to%20action%20report.pdf.

Thomson, G. 2008. A short overview of regional positions on foot-and-mouth disease control in southern Africa, *Transboundary animal disease and market access: future options for the beef industry in southern Africa, Working Paper 2,* Institute of Development Studies at the University of Sussex, Brighton, UK.

Umali, D.L., Feder, G., de Haan, C. 1994. Animal Health Services: Finding the Balance between Public and Private Delivery. World Bank Research Observer 9(1), 71–96.

Vosloo, W. Bastos, A.D.S., Sangare, O., Hargraves, S.K., Thomson, G.R., 2002. Review of the status and control of foot-and-mouth disease in sub-Saharan Africa. Rev. sci. tech. Off. int. Epiz. 21(3),437–449.

Winter-Nelson, A., Rich, K.M. 2008. Mad Cows and Sick Birds: Financing International Responses to Animal Disease in Developing Countries, *Development Policy Review* 26 (2), 211–226.

World Health Organization (WHO), 2009. Integrated Control of Neglected Zoonotic Diseases in Africa: Applying the "One Health" Concept. Report of a Joint WHO/EU/ILRI/DBL/FAO/OIE/AU meeting, ILRI Headquarters, Nairobi, Kenya, 13–15 November 2007. WHO, Geneva.

Chapter 17
Promoting Rural Livelihoods and Public Health Through Poultry Contracting: Evidence from Thailand

Samuel Heft-Neal, David Roland-Holst, Songsak Sriboonchitta, Anaspree Chaiwan, and Joachim Otte

Introduction: The Twin Challenges of Public Health and Rural Poverty

Highly Pathogenic Avian Influenza (HPAI) first emerged in Southeast Asia in 2003–2004. Initially, containment policies ranged from focusing on mass culling (Thailand) and vaccination (Vietnam) to the elimination of all wet markets (Hong Kong). Although these measures were applied with varied success, it has become clear that a new generation of policies is necessary to address the infrequent, but continued, outbreaks of an apparently endemic disease. The nature of these circumstances require that the new generation of policies focus on long term adjustment and take into account acceptable risk levels, farmer livelihoods, and financial sustainability. It is within this context that we look at geographical potential for medium scale contract farming in the informal poultry sector in Thailand.

Optimal utilization of limited resources and financial sustainability are increasingly important considerations. Promoting privately financed schemes that can improve production quality, such as contracting mechanisms, producer cooperatives, and certification programs, are potential approaches to achieving this sustainability. The long history of successful contracting schemes in the formal Thai poultry sector suggests that contracting may be a viable approach to quality improvement for some producers in the informal poultry sector as well.

S. Heft-Neal (✉) • D. Roland-Holst
Department of Agricultural and Resource Economics, University of California, Berkeley, CA, USA
e-mail: sheftneal@berkeley.edu

S. Sriboonchitta • A. Chaiwan
Department of Economics, Chiang Mai University, Chiang Mai, Thailand

J. Otte
Food and Agriculture Organization of the United Nations, Rome, Italy

D. Zilberman et al. (eds.), *Health and Animal Agriculture in Developing Countries*, Natural Resource Management and Policy 36, DOI 10.1007/978-1-4419-7077-0_17,

The purpose of this research activity is to identify specific areas in Thailand where contracting schemes are most likely to be viable in the informal sector. Integrating data from small-sample, highly detailed poultry producer surveys, with data from the national Socio-Economic Survey (SES), we identify regions with characteristics similar to areas where we have observed successful contracting programs.

We approach the problem of identifying these areas in three steps:

- Identify farmers in the small sample survey that are most likely to succeed as contract-farms.
- Establish criteria for successful contract farming based on characteristics of this subgroup.
- Use the relevant attributes identified in the small sample survey to identify households in the large sample survey where contracting is most likely to be viable.

The next section contextualizes the problem by describing the informal poultry sector in Thailand. Section "Types of Market Transactions" discusses the nature of market access barriers and their relationship to poverty. This is followed in Section "Volume of Transactions" by a discussion of contracting as a means of using private agency to overcome access barriers and other market failures that have made extensive rural poverty a chronic condition in Thailand and other countries. By using mechanisms of this kind, public policy can promote more constructive engagement between the urban enterprise sector and rural majorities, using enhanced market access to achieve self-directed poverty reduction. While the state plays a facilitating role in this process, its primary impetus and sustenance comes from private enterprises and individual farmers, avoiding the need for large, open-ended fiscal commitments, and incentive problems associated with public transfer schemes. A description of the survey data follows. Subsequently, criteria for identification of eligible farmers are discussed and identification of viable regions is extended from the small sample survey to the national survey. The final section presents our conclusions.

The Informal Poultry Sector

The next section draws on the authors' previous work in the region to provide an overview of the roles and activities of the primary actors within the informal poultry sector. First, we describe the production systems most commonly utilized by smallholder farmers. Second, we describe other actors involved with distribution and discuss their roles within the informal market chain.

Production Systems

Native chicken breeds, primarily utilized by smallholders, are ideal for informal production methods, in large part because they are not capital intensive. Hens on farm are used for restocking the flock and may provide some eggs for consumption (native breeds have low hatchability rates).

Most birds scavenge for naturally occurring feed sources (worms, seeds, etc.), rice byproducts (polished paddy rice and broken rice) and for household food scraps. Some farmers with larger flocks provide their birds with nominal inputs of feed, though rarely more than one feeding per day. Given the expense of commercial feed, and the native chicken's ability to scavenge, normally it is not cost-effective to provide the birds with commercial feeds. However, when commercial feed is used, often with crossbred chicken, it is usually supplemented with scavenging. Similarly, chickens can drink standing water or crop runoff, in addition to any extra water that is available. Simple basins are often used to collect the water for chickens to drink.

Our previous studies of backyard farmers in the north and northeast of Thailand (Heft-Neal et al. 2009) found that biosecurity investments were low. Sixty-five percent of backyard farms did not vaccinate their flock. Similarly, we found that the same percentage of farms did not use medicine to protect their flocks from disease. When used, pharmaceuticals can be purchased from local shops or acquired directly from the Department of Livestock. Approximately 30% of farms used housing to contain their chickens. Although flock containment can protect chickens from predators and decrease exposure to diseases, it hinders a chicken's scavenging ability and consequently increases the cost of rearing poultry. If housing is used it can be constructed out of inexpensive materials such as locally available wood and chicken wire. Closed housing systems are not used.

In addition to housing and the use of pharmaceuticals, disposal of animal waste is an important aspect of biosecurity. Traditionally, smallholder farmers have increased the viability of independent farms by maintaining integrated chicken and fish farms, where chicken feces are used as fish feed (Porn-Amart, 2003). Industrial farms can sell chicken dung to be used in fertilizer. However, in response to the HPAI outbreaks, Farm Standards were passed, which prohibits unregulated use and sale of chicken feces because of the risks associated (Sudsawasd and Wisarn 2008). While large farms have the resources and the incentives to meet the requirements to sell the large amounts of dung the farms generate, smallholders have neither and thus are less likely to sell dung. Instead, it is more viable for small farms to ignore government warnings and use dung for fish farms or fertilizer, or dispose of it. The outcome depends on many factors including the size of flock (amount of dung generated), presence of crops, and willingness to risk undertaking prohibited activities (Fig. 17.1).

While there tends to be low usage of pharmaceuticals on backyard farms, most vaccines and medicines are relatively inexpensive running from 1 to 1.5 baht per dose. Instead there are other reasons for not vaccinating flocks. Some common reasons that farmers cited include that vaccinations are not easily available or that small flock sizes mean that chickens are less economically important than other farming activities and thus do not warrant investment (of money or time). Moreover, many people feel that indigenous breeds of poultry are disease resistant and thus vaccination is unnecessary.

Enclosures are also relatively inexpensive to construct; however, they may be associated with increased feeding costs because caging chickens prevents birds

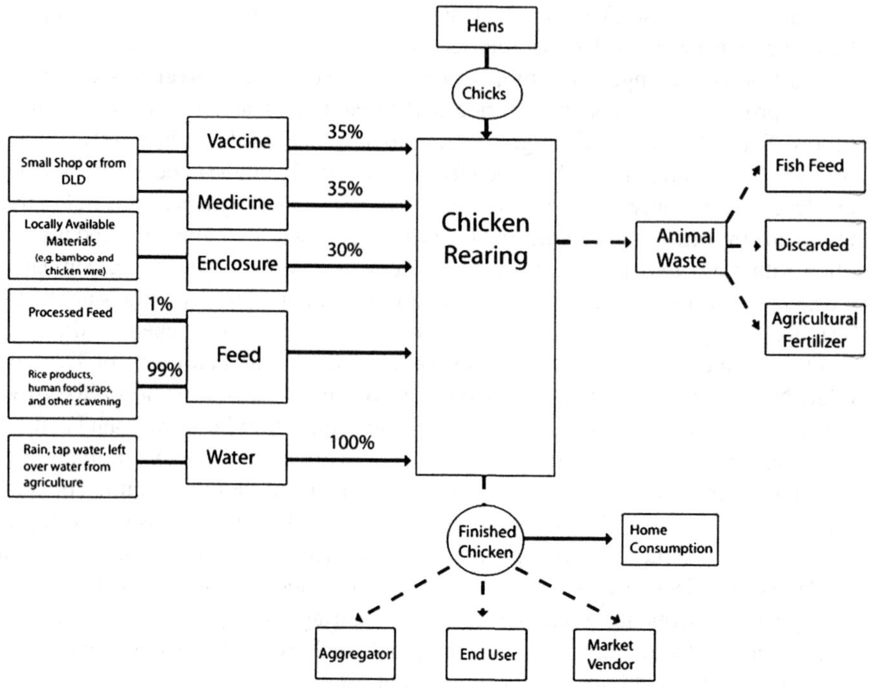

Fig. 17.1 Resource utilization among smallholder farmers

from scavenging for all of their food. Water and chicks are essentially costless and electricity is unnecessary.

The outputs on small farms are a small amount of animal waste, finished chicken, and birds that die from a reason other than slaughter. Farmers expect and accept high mortality rates.

Market Transactions

The market chain is composed of many actors playing varying roles in connecting producers and consumers. Generally, between producers and consumers, several transactions take place and much information is lost. The following section characterizes these transactions and the actors that facilitate them.

Types of Market Transactions

The primary actors in the informal market chain are farmers, aggregators, and market vendors. Some actors may take on more than one role within the system. For example, aggregators may collect birds from farmers and sell them to several market vendors but also operate their own market spaces where they sell directly to consumers. The percentages of transactions that travel through each actor depend on many factors including regional characteristics, the scale of the participating actors, and seasonality. However, average percentages reported by actors included in our survey from Chiang Mai, KhonKaen, and Nakhon Phanom provinces are presented in Fig. 2.2.

Volume of Transactions

Figure 17.2 looks at flows through the supply chain according to percentage of sales. However, to gain a fuller perspective of how the market chain functions, this next section looks at the volumes of sales.

End users are the most common buyers for every seller (i.e., farmers, aggregators, market vendors); however, individual transactions tend to be very small. Figures 17.3–17.5 highlight both the average number of chickens sold in each transaction type and the average number of people a single actor trades with. For example, an aggregator may buy 5–15 chickens from each of 20 farmers in a month and sell those 200–300 birds to 15 different vendors, 20 restaurants, and 30 end-users with decreasing transaction size. Compare this to farmers who sell to four or five sources per year and it becomes apparent that different market actors have varying degrees of familiarity and market power in negotiating transactions. The farmer depicted in Fig. 2.2 raises, on average, less than 50 chickens. Among farmers of larger scale (50–200 birds) the number of transactions tends to be only slightly higher while the size of transactions tends to be significantly larger.

The natural timescale for each actor is also different. Small-scale farmers tend to view their production scales as an annual undertaking, selling birds only once every few months, while traders think in terms of monthly sales and income. Market vendors, however, work on a daily scale. While aggregators also tend to trade every day, daily volumes of trade tend to vary while monthly volumes are more consistent, although also seasonal.

A farmer may sell 20–30 birds per year involving occasional sales to neighbors as well as repeat transactions with the same aggregators and vendors once every few months.

Aggregators included in our survey traded an average of 275 birds per month. These birds were purchased from 20 different farmers, many of whom the aggregator had bought from in previous months. Aggregators sold to an average of 75 unique sources per month, largely made up of small transactions with end-users, as well as some larger transactions with shops and vendors.

Market vendors included in our survey sold an average of 15 indigenous birds per day. However, these vendors were also the only actors in this supply chain to also be integrated into the broiler supply chain as many vendors sold indigenous chicken to supplement their larger supplies of broilers. Vendors tended to buy from

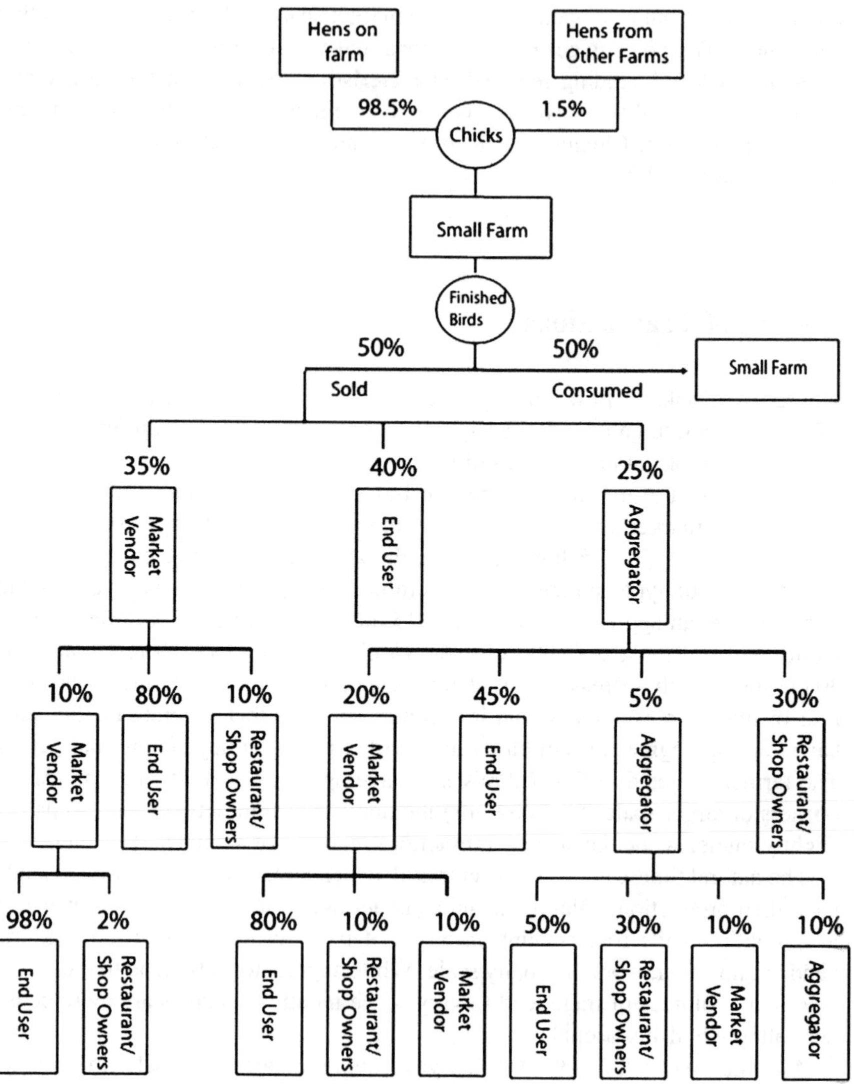

Fig. 17.2 Percentage of transactions along the supply chain. *Source*: Heft-Neal et al. (2009)

one aggregator and often a few farmers. Some vendors raised their own chicken as well. Vendor sales went primarily to end-users with some restaurants and shops buying a few birds at a time. Many of these transactions were repeated daily (with aggregators and restaurants), weekly or bi-weekly (with end-users), and every few months (with farms).

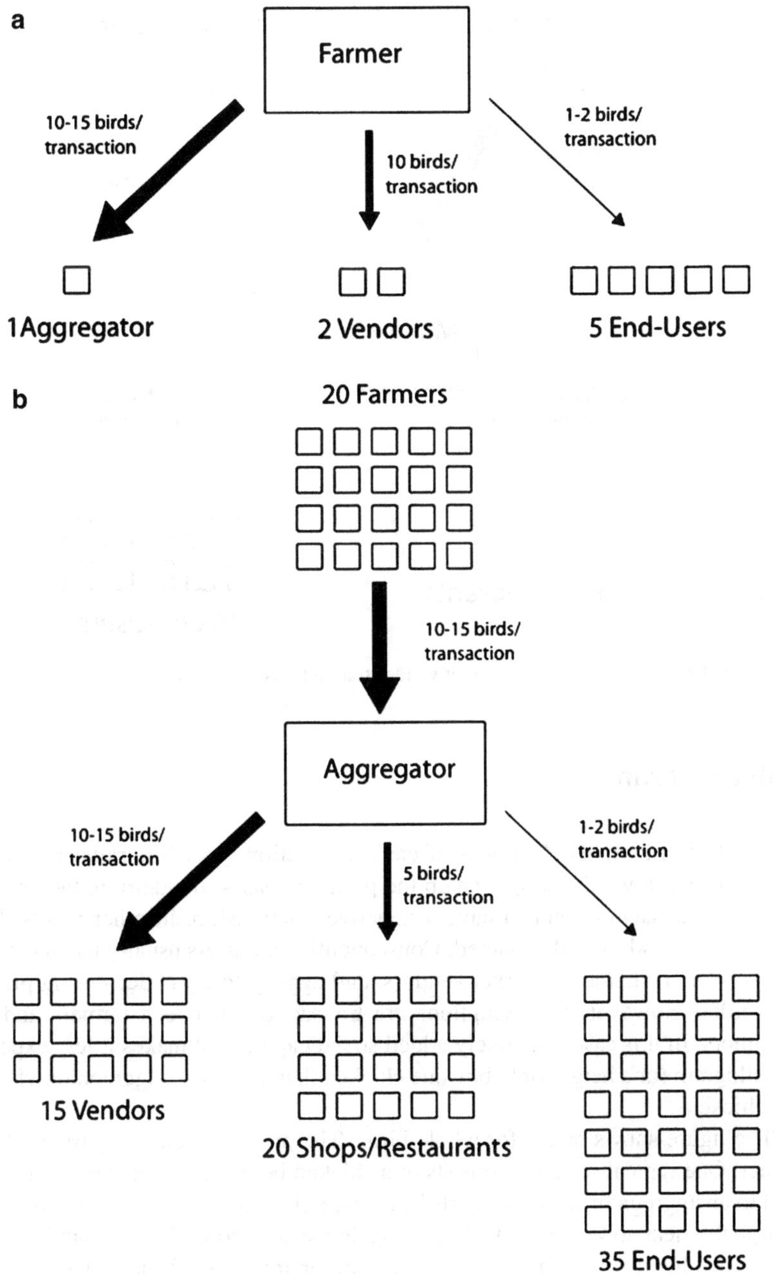

Fig. 17.3 (**a**) Frequency and size of aggregator transactions (per year). (**b**) Frequency and size of aggregator transactions (per month)

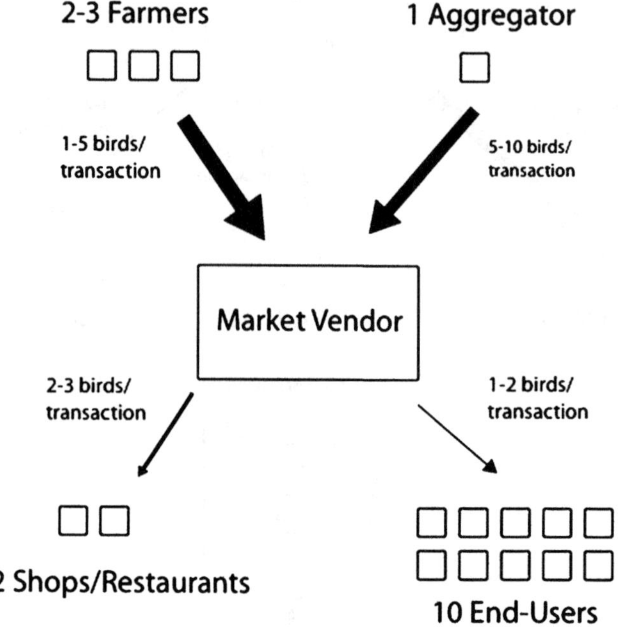

Fig. 17.4 Frequency and size of market vendor transactions (per day)

Value of Transactions

Figure 17.5 highlights the prices of each transaction type. Naturally, as products move further down the supply chain the price increases. In addition, the price and size of a transaction tend to have a negative relationship. In other words, larger transactions tend to be discounted. Consequently, end users usually pay the highest prices while restaurants, market vendors, and aggregators pay decreasing prices in that order. However, the exceptions are transactions between farmers and their neighbors. In this case, it is likely added social capital that makes it less expensive per kilogram for a neighbor to buy one chicken than it is for an aggregator who buys ten chickens.

The figure shows prices for whole birds. The price of premium parts is slightly higher. One typical scenario consists of a chicken being grown on a backyard farm, sold to an aggregator who transports birds to market and sells them to vendors, who then sells meat to an individual to bring home and cook for her family. In this scenario, illustrated in the diagram, the farmer receives 62 baht, the aggregator 10 baht, and the vendor 18 baht. The end-user pays a price of 90 baht/kg, a 50% increase above the farm gate price. If the end-user buys premium parts then the price is likely over 100 baht/kg with the extra profit going to the vendor.

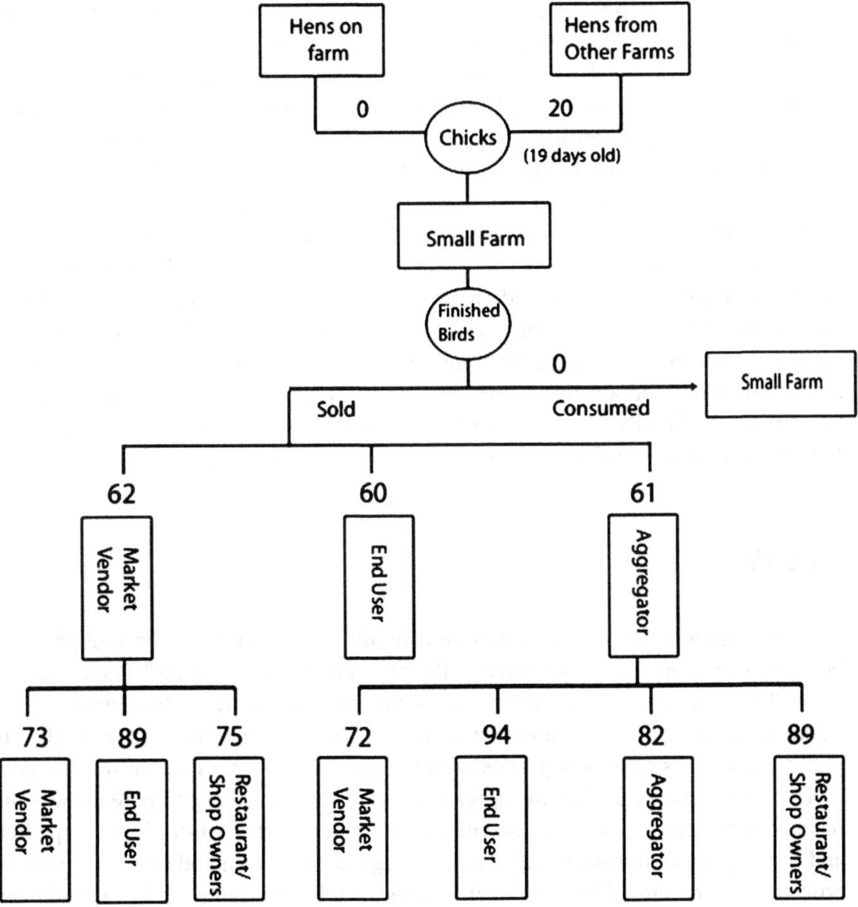

Fig. 17.5 Average price along the market chain for indigenous birds. *Source*: Heft-Neal et al. (2009)

Market Access and Poverty Reduction

Given the existing marketing structure, there is significant room to improve the terms of trade among the most vulnerable actors and to extend the system to include currently omitted actors. Greater market access can be a potent catalyst for poverty alleviation in the rural communities of developing economies. This is particularly true for more remote areas that lie outside the networks of transport and other infrastructure that link urban areas. These areas are also more isolated from administrative and informal linkages that could facilitate market access, extension service, and technology diffusion. Finally, lack of market participation compounds itself by limiting capital accumulation and the potential to invest for expanded production capacity. Likewise, external reform and greater outward orientation at the national level can be a catalyst for growth, leveraging external markets to expand productive capacity, and resource use in the domestic economy. Market access is a very high priority for poverty reduction generally, and our research evaluates access conditions so that barriers to entry can more easily be overcome, allowing the poor to participate in self-directed poverty reduction.

Transactions costs are a primary determinant of economic behavior, and they exert important influence on market participation decisions. For the rural poor, who can be isolated and constrained in transport choices, logistical costs limit both supply and demand interactions with larger markets. Because of socioeconomic factors (language, education, etc.) and technology limitations, information available about market opportunities may be very limited, again undermining incentives to participate. Finally, a broad spectrum of formal and informal institutions can limit participation in urban markets by isolated rural groups.

Logistics

Poor farmers who take their agricultural products to market can be thought of as being in the ice business. The value of their product begins to "melt" as soon as they leave the farm gate. Perishability means that they can only realize revenue on a fraction of their product, and the market value of what is sold can also be undermined as its quality degrades with transport, exposure, and waiting for sale.

For these reasons, logistical support can significantly enhance the realized value of farm produce and increase incentives for market participation. More expedient and less expensive transport, cold chain storage networks, and other innovations are now taken for granted by producer groups in OECD countries, but remain rare in developing countries where infrastructure and investment barriers are large compared with real and potential farm incomes.

In the case of livestock, the logistic issue is complicated by special characteristics of animals. Some have their own motility, which trades off perishability against convenience of shipping inanimate carcasses or processed animal products. The complexity of this decision, with many contingent factors such as processing and vehicle technology, often induces smallholders to relinquish their animals to intermediaries at the farm gate. This limits their bargaining power, value added capture, and incentives to improve quality.

Information

Given their geographic, social, and economic isolation, it is hardly surprising that rural smallholders have limited information about downstream market conditions. The intermediary relationships they encounter do little to improve this, as aggregators and other middlemen bargain with them individually in the countryside and offer little insight regarding downstream market conditions, including accurate information on prices, product variety, and consumer taste. Mobile telephony has been improving this situation for some producers of commodity crops, which have standardized prices and quality characteristics, but there is little information available to smallholders who are in a position to develop differentiated and higher value local varieties.

Institutions

Private institutions, both formal and informal, can be guarantors of successful agricultural development. Informal institutions have long managed common property resources for more sustainable use, and modern producer cooperatives provide important economies of processing scale, communications, and standards for product quality, all of which increase producer net incomes. Agricultural policies, whether targeted at livestock or other products and practices, need to take explicit account of the real and potential role of private institutions in facilitating pro-poor development.

Terms of Market Participation

From an economist's perspective, market participation is the sum of all activities related to formal sector income and expenditure, in the same sense as the definition of GDP itself. When seeking more microeconomic insight, however, detailed institutional characteristics become important. For example, smallholders often have very limited participation because they face local intermediates on both sides (income and expenditure) of their formal sector activities. On the income side, farmers may sell to aggregators and distributors who exploit their limited market access and disengage them from downstream consumers.

The intermediary problem limits smallholder progress down the food value chain and is thus another barrier to market access. It can have other serious consequences, however, particularly for product quality (Ifft et al. 2007). The product quality problem has two parts, market power and moral hazard. Intermediaries use their proprietary information and market access to increase market power. This means monopsony power for buying intermediaries and monopoly power for resellers, and both contribute to a wedge of rents that could be shared by smallholder and consumer households in a more competitive environment. Moreover, removing smallholders from consumer relationships destroys a primary incentive to invest in product quality, the commensurate premium that would accrue to the producer could they be identified. Simply put, direct producer–consumer links foster a virtuous cycle of quality and revenue improvement, but this is wiped out when intermediaries interpose themselves and mask the identities of producers and consumers.

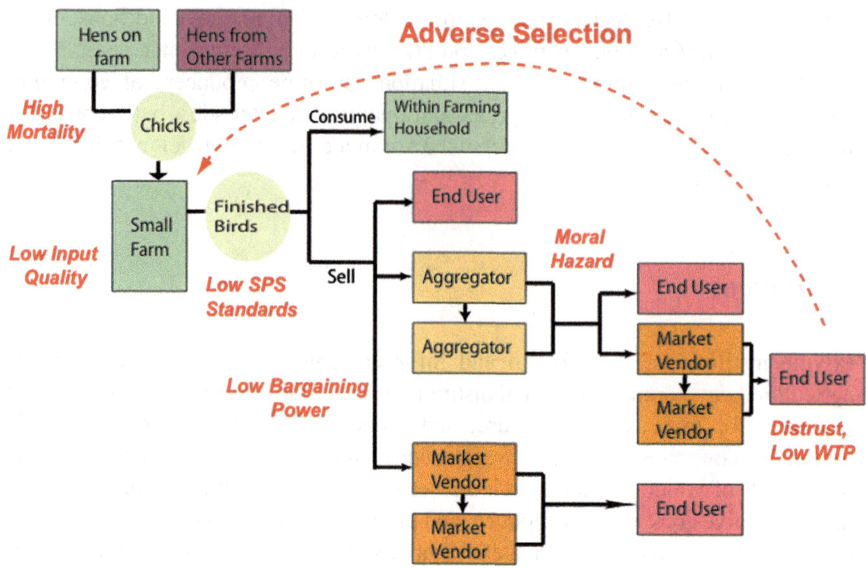

Fig. 17.6 Market failures in the informal poultry sector

Intermediary activities increase moral hazard in rural–urban market systems. Because they aggregate outputs from many producers, usually destroying origin information in the process, intermediaries reduce buyer's capacity to assess product quality/risk and thereby contribute to adverse selection (Fig. 17.6). Markets with identifiable origin information, by contrast, provide strong reputation incentives for producers to improve product quality/safety, and consumers can appropriately reward this.

Contracting as a Means of Overcoming Market Access Barriers and Information Failures

Contracts are an effective primary means of transferring property rights or resource services, permanently or temporarily, and thus have the ability to extend the process of income creation across complex networks. When contracts are transparent and enforceable, they contribute to extensive multiplier linkages, across diverse constituencies, promoting growth and economic participation, interdependence and risk reduction. All these factors improve the likelihood of sustainable and inclusive economic activity.

When designed effectively, contracting provides a direct link between producers and retailers (Fig. 17.7). Along the supply chain, this relationship can provide producers with the benefits of continuous market access and the opportunity to establish a reputation for quality. Along the value chain, retailers gain the ability

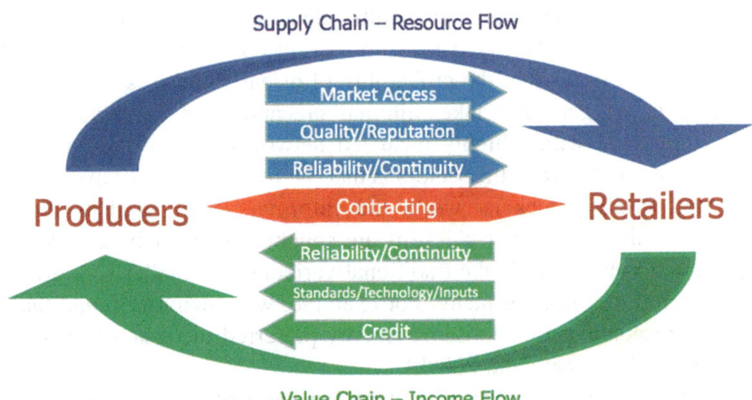

Fig. 17.7 The role of contracting in overcoming market failures

to impose quality standards on producers and may be in the position to provide credit to facilitate producers' adoption of these standards. These factors increase producer incentives to invest in product quality thereby increasing consumer safety and raising producer and retailer profit potential.

Because of their complex behavioral properties, rights are an important area for policy makers, particularly those in countries making transitions from informal to formal, rural to urban, civil society. To a significant extent, economic rights are a hallmark of modern market economies because their incentive properties animate the forces of entrepreneurship and provide the "soft infrastructure" of contractual instruments needed for more efficient specialization, forward looking investment, and risk management. In the livestock sector, policies regarding ownership have far reaching implications, and in recognition of this fact we examine Thailand's livestock sector for growth potential that reconciles the needs of both established and emerging economic interests.

In recent decades, Thailand has transformed into one of the world's major poultry exporters. This transition to a primarily formal sector will surely continue, but abrupt changes could destabilize the livelihoods among the economically vulnerable rural majority (NaRanong 2007). As a middle way during this transition, contracting offers a socially constructive approach to reduce HPAI risk while steadily improving economic conditions for rural farmers (Taenkaew 2001). Moreover, these policies can be localized to the diverse realities of Thailand, but coordinated to avert distortions and improve efficiency. Market oriented policies offer vital opportunities for private cost sharing and self-directed livelihood improvement. Governments can support these prorural supply networks by facilitating an environment congenial to enterprise development. It is within this context that we propose contracting be considered in particular areas as a mechanism to develop rural enterprise and reduce HPAI risk.

It has been debated whether large-scale industrial farms or small-scale "backyard" farms are at greater risk of spreading HPAI (Otte et al. 2007). Most large-scale industrial farms in Thailand utilize world-class levels of biosecurity.

However, they also maintain high poultry densities and expansive distribution networks. In other words, the risk of an outbreak is low; however, conditional upon an outbreak occurring, the probability of disease spread is relatively higher. Small-scale farms generally take minimal precautions but maintain low flock densities and trade within highly localized networks. In addition, an important general distinction between the sectors is that the formal sector primarily produces foreign-bred broiler chickens while the informal sector produces traditional varieties of birds. The broiler chickens are sought out for their fast growth rates and high meat volume while the traditional varieties are smaller and grow slower but have a preferred taste to many. For example, we previously found that 2/3 of households responding to a consumer survey preferred the taste of the traditional bird varieties (Heft-Neal et al. 2008b).

The formal sector, both vertically integrated and contract farms, are responsible for approximately 98% of all birds produced in Thailand. However, these farms account for only 2% of all flocks in the country (DLD 2009). Consequently, the fact that most HPAI outbreaks have occurred on "backyard" farms may not reflect higher levels of risk as much as the overwhelming number of farms of this type. It is within this context that the debate between risk factors continues. Researchers lack consensus on comparative risk levels between the smallest and largest farms, while medium-scale farms in the informal sector with relatively large flock sizes, high poultry densities, and minimal biosecurity seem to pose the greatest risk (Otte et al. 2006). An efficient utilization of resources would target risk reduction among this group. Contracting is one potential approach to achieve this targeted risk reduction while promoting rural livelihoods.

In Thailand, like the rest of the Mekong Region, poultry producers in the informal sector face a range of access barriers and information failures that result in low levels of market participation and product value/income. Despite this, consumers continue to express strong willingness to pay for traditional varieties and safety (Heft-Neal et al. 2008b). Contracting may offer an institutional mechanism to overcome these hurdles and strengthen informal sector incentives to invest for higher value production.

There is a long precedent of successful contracting agreements in the formal poultry sector (Singh 2005). CP Foods has set the standard in this area, along with other companies such as Betagro and Saha Farms. These companies have primarily utilized two distinct arrangements: total contracts and procurement contracts. With total contracts, the farmer owns land and labor but not birds, and the company provides all inputs. The farmer receives a small compensation for each bird produced and may be penalized for birds lost. Procurement contracts, however, entail the farmer purchasing birds and other inputs at previously agreed upon prices. The company provides technical support and the farmer receives a higher price per bird upon completion of the production cycle. Interviews with contract farmers in the north and northeast revealed a farmer preference for procurement contracts because they are potentially more profitable for the farmer. However, they require more capital initially to purchase inputs. Both contract types require high levels of initial investments in infrastructure.

Procurement contracts, with less intensive production technology requirements, may be ideal for the informal sector. A substantial reduction in risk levels could be achieved without the expensive task of conforming to international export standards required in the formal sector. Moreover, procurement contracts have less logistical requirements for the contractor and the profit potential for the farmer is greater. Investment requirements are likely to prohibit the poorest household from participation without assistance. In fact, medium-sized enterprise development plays an important role in rural development, and it is the medium-sized enterprises, which seem to pose the greatest disease risk.

The following section identifies the characteristics related to successful contracting as well as the regions in Thailand where these characteristics are most common.

Survey Data

The data utilized come from two household surveys. The first is the 2006 SES conducted by the National Statistics Office of Thailand. The second is a component of poultry market chain surveys carried out in the north and northeastern regions of Thailand by the authors in 2008.

Socio-Economic Household Survey

The 2006 SES was a national two-stage stratified sample survey. In total 51,970 households were interviewed across Thailand between January and December 2006. Provinces were used for stratification, 76 strata in total, and each was divided into municipal and nonmunicipal areas (i.e., urban-rural). In the first stage, primary sampling units (PSU) consisting of several block subsections of villages were selected independently in each province using probability proportional to size (PPS) sampling. In the second stage, every household within the selected PSU was listed and then individual households were chosen – 10 in municipal districts and 15 in nonmunicipal districts. The questionnaire was organized into four primary modules focusing on household characteristics, incomes, expenditures, and employment activities.

PPLPI Poultry Market Chain Surveys

The PPLPI data come from resource flow surveys carried out from January to June 2008 (Heft-Neal et al. 2008a). The sample frame consisted of three provinces: Chiang Mai in the north, KhonKaen and Nakhon Phanom in the northeast.

The sampling method was modeled after the SES methodology. The survey set was made up of four components, each with their own questionnaire; market vendors, aggregators, independent farmers, and contract farmers. In all, more than 1,600 resource flow surveys were carried out. For the purposes of this research activity, we utilize data from the farmer surveys. Of the 1,500 plus farmer observations, 1,457 were independent farmers and 63 were contract farmers. The questionnaires consisted of four sections: household background, poultry incomes, production methods, and trading relationships. In addition, the contract farmer questionnaires included an additional component that detailed the investments made prior to entering contractual relationships with companies.

Identification of Eligible Farmers

Utilizing data from our small sample survey of independent farmers, we identify the medium-scale farmers that raise at least 50 chickens *solely for the purpose of sale*. We omit farmers who raise any poultry for the purpose of consumption because it suggests that poultry production is less of a primary economic activity. In addition, we omit farmers that raise less than 50 birds because of the scale necessary for a contracting scheme. In the future, these requirements can be relaxed. For example, farmer cooperatives could be used to scale up production levels to achieve the necessary scale to make contracting profitable.

Having identified potentially eligible farmers for contracting, we look at identification criteria such as income from poultry, total income, land size, flock size, and credit availability. Finally, we identify regions where large number of observations in the SES survey meets the identification criteria to predict which regions are the most eligible for contract extension activities.

Identifying Relevant Attributes in the Small Sample

From the PPLPI survey data, we are able to identify 162 eligible farmers from our sample of 1,457 farmers in the informal sector that meet the criteria described above. Descriptive statistics of the identified farmers, as well as the remaining independent farmers and the already contracted farmers are listed in Table 6.1.

There is a large discrepancy in scale between the farmers identified in the informal sector and the contract farmers in the formal sector. Contract farmers in the formal sector raise broiler chickens in cages with high densities while farmers in the informal sector raise traditional varieties with free-range production methods. A successful contracting scheme in the informal sector would likely need to achieve a scale in between those above to make infrastructure investment worthwhile while maintaining free-range production methods.

The primary factor that distinguishes the farmers suitable for contracting from the rest of the informal sector is that poultry production is a serious economic activity. Although the average farmer in this group only derives 10% of their income from poultry production, flock sizes of 100 birds require time and energy investments that signal these households raise poultry for more than a hobby. Moreover, total capacity suggests that there is ample room for scaling up production levels if a farmer decides to invest in improved infrastructure. Consequently, these are the households that are the most eligible for participating in contracting programs without additional assistance.

This section discussed qualifications for eligible farms in the small sample. The next section uses the SES survey data to identify regions with farmers that possess descriptive statistics similar to the eligible farmers described in Table 6.1.

Identifying Eligible Farmers in the Large Sample

To identify eligible farmers in the SES survey, we develop parallel criteria for eligibility. Farmers, given they produce poultry, are identified using four criteria:

- Percentage of agricultural income from livestock
- Percentage of total income from agricultural activities
- Total cash income
- Land area for agricultural use

Criteria 1 and 2 proxy for poultry income, which is necessary because agricultural income is aggregated in the SES data. Criteria 3 is comparable to total cash income in Table 17.1 and criteria 4 is used to proxy for flock size and total capacity since these data are not available in the SES.

To focus on primarily agrarian households, we require that at least 80% of total income come from agricultural production. To meet the criteria established with the PPLLPI data, we also require that *livestock* income be at least 6,000 baht per month and account for at least 11% of all agricultural income. (Note: poultry income is not

Table 17.1 Summary statistics by eligibility group, PPLPI survey

Group	Poultry income (baht/year)[a]	Total income (baht/year)[a]	% Income from poultry	Flock size	Total capacity
Independent farmers – eligible	6,000	80,000	11	105 birds	190 birds
Independent farmers – other	1,000	50,500	4	35 birds	55 birds
Contract framers (broiler farms)	180,000	220,000	80	6,700 birds	9,000 birds

[a] Poultry income and total income are not equivalent to profits, but represent cash income generated (not costs)

available, only income from all livestock as well as a dummy variable for poultry production).

In addition, we select households that generate between 50,000 and 150,000 baht per year in total cash income, so that they would fall within the 95% interval of households deemed eligible by the PPLPI sample. Intuitively, we would like to initially target farmers that are relatively poor, yet have the means necessary to make investments in production technology. Subsequently, we can relax these requirements if additional assistance is provided to help households overcome market failures. Finally, since there are no data for flock size, we use land as a proxy and estimate the PPLPI data criteria in terms of agricultural land used for livestock.

Using these criteria we first select the most eligible households. These households are the most likely to be viable contract farmers without additional assistance. Subsequently, we relax the income requirements to select households that would be eligible with additional financial assistance, most likely in the form of credit. Finally, we relax the flock size (land) requirement to highlight households that have many of the same traits as the most eligible households, but would need to pool resources with others to meet minimum scale requirements for contracting.

Using these selections, we evaluate the eligibility of regions by the number of households that meet the criteria within a given PSU. The provinces with the most selections are listed in the results section.

Limitations

There are several obvious limitations to the approach taken here. The first is that collecting detailed agricultural production data is not the purpose of the SES. Consequently, most agricultural data in the SES is aggregated. We are unable to disaggregate poultry production data from other livestock production data. This puts undue weight on regions with high levels of nonpoultry livestock production. Ideally, we would incorporate data from a detailed agricultural survey carried out at the national level. However, given the available data, we believe this is a "best-guess" estimate of regions that would be viable for pilot programs.

Results

Applying these criteria to the SES data, we find 654 "most eligible" households out of the nearly 45,000 total observations (having dropped observations with missing data). These households would be the most suitable to enter contracts without any additional assistance. After relaxing the minimum income requirement, an additional 271 households become eligible, raising the total number of viable households to 925 if financial assistance is provided. Finally, we relax the minimum

Table 17.2 Summary of eligibility categories, SES

Eligibility category	Number of observations	Accumulative observations	Percent of observations	Monthly per cap income
1	654	654	1.5	5,000
2	271	925	2.1	1,200
3	1,405	2,330	5.2	1,600

1 eligible without assistance; *2* eligible with financial assistance; *3* eligible for co-operative farming

Table 17.3 Provinces with most eligible households (no assistance)

Province	Region	Eligible SES observations	Millions of native chicken[a]	Millions of broiler chicken[a]
Roi Et	Northeast	34	1.9	0.5
Amnat Charoen	Northeast	29	0.5	0.4
Sakon Nakhon	Northeast	26	2.3	0.2
Chaiyaphum	Northeast	23	1.4	3.4
Ubon Ratchathani	Northeast	22	2.4	3.7
Kanchanaburi	Central	22	0.8	7.9
Phatthalung	South	22	0.6	2.1
Ratchaburi	Central	19	0.4	5.8
Tak	North	18	0.5	0.0
Kalasin	Northeast	18	1.1	0.1
Maha Sarakham	Northeast	16	1.1	0.3
Phitsanulok	North	15	1.0	0.5
Nakhon Si Thammarat	South	15	1.9	1.9
Phrae	North	15	0.8	0.1
Uthai Thani	North	15	0.3	0.8

[a] Number of chickens raised in province (rounded to nearest 100,000 birds). DLD (2009)

farm size criteria to include households that would be eligible if resources were pooled to meet minimum scale requirements for viable contracting. In total, we identify 2,330 households that would be eligible with some form of assistance (Table 17.2).

Most of the provinces with the greatest number of eligible households are in the northeastern and northern regions of Thailand. In fact, the criteria select only two provinces each from the central and southern regions, respectively. These results are not surprising. Households in the north and northeastern regions tend to rely more heavily on agricultural production for their livelihoods, including livestock. Consequently, supporting rural enterprise development in these regions is a valuable use of resources.

Tables 17.3–17.5 list the provinces with the largest number of observations in each category. Figures 17.8–17.10 highlight these provinces on a map. Once a household is classified as eligible, it is not counted in subsequent measures. For example, a household in Roi-Et that would be eligible without financial assistance would not be counted as a household that would be eligible with financial assistance.

Table 17.4 Provinces with most eligible households (financial assistance)

Province	Region	Eligible SES observations	Millions of native chicken[a]	Millions of broiler chicken[a]
Sakon Nakhon	Northeast	21	2.3	0.2
Roi Et	Northeast	19	1.9	0.5
Surin	Northeast	15	2.1	1.0
Buri Ram	Northeast	15	4.3	4.0
Phayao	North	15	1.4	0.1

[a] Number of chickens raised in province (rounded to nearest 100,000 birds). DLD (2009)

Table 17.5 Provinces with most eligible households (co-operative farming)

Province	Region	Eligible SES observations	Millions of native chicken[a]	Millions of broiler chicken[a]
Roi Et	Northeast	100	1.9	0.5
Sakon Nakhon	Northeast	48	2.3	0.2
Surin	Northeast	47	2.1	1.0
Si Sa Ket	Northeast	41	1.8	0.4
Ubon Ratchathani	Northeast	41	2.4	3.7
Yasothon	Northeast	39	1.0	0.5
Nong Bua Lam Phu	Northeast	39	0.6	0.3
Phayao	North	39	1.4	0.1
Chiang Rai	North	39	1.8	0.5
Uthai Thani	North	37	0.3	0.8
Phrae	North	36	0.8	0.1
Maha Sarakham	Northeast	35	1.1	0.3
Kalasin	Northeast	33	1.1	0.1
Phitsanulok	North	33	1.0	0.5
Ratchaburi	Central	32	0.4	5.8
Buri Ram	Northeast	31	4.3	4.0
Kanchanaburi	Central	30	0.8	7.9

[a] Number of chickens raised in province (rounded to nearest 100,000 birds). DLD (2009)

Despite the requirement of unique observations meeting each criteria, Roi Et and Sakon Nakhon provinces have among the most observations in every category. Consequently, according to the stated criteria, these two northeastern provinces are the most viable for any form of contracting program. Other northeastern provinces ranking highly include Surin, Buri Ram, and Ubon Ratchathani. In the northern region, Phayao and Phrae rank highly in two of the three categories, respectively.

In addition to the number of households meeting the eligibility criteria, the tables report provincial native and broiler chicken production from 2009. Roi Et and Sakon Nakhon, the most eligible provinces, each reported nearly two million native chickens produced in 2009, and less than 500,000 broiler chickens. The size and importance, relative to the formal sector, of the native chicken production sector in these provinces supports the idea that contracting is most viable in these areas. The relative size of the two sectors in each province provides insight into the nature of the local livestock economy. Provinces with large native poultry sectors, relative

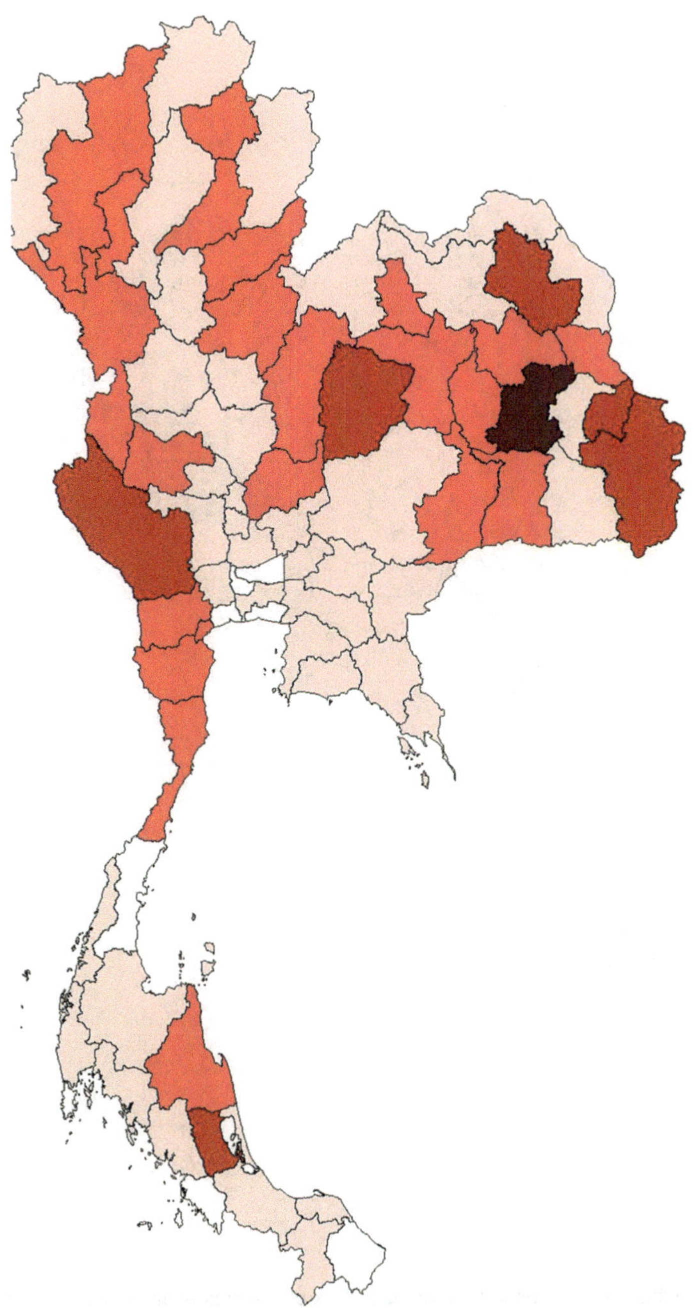

Fig. 17.8 Map of Provinces with most eligible households (no assistance)

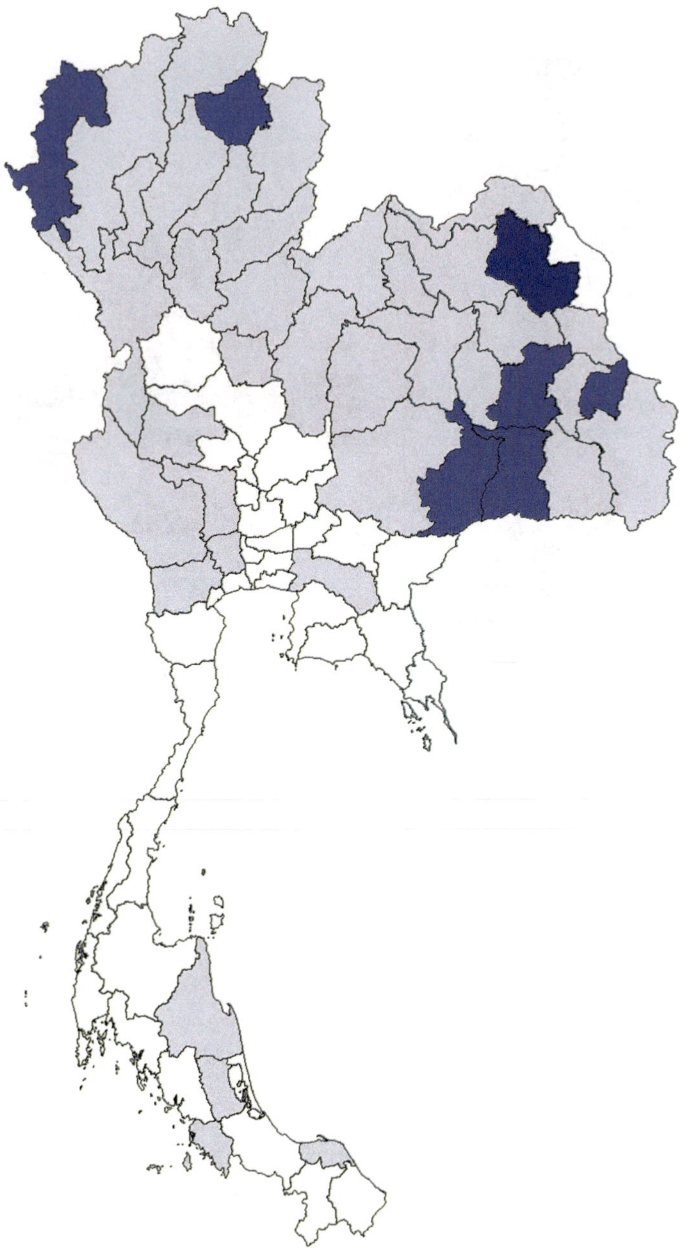

Fig. 17.9 Map of provinces with most eligible households (financial assistance)

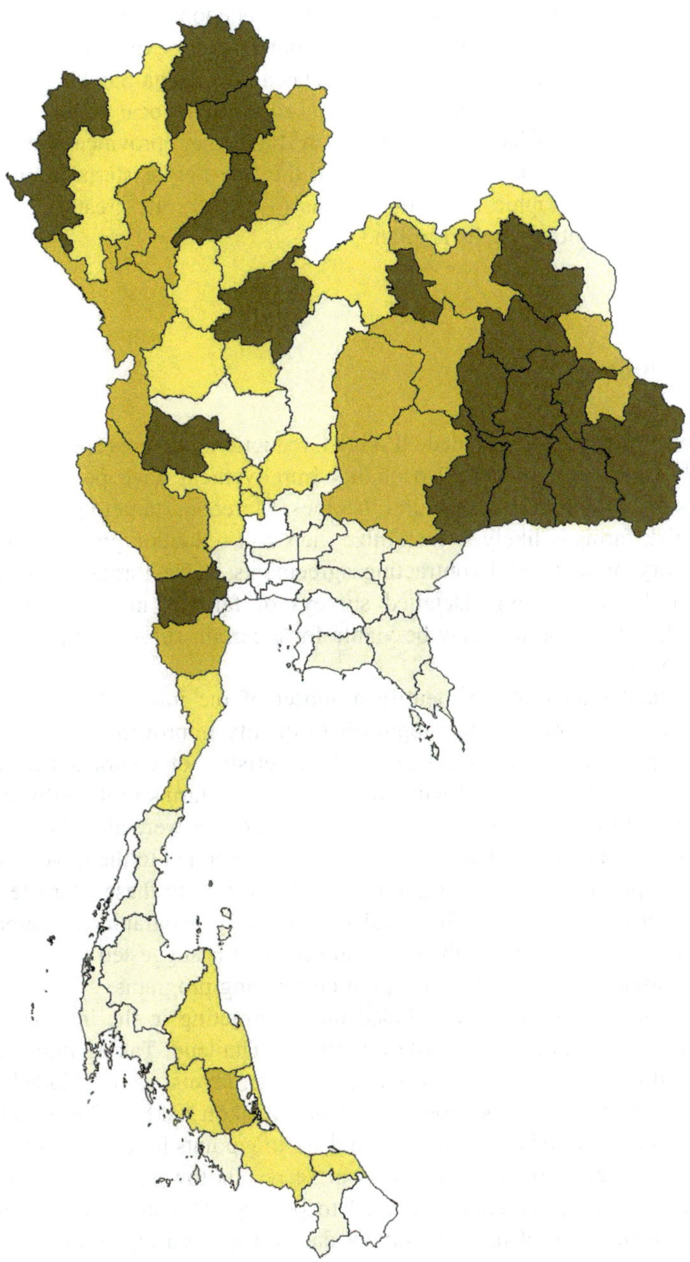

Fig. 17.10 Map of provinces with most eligible households (pooling resources)

to broiler sectors, are more likely to have local economies based on the informal sector and would thus seem to have greater potential for expansion of contracting.

The native chicken sectors in Surin, Kalasin, and Maha Sarakham, also in the northeast, appear to have similarly dominant roles in the local poultry sector while the broiler sectors in Chaiyaphum and Ubon Ratchathani provinces dominate even their large native chicken sectors. In fact, the five northeastern provinces listed above form a geographic block that seems to provide the greatest potential for native chicken contracting in Thailand.

Conclusions

In the long term, publicly funded HPAI risk reduction measures are not sustainable. Privately financed approaches to quality improvement have potential to reduce outbreak risks while increasing rural incomes. Moreover, targeting these programs to high-risk farms is likely to maximize their risk reduction potential. Thailand's long history of successful contracting agreements in the formal sector has set a precedent for contracting. Detailed surveys of farmers in the informal sector suggest that this approach may be viable for a certain subsection of the informal sector as well.

This chapter attempts to identify a subset of the informal sector for which contracting would be a viable approach to quality improvement. Using detailed small sample survey data, production characteristics of contracted farms in the formal sector were compared with characteristics of farms in the informal sector. Farms in the informal sector with similar characteristics were identified and a set of criteria were established. These criteria were then applied to the large sample SES survey and provinces were identified where farmers with these characteristics are most numerous. The approach intended to identify geographical regions where contracting is most viable in the informal sector. It is suggested that these regions would be logical starting points for pilot contracting programs.

Using this approach, we concluded that contracting in the informal sector is likely most viable in the northeastern region of Thailand. Taking into account the established criteria, as well as current production levels, we concluded that sites within the geographical block formed by Roi Et, Sakon Nakhon, Surin, Kalasin, and Maha Sarakham would be the most natural starting points for contracting programs.

However, data limitations may bias these results toward provinces with other types of livestock production unrelated to poultry. This approach would benefit from the availability of national survey data focused on agricultural production. Nonetheless, we argue that this chapter presents a reasonable starting point for targeting regions where contracting in the informal sector is more likely to be viable for potential pilot programs that strive to achieve HPAI risk reduction in conjunction with rural development.

References

Department of Livestock Development (2009): http://www.dld.go.th/ict/th/index.php?option=com\content\&view=article\&id=89:-2552\&catid=74:2009-11-01-07-43-07\&Itemid=60

Heft-Neal, S., J. Otte, W. Pupphavessa, D. Roland-Holst, S. Sudsawasd, and D. Zilberman (2008a). *Supply Chain Auditing for Poultry Production in Thailand.* RR Nr. 08–09, FAO, Rome, September.

Heft-Neal, S., J. Otte, D. Roland-Holst (2008b) *Poultry Market Surveys for Thailand.* Mekong Working Paper No. 12, FAO, Rome.

Heft-Neal, S., J. Otte, and D. Roland-Holst (2009). *Assessment of Smallholder Indigenous Poultry Producer Viability in Thailand.* Mekong Team Working Paper No. 8. FAO, Rome.

Ifft, J., J. Otte, D. Roland-Holst, and D. Zilberman (2007). *Demand-Oriented Approaches to HPAI Risk Management.* Pro-Poor Livestock Policy Initiative Research Report Nr. 07–14; October 2007.

NaRanong, V. (2007). *Structural Changes in Thailand's Poultry Sector and its Social Implications.* Thailand Development Research Institute. Bangkok, Thailand.

Otte, J., D. Pfeiffer, T. Tiensin, L. Price, and E. Silbergeld (2006). *Evidence-based Policy for Controlling HPAI in Poultry: Bio-security Revisited.* Pro-Poor Livestock Policy Initiative Research Report 249.

Otte J, Pfeiffer D, Tiensin T, Price L and Silbergeld E (2007). *Highly pathogenic avian influenza risk, biosecurity and smallholder adversity.* Livestock Research for Rural Development. Volume 19, Article 102.

Porn-Amart, T. (2003). *The Success of Native Chicken Raising as a Minor Occupation in Sansai District, Chiang Mai Province.* Master Thesis, Faculty of Agriculture, Chiang Mai University, Chiang Mai, Thailand (in Thai).

Singh, S. (2005). *Role of the state in contract farming in Thailand: experience and lessons.* ASEAN Economic Bulletin. Monday, August 2 2005.

Sudsawasd, S. and P. Wisarn, (2008). *Structural Transition in Thailand's Poultry Sector.* Thailand Development Research Institute.

Chapter 18
Micro Contracting and the Smallholder Poultry Supply Chain in Lao PDR

Drew Behnke, David Roland-Holst, and Joachim Otte

Introduction

The population of Lao PDR remains a predominately rural population and engaged in subsistence agriculture. For this reason, livestock generally and poultry in particular can be instrumental to sustaining and improving livelihoods. However, these livelihood potentials are far from being realized, because a myriad of market access challenges and information/incentive failures deter participation and undermine value creation across low income supply chains. These include, but are not limited to, transport costs, search costs, the nature of sales (often in times of emergency), information asymmetries, and agency problems that undermine investment incentives and reduce the likelihood of smallholders effective market participation.

Survey work conducted in Lao PDR can be used to support evidence-based solutions to these market failures. Market chain questionnaires were implemented in five provinces from December 2009 to July 2010 and include detailed producer, trader, and vendor surveys. Data from questionnaires provide a supply chain audit demonstrating how the smallholder supply chain functions and where failures occur.

Furthermore, information was collected on the role of informal contracting systems. In industrial poultry supply chains, contracting schemes have proven successful in allowing producers to link directly to vendors, increasing both the quality and value of products. However, in a smallholder system, conventional contracts are not logical given their complexities, high commitments, and rigid

D. Behnke (✉)
Department of Economics, University of California, Santa Barbara
e-mail: dbehnke@umail.ucsb.edu

D. Roland-Holst
Department of Agricultural and Resource Economics, University of California, Berkeley, CA, USA

J. Otte
Food and Agriculture Organization of the United Nations, Rome, Italy

D. Zilberman et al. (eds.), *Health and Animal Agriculture in Developing Countries*, Natural Resource Management and Policy 36, DOI 10.1007/978-1-4419-7077-0_18,

structures. Yet adapting a contract system to smallholder systems remains attractive because it could help reduce the inefficiencies found in the poultry supply chain. Here lies the role of micro-contracts; types of informal agreements that allow more flexibility than a formal contract system yet allow producers to link directly to vendors. Micro-contracting presents a potential solution to the market failures that arise in the smallholder poultry supply chain and are worthy of further examination.

Background

Lao PDR Poultry Sector

Smallholder production in Lao PDR is dominant representing 94% of all agricultural outputs (Wilson 2007a). Almost nearly as widespread is poultry production, with 90% of households keeping at least one species of bird (Wilson 2007b). Lao PDR has an estimated 20.5 million birds in its national flock, composed of approximately 77% chickens and 22% ducks (Table 18.1 and Fig. 18.1). Looking at the national poultry flock by type of bird demonstrates that local birds, especially local chickens, represent the vast majority of birds. However, it should be noted that number of commercial birds are under-represented, as the estimates on the national flock by types of birds are drawn from a household survey and large commercial farmers were not represented in the data. To gauge the size of the commercial production system, we must therefore look at the distribution of the poultry population by production system, which can be seen in Fig. 18.2.

Table 18.1 Poultry population and density by province, 2007

	Poultry population (thousands of birds)	% of total	Poultry density (birds per sq km)	Birds per person
Lao PDR	20,453	100	86.4	3.5
Luang Namtha	324	1.6	34.7	2.1
Phongsali	532	2.6	32.7	3.1
Oudomxay	834	4.1	54.3	3.0
Vientiane Capital	808	4.0	206.1	1.1
Savannakhet	2,007	9.8	92.2	2.3

Source: MAF (2008) (survey provinces listed)

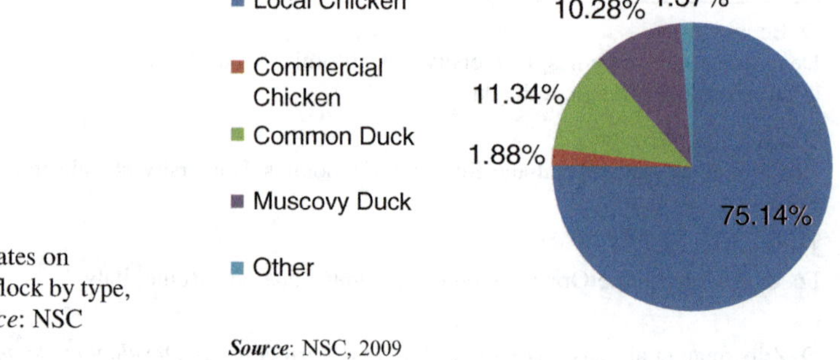

Fig. 18.1 Estimates on national poultry flock by type, 2007/2008. *Source*: NSC (2009)

Fig. 18.2 Poultry population distribution by production system, 2006. *Source*: NSC (2008)

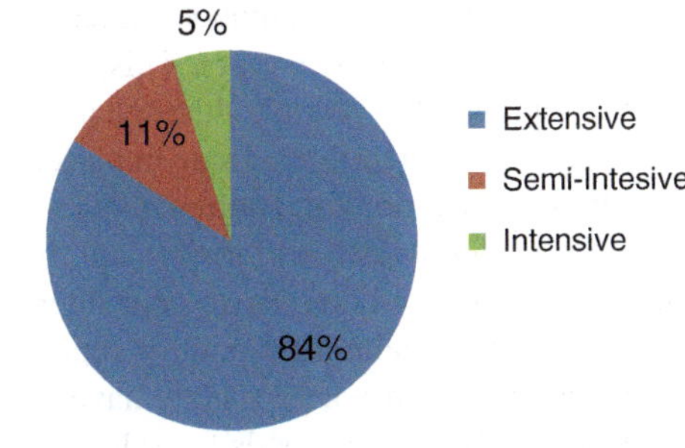

Source: NSC, 2008

Poultry Production Systems and Supply Chain

Officially, the Department of Livestock and Fisheries only separates production between backyard and commercial systems and reports there are some 800,000 birds raised in a commercial system, which represents approximately 3.8% of the national flock (DLF 2009). However, it is more accurate to disaggregate production systems further into one of the following three categories: (1) traditional, small-scale, extensive backyard poultry production, (2) semi-intensive, small- to medium-scale, market-oriented, commercial poultry production, and (3) intensive, large-scale, industrially-integrated poultry production. As the DLF lumps semi-intensive production with traditional backyard systems, only estimates are available on the distribution of birds by this classification of production systems, which are found in Fig. 18.2.

Extensive Production

Traditional backyard systems are the primary production method and account for approximately 84% of the national poultry flock. This production method is common throughout the country in both urban and rural areas, and it is especially favored amongst the poor due to its low inputs. In this system, birds scavenge for food during the day and sleep at night in trees, underneath houses, in natural sheds, or in rudimentary enclosures. Therefore, the costs of inputs are extremely low and are often only the small cash expense for day old chicks (DOCs). However, once birds become mature, there is no need to purchase DOCs because they can be self-supplied through hatchlings and thus the system is self-replacing. Flock sizes are typically small (under 100 birds), and different bird species are raised in conjunction with each other. The majority of small-scale poultry production is intended for household consumption and only after satisfying family nutrition requirements are birds sold or given as gifts. When sales do occur, producers face a variety of options. First, they can simply sell their products themselves in effect acting as vendors. This commonly occurs if the producer regularly operates as a vendor or if they sell products to neighbors. Second, a producer may sell their products to vendors directly in the market. This often requires a preestablished relationship or agreement with the vendor, although in some cases producers may bring birds to the market with no prearranged sale in place. Finally, a producer can sell their products to a trader or aggregator. Aggregators rely on networks within different villages and buy birds from numerous farmers and then transport the birds to the market where they are either sold directly to consumers or to vendors. Aggregators are a unique feature of the smallholder poultry supply chain and they owe their existence to a production system where traders have to buy from numerous sources to obtain a satisfactory bundle of goods for sale.

Semi-Intensive Production

Semi-intensive production systems are often lumped into traditional backyard systems, but with flock sizes of 50–1,000 birds and intensive qualities it should be considered a unique production system. Semi-intensive production accounts for an estimated 11% of Lao PDR's poultry population. Housing is provided for birds either in permanent or makeshift enclosures, with both feed and water being provided in plates, trays, and/or bowls. The level of intensiveness can vary widely and often gardens, backyards, or vacant pieces of land are fenced-in to house birds at night after scavenging for feed during the day. Bio-security measures, albeit limited, represent another key difference between traditional backyard systems as sanitation, treatment, and management are given more priority to limit disease outbreaks. Birds are more frequently sourced from external channels, although the majority of restocking remains self-replacing from the flock. Production is usually focused on poultry meat or eggs. Compared with traditional backyard production systems, these outputs have higher rates of marketing and more formal marketing channels.

Intensive Production

While the lines are often blurred between traditional and semi-intensive production, intensive production demonstrates a clear divide between poultry production systems. Intensive production operations generally have flocks of 1,000–10,000 birds and account for approximately 5% of the national poultry population. These operations are characterized by advanced production infrastructures designed for commercialization that include elaborate housing, feeding, and drinking systems. Many of these farms have contracts with CP, a large Thai conglomerate, and source energy dense commercial feed and replacement chicks directly from Thailand. Furthermore, infrastructure and technical knowledge is also sourced from Thailand, and these farms operate very similarly to CP contract farms in Thailand. CP is by far the strongest branded commercial poultry product, even with unique Lao packaging. CP products are commonly found in traditional wet markets amongst vendors of indigenous products. In total, the DLF reports there are 193 commercial farms split between broiler chickens (4), layer chickens (112), layer ducks (52), and quail (25). Most of these operations are found in or around the large urban areas of Lao PDR, with the majority (61%) located in the Vientiane Capital (DLF 2009).

Smallholder Poultry Supply Chains: Evidence from Lao PDR

Throughout the GMS, the smallholder poultry supply chain is nearly indistinguishable. In general, it can be described as the marketing and sales of poultry products between three actors: producers, traders, and vendors. Questionnaires were specifically tailored to capture the differences between these actors and reveal how each group contributes to the marketing of poultry products.

Producer: Rates of Marketing

Poultry farmer surveys consisted of three distinct questionnaires: a smallholder chicken and duck producer questionnaire, a largescale chicken producer questionnaire, and a largescale duck questionnaire. The largescale questionnaires are intended for semi-intensive producers households and are applicable for both meat and egg production.

Before birds can enter the supply chain, we must first examine the producer's decision to market their products. Looking at small-scale production first, it can be seen how limited sales of poultry products are. Although 41.2% of respondents sold at least one type of bird, the vast majority of sales occur infrequently. Only 1 of the 1,050 small-scale producers reported selling birds weekly, and just 4% reported selling birds monthly. Egg sales amongst smallholders are even more limited with only 1.8% of respondents reporting that eggs are ever sold. Just like birds, the vast majority of the sales are infrequent with 1.4% of producers selling eggs less than once a month. Disaggregating between chicken and duck shows that chickens are sold more frequently than ducks. Only 28.4% of small-scale producers reported selling ducks compared to 36.6% for chickens. In regards to egg sales, duck eggs were marketed slightly more with 1.4% of producers selling duck eggs compared to 1.1% for chicken eggs (Table 18.2).

Looking at the difference between rural and urban areas, we see that marketing is slightly more common amongst urban small-scale producers. In urban areas, 43.3% of respondents sold at least one type of bird compared with only 40.9% in rural areas. Again separating between chickens and ducks reveals that chickens are marketed more frequently. For chickens, 37.1% of urban small-scale producers sold birds compared with 36.6% for rural producers. For ducks, the discrepancy between urban and rural is smaller with 28.9% of urban producers selling birds vs. 28.4% for rural.

Table 18.2 Frequency of sales, small-scale producers

	Chicken (%)		Duck (%)	
	Birds	Eggs	Birds	Eggs
Never	63.43	98.86	71.62	98.57
Weekly	0.10	0.10	0.10	0
Monthly	3.62	0.19	3.14	0.19
Less than once a month	32.86	0.86	25.14	1.24

Large producers have much higher levels of marketing and monthly sales are common. Large chicken producers that raise birds for meat sell an average of seven birds per month, although there is a great variation of sales within months and between months. For example, sales within a single month ranged anywhere from 0 to 170 birds and between month averages ranged from 2 to 18 birds per month. The same is true for large duck meat producers who sell an average of 12 birds per month, but with large disparities within and between months. Although some large producers sell birds every month, it is more common for producers to sell birds periodically, and thus there are clusters of sales explaining the large variations. Furthermore, sales have a distinct temporal nature with the highest sales months occurring during the festival seasons especially in April and November.

With regards to egg production, large producers had larger flock sizes to provide a marketable quantity of eggs. In these systems, flock sizes of couple thousand birds are common and although bird sales are scarce, when they do occur the numbers were generally quite large. For example, one large chicken egg farm with a flock size of 5,800 layers had two sales in the previous year: 2,000 birds in June and 3,000 in December. This reflects that the marketing of birds in an egg production system occurs when birds are no longer actively laying, and thus the sales are infrequent and typically large in size. In regards to egg sales, there is a great variation in the data due to the difference of flock sizes and thus the average amount of eggs sold per month is nondescriptive. A much more telling statistic is the average number of eggs laid per bird. For chickens, active layers laid 16 eggs on average over the course of the previous month while active layer ducks averaged 14 eggs.

Producer: Marketing Channels

When a small-scale producer does sell their products, they reported selling them through four different channels: aggregators, market vendors, households and other farmers, or to restaurants and food vendors directly. Enumerators asked respondents to list what percent of birds and eggs were sold through these channels and the averages are listed in Table 18.3. The Vientiane producer data reveals that aggregators represent the single largest group of sales. However, most small-scale producers arrange the majority of sales on their own, either selling directly to end-users or to

Table 18.3 Marketing channels, percentage of sales, small-scale producers

	Chicken (%)		Duck (%)	
	Birds	Eggs	Birds	Eggs
Aggregator	42.96	8.33	46.12	6.25
Market vendor	20.26	26.67	18.55	56.25
Households/other farmers	36.46	65	34.63	37.5
Restaurants/food vendors	0.33	0.86	0.67	0

vendors. This is especially true for egg sales, where only two farmers reported selling eggs to traders.

Looking at the percentage of sales averages masks some other unique features of the data. All of the averages have large standard deviations as sales range from 0 to 100% and typically a producer sells all of their products (or none) through one of the four channels. Thus the most common responses are 0 and 100% and looking at the data this way reveals further information about how products are sold. Aggregators are the most common channel for exclusive sales with 33.5 and 35.3% of chicken and duck meat producers selling all their products to traders. When producers sell to vendors or end users, it is more common for them to do so in conjunction of other sales. For example, only 9 and 23% of bird sales went exclusively to market vendors and households, respectively.

For large producers, sales to aggregators and vendors are more common. This reflects the fact that production is commercially oriented and is intended for consumers at markets. Thus large producers are more likely to rely on traders or arrange sales directly with vendors. This is in contrast to small-scale production where sales to neighboring households are common as producers sell to people within their village when extra income is needed. Sales to hatcheries and integrators also appear in large production, but the average percentage of sales for these two categories are particularly misleading as only one producer each reported selling through these channels and their sales percentage were much higher than the average reflects. For example, the producer who sold eggs to the hatchery sold 80% of their eggs through this channel while the producer that sold birds to the company sold 100% of their birds (Table 18.4).

Aggregator

Moving down the supply chain to aggregators, survey data can further illustrate the role aggregator's play in smallholder poultry marketing. Aggregators typically trade with numerous sources and a trader may routinely visit several different villages and purchase birds from many different farmers within the villages. This is especially true for traders who source from small farmers as they can only buy a few birds from each source. Thus they must aggregate birds from many sources to obtain a marketable quantity of goods for sale. On average, aggregators sourced from 13 different sources over the course of a month, ranging anywhere from 1 to

Table 18.4 Marketing channels, percentage of sales, large producers

	Large chicken (%)		Large duck (%)	
	Birds	Eggs	Birds	Eggs
Aggregator	49.69	69.52	49.67	45.83
Market vendor	25.41	23.55	21.09	25.83
Households/other farmers	22.35	3.39	24.46	28.83
Restaurants/food vendors	0.51	0.97	4.78	0
Hatchery	0	2.58	0	0
Company	2.04	0	0	0

Table 18.5 Average number of birds/eggs traded per week, aggregator

	Chicken		Duck	
	Purchased	Sold	Purchased	Sold
Meat	150.88	64.67	134.6	47.58
Active layer	27	15.29	30	21
Spent layer	31.5	18.2	29	30.2
Males (breeding)	15	16.6	24.83	11.5
Chicks/ducklings	215	70	80	17.5
Eggs	10	10	8	40

108 sources. In general, traders who source from few sources often trade with large farms and those that have numerous sources work with small-scale producers.

The traders our enumerators encountered sourced chickens exclusively from small farms (less than 100 birds) and other traders. The same was virtually true for ducks, except one aggregator reported sourcing 30% of ducks from a large farm. Farms with less than 50 birds were the single largest source for the traders with approximately 63% of purchases coming from these farms.

With regards to the size of the trading operations, aggregators purchase an average of 151 chicken and 135 duck broilers a week. These birds are the most commonly purchased type, although some traders reported buying layers, males for breeding, chicks/ducklings, and eggs as well. For traders who sell the other types of products, the averages are listed in Table 18.5.

As Table 18.5 demonstrates, the reported buying and selling margins do not always add up. This could occur for a variety of reasons. First, aggregators may purchase more birds than they sell and keep the birds for later dates. It is common for aggregators to keep birds overnight and 50% said they routinely do, keeping birds for 1–7 days at a time. Second, aggregators may want to underemphasize the amount of birds they sell in fear that if their revenues are too large they may be subject to tax and inspection. Although enumerators stated they were working with the National University of Laos and the FAO, many respondents can still be skeptical. Finally, there is likely a large amount of human error both from the respondents and enumerators. In theory, a good enumerator should catch any large

Table 18.6 Marketing channels, percentage of sales, aggregators

	Chicken (%)	Ducks (%)
Market vendors	70.38	76.19
Consumers	24.23	18.57
Restaurants/shops	1.15	0.95
Other traders	4.23	4.29

Table 18.7 Marketing channels, percentage of purchase, market vendors

	Chicken (%)	Ducks (%)
Trader (buy at market)	36.77	39.72
Trader (buy at home)	9.79	23.06
Directly from Farm	13.54	19.44
Other market vendors	5.73	11.11
Own flock	0.21	1.11
Company (CP)	33.96	5.56

discrepancies between the reported buying and selling quantities, but in practice this is not always the case. Additionally, many aggregators may not readily know these numbers off the top of their heads and simply provide loose estimates.

Aggregators most commonly sell their products to market vendors and consumers, and these two groups account for 94.61 and 94.76% of chicken and duck sales, respectively. Vendors constitute the majority of these sales with over 70% of sales for both chickens and ducks (Table 18.6). Vendors are also the channel where exclusive sales are most common and 46.15 and 57.14% of aggregators reported selling all their chickens and ducks, respectively, to vendors alone.

Market Vendors

Vendors represent the final link of the supply chain and again survey data provides information on how these actors contribute to the marketing of poultry. Market vendors in Vientiane purchase their birds from a variety of sources, which are listed in Table 18.7. Purchases from aggregators are the largest source of birds for vendors, where birds are either purchased at the market or the vendor's home. Purchasing birds directly from a farm was also common, especially for duck vendors. This occurs either when a vendor buys form several farmers in a village (in effect acting as an aggregator) or farmers deliver birds to the market.

For chicken vendors, purchasing poultry products from a company (specifically CP) was also a popular channel of purchases. These products are produced domestically and CP delivers products directly to the market for the vendors to sell. Vendors of CP products were most frequently encountered in Vientiane with 31% of the vendors reporting they sourced 100% of their products from the company.

Table 18.8 Average number of birds/eggs traded per day, market vendors

	Chicken		Duck	
	Purchased	Sold	Purchased	Sold
Commercial	37.61	34.38	22.5	21.5
Local	19.58	15.7	15	8.73
Eggs	843.16	612	625	472.5

Table 18.9 Marketing channels, percentage of sales, market vendors

	Chicken (%)	Duck (%)
Consumers for HH use	64.79	61.67
Restaurants/shops	9.69	10.67
Traders	5	4.67
Other vendors	9.48	7.33
Farmers	4.06	2
Unsure who buyer is	6.46	12.33

On average, vendors of commercial birds purchase larger volumes than vendors of local birds. Commercial chicken vendors purchase an average of 22 kg or 38 birds per day, vs. 18 kg or 20 birds a day for local chickens. Commercial duck vendors are less common than chickens, but their volumes are larger for the most part. Commercial duck vendors purchased an average of 13 kg of meat or 23 birds per day. Local duck meat vendors purchased an average 14 kg a day or 15 birds a day (Table 18.8). Vendors were asked to think about their volumes in either kilograms or heads per day so that the most natural response would be recorded. Responses were in favor of heads per day (62.5% of responses), although most vendors price their products by kilogram. This is in contrast to other regions of the country where birds are sold per head only, and most vendors never think of their products by weight.

With regards to sales volumes, vendors typically sell all of their purchases or have a few leftover products. Vendors most commonly keep their unsold products for sale the next day (97% reported doing this). Most sales are to consumers for household use, followed by restaurants/shops, and other market vendors (Table 18.9).

Micro-Contracting: Evidence from Lao PDR

The supply chain data effectively demonstrates that sales of poultry products among smallholders are limited and when they do occur intermediary agents have a large role in the marketing of these goods. It is the prevalence of these intermediary agents that create market failures in the supply chain resulting in reduced product quality, sales, and income for producers.

Asymmetric information is particularly damaging for smallholder producers because it creates problems of moral hazard and adverse selection in the supply chain. Producers have no incentive to invest in the quality of their products as birds are often mixed with others during transportation and sales. This mixing of birds means there is no way for producers to signal products of higher value, and thus producers must default to low value products. On the opposite end, vendors and consumers have asymmetric information about products and are unable to tell high quality products from low quality. Naturally, purchasers of poultry products will expect all birds to be of the standard low value variety, and without a signaling mechanism they have no reliable method of determining otherwise. Not only does moral hazard and adverse selection limit the value of producer's products, but also it has ramifications for animal disease as well. Producers have a disincentive to increase biosecurity levels, because again there is no way for them to signal that birds are disease free and healthy. These access barriers and market failures significantly undermine smallholder's potential for income, product diversification, and the savings/investment decisions needed for local economic growth and modernization.

Just as microcredit arose to solve the problem of capital constraints from information failures in financial markets, we see micro-contracts as tool for reducing market constraints in the smallholder supply chain.

Definition: Micro-Contracting

Unlike other GMS countries where a traditional contracting system is common, Lao PDR has a nascent or nonexistent system. Essentially the only outlet for a contract comes from the Thai conglomerate, CP, and these only exist for large, industrial farms. Small-scale producers must rely on loose, oral, informal agreements with aggregators and vendors, and we define micro-contracts as any agreements between two performing parties that fall under one of the following categories: a predetermined time of purchase, a predetermined price, a predetermined quantity, providing a price discount, prepayment for future products, giving credit to the seller, providing a guaranteed market, or gathering products at the farm. These agreements currently exist only in an informal manner, yet we believe by taking these informal agreements and using them to emulate a traditional contract system, the problems associated with the smallholder poultry supply chain can be alleviated.

Producer

In the current state of the smallholder poultry, supply chain micro-contracts are extremely limited. Only about 5% of small-scale producers reported having any sort of informal or oral agreement with those they sold to. For these producers, sales to

Table 18.10 Percentage of producers with informal agreement for sales

Marketing channel	Smallscale (%)	Large chicken (%)	Large duck (%)
Aggregator	5.43	20	25
Market vendor	5.24	22.50	21.43
Households/other farmers	6.19	13.75	14.29
Restaurants/food vendors	0.19	7.5	7.14

Table 18.11 Items covered by informal agreements, percentage of producers

	Smallscale (%)	Large chicken (%)	Large duck (%)
Time of sale	25.69	30.43	40
Price	44.04	39.13	40
Quantity	22.94	39.13	30
Price discount	14.68	13.04	30
Product prepayment	0.92	8.7	0
Credit to seller	7.34	N/A	N/A
Guaranteed market	2.75	8.7	10
Provide inputs	0.92	4.35	10

households and other farmers were slightly more common to have an informal agreement, while aggregator and vendor sales were nearly even. Agreements with restaurants and food vendors were essentially nonexistent with only two small-scale producers reporting having a system in place (Table 18.10).

For large producers, micro-contracts are much more common and informal agreements are more common with aggregators and vendors than end users. This should be expected given sales to these actors are more common. Approximately 20–25% of large producers reported having informal agreements in place with traders and vendors vs. only 14% for households and other farmers (Table 18.10). Although it is tempting to conclude that large producers achieved larger flock sizes and levels of marketing because of the prevalence of micro-contracts, this cannot be confirmed due uncertain causality. It is also possible that larger flock sizes allow producers to more frequently engage in informal agreements

Producers with informal agreements were asked what they entail and the majority of agreements cover a predetermined time of purchase, price, and quantity. In particular, a predetermined price was the single most common type of agreement for all producers. Agreements on the timing of sale and the quantity sold were more frequent amongst large producers, which can be expected due to the larger volumes of sales. Price discounts were the next most common type of agreement. Price discounts can be given to buyers to create incentives for future purchases. Furthermore, there is a large bargaining culture in Lao PDR and discounts on most purchased goods are common. Credits, either to the producer through prepayment for future products or to the seller by only paying for goods that are sold, were relatively uncommon. Producer credit only appeared for large chicken producers and seller credit was only seen amongst smallholders (Table 18.11).

Table 18.12 Interest in informal agreements, producers without any (percentage)

	Smallscale (%)	Large chicken (%)	Large duck (%)
Time of sale	10.31	29.63	25
Price	14.22	40.74	50
Quantity	9.88	29.63	25
Price discount	4.23	16.67	6.25
Product prepayment	3.37	22.22	18.75
Credit to seller	6.51	N/A	N/A
Guaranteed market	10.42	38.89	50
Provide inputs	5.97	16.67	18.75

Table 18.13 Reasons for not wanting to participate in informal agreements, producers

	Smallscale (%)	Large chicken (%)	Large duck (%)
Contract system is unreliable	14.31	31.82	14.29
Prefer to work independently	62.24	59.09	71.43
Do not have a channel that offers a micro-contract	8.41	13.64	14.29
Do not know who to make a micro-contract with	9	18.18	0

Enumerators also asked producers without an informal agreement for sales what (if any) agreements they would be interested in. Table 18.12 reveals that small-scale producers are generally less interested in participating in any sort of informal agreements than large producers are. For small producers, the most attractive agreements would be a predetermined price, time of sale, and quantity as well as guaranteed market. For large producers, the most desirable agreements were predetermined prices and having a guaranteed market. Overall these percentages are quite low for small-scale producers because it includes producers who rarely or never sell and are thus less likely to be interested in forming any agreements.

For producers that did not have an interest in informal agreements, enumerators asked respondents to list their reasons. The most common explanation was that the producers preferred to work independently and 62.23% of all producers listed this as a reason. Next most common was the belief that the contract system is unreliable followed by producers having no person that would agree to an informal agreement. Search costs were the least common reason given, with no large duck producers citing this as a reason (Table 18.13).

It is of particular interest to examine how rates of marketing differ between those with micro-contracting systems and those without. Among small-scale producers, poultry sales are more frequent for those with informal agreements with aggregators and vendors. For end users, marketing levels are similar to those with agreements and those without (Table 18.14). This can be expected because sales to neighbors and households are more informal channels while aggregators and vendors would have the most potential for informal agreements.

Table 18.14 Informal agreements and bird marketing levels, smallscale

	Aggregators (%)		Vendors (%)		End users (%)	
	Agreements	No agreements	Agreements	No agreements	Agreements	No agreements
No sales	20.2	27.7	19.1	28	30.8	31.2
Monthly sales	11.4	7.9	6.4	6.9	6.2	5.2
Sales less than once a month	64.4	64.5	74.5	65.1	63.1	63.7

Table 18.15 Percentage of aggregators with informal agreement for purchases and sales

Marketing channel	Purchases (%)	Sales (%)
Farmers	11.11	0
Other aggregators	11.11	3.7
Market vendors	14.81	25.93
Consumers	N/A	7.41
Restaurants/shops	N/A	3.7

The survey data reveals that marketing levels are higher for producers with informal agreements, but we cannot make statements about the impact of micro-contracts on marketing rates due to the issues of causality. Again it could be possible that higher levels of marketing lead to informal agreements.

Aggregator

The fundamental aggregator business model is to capitalize on the buying and selling margins of poultry products. Thus, it should be expected that micro-contracts are especially appealing to aggregators as it can help formalize and regulate the supply and sales of their products. The data confirms these expectations, as aggregators are more likely to engage in informal agreements than smallholders are. Informal agreements were formed for both poultry product purchases and sales and market vendors were the most common group that vendors had agreements with. Aggregators also formed agreements with farmers and other traders as well, and for these groups, purchasing agreements were much more frequent. Aggregators also reported having informal agreements with consumers, and this group represented the second most common source of micro-contracts for sales (Table 18.15).

The most common form of agreements formed by aggregators was establishing a predetermined price and are utilized because establishing clear buying and selling prices before sales occur can ensure profits for aggregators. Next most common are agreements on timing of sales and quantities, which is consistent with the producer data. Price discounts also appear, but are favored less than the other agreements. Credit in any form was not used by any of the aggregators (Table 18.16).

Table 18.16 Items covered by informal agreements, percentage of aggregators

Time of sale	63.63%
Price	81.82%
Quantity	54.55%
Price discount	36.36%
Product prepayment	0%
Credit from producer	0%
Credit to seller	0%
Guaranteed market	18.18%

Table 18.17 Interest in informal agreements, aggregator (percentage)

Time of sale	33.33%
Price	53.33%
Quantity	40%
Price discount	40%
Product prepayment	13.33%
Credit from producer	6.67%
Credit to seller	6.67%
Guaranteed market	33.33%

Table 18.18 Reasons for not wanting to participate in informal agreements

Contract system is unreliable	20%
Prefer to work independently	100%
Do not have a channel that offers a micro-contract	0%
Do not know who to make a micro-contract with	0%

Just like the producer questionnaire, enumerators asked enumerators without micro-contracts what informal agreements they would be interested in forming. Establishing a predetermined price was the most desired informal agreement, followed by predetermined quantity and receiving a price discount on purchases. Some aggregators expressed an interest in forming credit agreements, with prepaying for future products being the most appealing (Table 18.17).

However, almost 19% had no interest in forming any of the informal agreements listed in Table 18.17. For these aggregators, the desire to work independently was the most common reason given, followed by the belief that a contracting system is unreliable (Table 18.18).

Market Vendors

Representing the final stage of the supply chain, market vendors have an essential role to play in the micro-contracting system. Survey data demonstrates that market vendors are most likely to have informal agreements for purchases with aggregators

Table 18.19 Percentage of market vendors with informal agreements for purchases

Farmers	8.16%
Aggregators	41.84%
Other vendors	8.16%
Company (CP)	20.41%

Table 18.20 Items covered by informal agreements, percentage of market vendors

Time of purchase	47.89%
Price	92.96%
Quantity	78.87%
Price discount	60.56%
Prepayment for future products	1.41%
Credit from trader/producer	1.41%
Guaranteed supply	0%

and 42% of respondents claimed they have agreements in place with this group, which is the largest percentage the entire Vientiane Capital data set. Next most common were informal agreements with CP, and 20% of vendors reported having agreements in place. Farmers and other vendors were the least common group that vendors formed agreements with and only 8% of vendors reported having agreements with these groups. Although agreements are the lowest with farmers and other vendors, it still a high percentage compared with other actors in the supply chain revealing how prevalent micro-contracts are amongst vendors (Table 18.19).

For market vendors, the most common type of informal agreements are establishing a predetermined price, a predetermined quantity, and receiving a price discount, respectively, which is more or less consistent with the other actors in the supply chain. Agreements on the timing of purchases are less common, which can possibly be explained by the fact that vendors routinely buy products daily and have an expected supply. Agreements on predetermined prices were the highest of all actors in the supply chain with almost 93% of vendors with informal agreements reported having them. This demonstrates how important having clear and consistent input prices are because it allows vendors to control their profit margins more effectively. Agreements on credit either in the form of prepayments, or only paying for products once they are sold were virtually unutilized (Table 18.20).

For market vendors without any informal agreements, vendors were equally likely to want a predetermined time of purchase, a predetermined price, and price discount. Establishing a predetermined quantity was also a popular desire. In general, interest in forming micro-contracts among market vendors was higher than all other actors in the supply chain. In fact, only one vendor expressed no interest in forming any informal agreements (Table 18.21).

Table 18.21 Interest in informal agreements for those without, percentage of market vendors

Time of purchase	50%
Price	50%
Quantity	40%
Price discount	50%
Prepayment for future products	0%
Credit from trader/producer	10%
Guaranteed supply	0%

Conclusion

Market chain data demonstrate the significant role aggregators have in the supply chain representing an essential link between producer and consumers. Aggregators are the most common channel for small-scale producers to sell to and likewise, aggregators are most likely to source from small farmers. Looking at sales, aggregators sell the vast majority of their products to market vendors and thus they provide an undeniable link between producers and end users.

Unfortunately, this creates market failures in the supply chain and it should come as no surprise that marketing rates amongst smallholders are so limited. Micro-contracts appear to be a potential solution to the problems of asymmetric information because they could provide a direct link between small-scale producer and vendors. However, our data show that small-scale producers are the least common group to form micro-contracts. Instead we see that intermediary actors most frequently form agreements. Aggregators and vendors receive their livelihoods from the sales of goods, which they do not produce and agreements help ensure their successful profit margins. Using the approach that intermediary agents take by creating agreements to establish prices, quantities and times of sales, it can be expected that smallholders could capture the profits from the margins that intermediary agents exploit.

We believe that micro-contracts can be an important tool for increasing market access rates and reducing market chain failures. Preliminary survey findings support these beliefs, but without testing the causality of micro-contracts on smallholder marketing, we cannot make absolute statements. This is one area where a pilot micro-contracting program would be especially useful. In the pilot project, both sample and control groups would be established were the sample group engages in micro-contracts with vendors and/or aggregators. Monitoring and comparing data between the two groups would allow us to determine the causality of informal agreements and marketing.

Another important aspect of a pilot micro-contracting project would be the formal establishment of informal agreements into micro-contracts. As it stands now, informal agreements are loose and unreliable and thus many producers will not use them. We hope to demonstrate that by formalizing micro-contracts, we can create a standard and reliable tool for producers that will help them enter high value urban markets and increase income from their products.

References

Burgos, S., J. Otte, and D. Roland-Holst (2008). "HPAI and Livelihoods in Lao People's Democratic Republic – A Review." FAO PPLPI Mekong Team Working Paper No. 5

DLF – Department of Livestock and Fisheries of Lao PDR (2009). Private Correspondence.

MAF – Ministry of Agriculture and Forestry of Lao PDR (2008). "Agricultural Statistics Year Book 2007." May 2008

NSC – National Statistical Center of Lao PDR (2008). "Statistical Yearbook 2007." June 2008

NSC – National Statistical Center of Lao PDR (2009). "Lao Expenditure and Consumption Survey 2008/09 (LECS 4)." October 2009

Wilson, R. T. (2007a). "Numbers, Ownership, Production, and Disease of Poultry in the Lao People's Democratic Republic." Worlds Poultry Science Journal 63: 655–663

Wilson, R. T. (2007b). "Status and Prospects for Livestock Production in the Lao People's Democratic Republic". Tropical Animal Health and Production 39: 443–452

Chapter 19
Poultry Sector Transition in Cambodia

Samuel Heft-Neal, David Roland-Holst, and Joachim Otte

Introduction

Across Southeast Asia, livestock sectors are in the midst of significant transitions. While these shifts were underway prior to the HPAI outbreaks which began in 2003, response and adjustment to these outbreaks have accelerated these trends in many countries. In light of these transitions underway, household surveys were undertaken in Cambodia to produce detailed assessments of poultry supply chain conditions in the Kampot and Siem Reap market casement areas. This was done with detailed and separate surveys at four levels, producers, traders, vendors, and consumers. The goal of this approach is to elucidate the properties of livestock networks as these are determined by production conditions, market access, contractual relationships, and overall market conditions.

Taken together, these results suggest that although there has been some shift out of the sector (most notably medium-large scale commercial chicken producers), most farmers have been completely unaffected by the outside world's preoccupation with HPAI. Our results can inform policy initiatives to improve incentives for higher poultry quality (including health status) and higher value added at each stage of the supply chain. As in other case studies of the Mekong region, this activity supports recommendations for sustainable market participation by smallholder poultry producers. These could include, but are not be limited to, programs for micro-credit and technology transfer, certified supply chains, and contract farming programs for bio-secure production of traditional bird varieties. In addition to reducing HPAI

S. Heft-Neal (✉) • D. Roland-Holst
Department of Agricultural and Resource Economics, University of California, Berkeley, CA, USA
e-mail: sheftneal@berkeley.edu

J. Otte
Food and Agriculture Organization of the United Nations, Rome, Italy

D. Zilberman et al. (eds.), *Health and Animal Agriculture in Developing Countries*, Natural Resource Management and Policy 36, DOI 10.1007/978-1-4419-7077-0_19,

risk and the economic vulnerability of rural poor farmers, these recommendations strive to increase product quality, safety, and revenue across the traditional bird supply chain.

This chapter serves to contextualize the activities that were carried out, present the findings, and suggest the policy implications that follow. The chapter begins with a section on the background of Cambodia's economy, the agricultural sector, and the livestock sector. Subsequently, the literature on HPAI in Cambodia is reviewed, with emphasis on past livelihood studies and, from the review, an appropriate approach to the project is developed. Next, survey results are presented, and we conclude with a discussion of policy implications.

Livestock Sector

Livestock accounts for 15% of agricultural GDP and serves an important role in many rural Cambodian households (National Institute of Statistics 2004). Buffalo and oxen are often kept for fieldwork activities while poultry and pork are commonly kept to supplement household diets and income. Local market chains serve the domestic market and consequently live bird traders are key marketing agents. Additionally, village and animal health agents serve as advisors to livestock production.

Poultry Sector

The 1990s saw annual increases of poultry production of more than 6% to nearly double total production over the course of the decade. However, since 2002 the number of chickens produced has declined each year besides 2004, when levels held steady. Meanwhile, duck production has continued to grow steadily and by 2007 Cambodia was producing more than 20 million head of poultry, 80% of which were chickens and 20% ducks (Burgos et al. 2008). Because of the short production cycles for chicken, each year there are more chickens slaughtered than there are live chicken stock. However, the opposite is true for ducks.

As incomes have continued to rise, poultry meat and egg consumption have been increasing in the past decade. By 2003, Cambodians were consuming more than 25,000 tons of chicken meat and 15,000 tons of eggs each year (AED 2005). The importance of livestock activities varies largely by region. Poultry densities are much higher in the southeastern and northwestern regions than in other parts of the country. Poultry densities are highly correlated to human densities.

The Cambodian poultry sector consists of three general types of production systems: traditional small-scale production, semi-intensive small/medium-scale commercial chicken/duck production, and medium/large-scale intensive industrial chicken/duck production. Traditional small-scale production is the most common system, accounting for more than 95% of poultry in Cambodia and involving

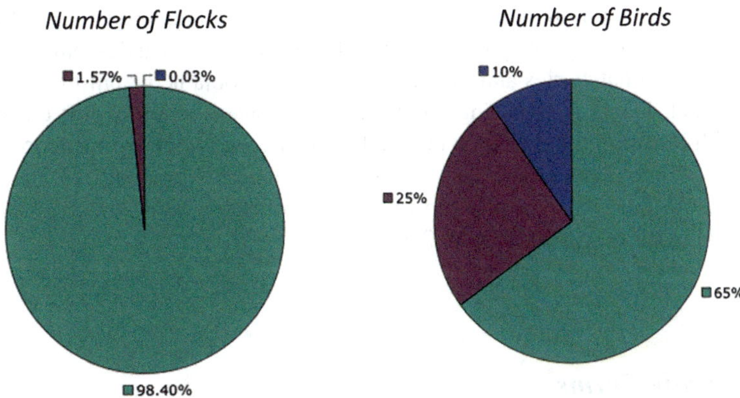

Fig. 19.1 Cambodian poultry production by production system. *Source*: Burgos et al. (2008)

2,000,000 households. In most rural households, raising poultry is one of the many activities in an extensive agriculture production system. More than half of all Cambodian households keep some poultry, including 60% of rural households and 25% of urban households. Among smallholder households raising poultry, 80% raise only chicken, 19% raise ducks and chicken, and only 1% raise solely ducks (VSF 2004) (Fig. 19.1).

The traditional system is characterized by small flocks and low inputs, allowing birds to scavenge for food and using hens for restocking the flock. Depending on the region, only 5–25% of chicken owners provide feed to supplement scavenging (CelAgrid 2007). Housing is very basic and mostly used for keeping birds safe from predators and thieves during the night. Some chickens spend their nights in trees. Birds produced in the traditional system are either consumed by the household, given or sold to neighbors/friends/family, or sold at the farm gate to traders.

Medium and large-scale commercial poultry production did not begin to develop in Cambodia until the mid to late 1990s. The entrance of CP into the Cambodian poultry sector played an important role in facilitating development. Not only did CP provide contracting opportunities, but also the company established hatcheries and feed factories that allowed some independent farmers to purchase these products locally rather than import them. Nonetheless, productions in these systems are resource intensive and often continue to utilize imported products from Thailand, China, and Vietnam (Burgos et al. 2008).

Commercial farms are major suppliers for large cities such as Phnom Penh, Battambang, and Siem Reap. However, less than half of Cambodian provinces have commercial poultry farms. The 2004 DAHP survey indicated 74 chicken layer farms, 108 broiler chicken farms, and 951 duck farms in Cambodia. The average size of a commercial farm is about 1,400 heads. Less developed production systems of this type rely both on naturally available feeds and manufactured animal feeds. Advanced systems rely solely on commercial feed. Broiler breed day-old-chicks are purchased from hatcheries and indigenous chicks are obtained from local markets or own-stock hatching. Housing varies from permanent to makeshift enclosures made with local primary building materials, such as mud bricks or bamboo, or tree branches. Compared to backyard farms, semi-intensive producers utilize more extensive bio-security measures; however, most of the system is still constructed from local materials (Burgos et al. 2008).

Small-Scale Farms

Small-scale farms are by far the most common farming system and are practiced all over Cambodia. Nationally, there are estimated to be more than 1.8 million flocks (MAFF 2007). However, poultry production for smallholders is only one of the many activities within agricultural livelihood systems. Small-scale poultry rearing is largely a seasonal venture. September to early February is an ideal fattening period because there is extra chicken feed such as rice grain left over from the rice harvest. Moreover, chickens can be fully fattened for sale during the high price period of Chinese New Year in February. There is often little home consumption in January as households prefer to sell their products during the high price period. After the hot season, in July and August, farmers begin to restock their flocks through hatching or obtaining chicks from outside sources (Seng 2007).

Broiler Farms

In 2004, DAHP recorded 108 chicken broiler farms with an average of more than 3,500 birds per farm. Broiler farms are mainly located in Kampong Speu, Siem Reap, Kandal, and Phnom Penh. Sixty eight of these farms are integrated with CP farms. As the integrator, CP provides DOCs, feedstuffs, pharmaceuticals, and technical advice. The farms provide infrastructure and labor, and assume part of the financial risk. All birds are output to CP owned slaughter houses and farm owners are paid by the head based on performance indicators. Market weight for broilers tends to be about 1.88 kg. Most farms operate on batch production cycles where broilers are all the same age (VSF 2004). DOCs can be provided by CP (integrated or private farms) as well as Medivet company and three other importers based in Phnom Penh. Prior to HPAI outbreaks, CP sold only 20% of its DOCs to private farms, however, after the outbreaks the proportion sold to private farms increased to 50% (VSF 2004).

Layer Farms

Layer farms produce eggs for human consumption and the manufacture of food products. In the 2004 DAHP survey, 74 chicken layer farms were recorded, 9 of which were contracted with CP. In addition, there are 57 pullet farms owned by CP. The average flock size was slightly more than 5,000 birds. Pullet raising farms provide replacements for unproductive layers, which are sold to be slaughtered for meat. Most layer breeds are imported because indigenous breeds are less productive (Burgos et al. 2008). Commercial egg producers compete with producers in neighboring countries, whose access to lower cost feedstuffs allow for cheaper production. One market survey found that 11% of chicken eggs and 36% of duck eggs came from neighboring countries (VSF 2004).The layer industry is geographically concentrated in Kandal and Kampong Speu, which collectively account for nearly 80% of the national layer population.

Duck Farms

In 2004, the DAHP survey found 951 duck farms with an average of 900 ducks per farm. Duck farms tend to be less standardized than broiler farms, involving about 30% broiler ducks and 70% layers (Burgos et al. 2008). Duck raising cycles generally coincide with rice production periods and peaks of demand. Initial investment is moderate (feed mixer, feed storage structure) and ducks are raised outdoor near a pond and in fields. High quality feed is often provided during the first 2 weeks of rearing and subsequently lower-quality feed is provided. Production cycles depend largely on local rice production cycles. Most ducklings are produced in Takeo province or imported from Vietnam. However, some eggs are imported from Thailand and Vietnam as well. October and November are the most popular months for purchasing ducklings. Duck laying cycles range from 4 to 24 months (Burgos et al. 2004).

Breeding/Incubator Farms

As of 2004, there was only one breeding farm in Cambodia. It is owned by CP and located in Kandal province. The farm produces day-old-chicks for broiler farms and pullet/layer farms. There are no duck or local chicken breeding farms; however, several provinces have hatcheries for production of local duck breed embryonated eggs or ducklings. Takeo is the most important one with 20–30 hatcheries (VSF 2004). Owners of hatcheries tend to have ties with local farmers and may offer credit and technical advice to trusted farmers. In Takeo province, many smallholder farmers purchase ducklings in May while commercial farmers purchase ducklings in October. The remaining months the hatcheries produce embryonated eggs (Seng 2007).

Wholesalers/Importers

Wholesale importers of chicken and animal feed import products from Thailand and Vietnam and distribute them in Cambodia. There are 3–4 major wholesales based in Phnom Penh and many regional agents operating at provincial scale. Operation volumes range from 80,000 to 175,000 chicks per month. Farmer orders are placed with the wholesalers in advance. The volume of feed imported by wholesalers depend on many factors including season and price, and range from 75 to 125 tons per month. The distribution network for national wholesalers consists of 20–25 trucking distributors throughout the provinces. In addition, there are several wholesalers that specialize in trading pharmaceuticals (VSF 2004).

Marketing

Many smallholders, in addition to home consumption, market some of their flock. Medium and large producers, on the contrary, generally sell nearly all of the birds they produce. Middlemen play a key role in bringing poultry and eggs from producers to markets. They transport poultry on bicycles, motorbikes, cars, and trucks. Traders can aggregate to different levels (i.e., village, district, province, etc.) and sell to local markets and higher administrative districts. Moreover, much of the poultry production is sold in the largest markets of Phnom Penh and Siem Reap. Market retailers exist at commune, district, and provincial levels. Consumers usually purchase live birds which are then slaughtered and prepared by the vendor at the market. While Cambodian products are not officially exported, there may be low levels of informal sale in neighboring countries during certain seasons. Imports from Thailand and Vietnam exist, but are difficult to estimate because much of the trade is informal.

HPAI in Cambodia

Cambodia has not been as affected by avian influenza as its neighbors; however, there have been 20 confirmed poultry cases and seven human deaths from HPAI since the disease was discovered in Cambodia in 2004 (Otte et al. 2008). During that time, more than 20,000 birds have been affected by the disease. There is a temporal pattern to the HPAI outbreaks. In the past 4 years, 70% of HPAI outbreaks in poultry have occurred between February and May while 25% have occurred between June and September, and 5% between October and January. Moreover, all seven human cases have occurred between February and May. In addition, there has been some confusion among farmers between Newcastle disease and HPAI. Part of this problem arises out of nomenclature (Hickler 2007). In addition, these diseases share similar temporal patterns, further confounding confusion. The most favorable season for raising poultry is July to December.

Three of the main challenges to developing the livestock sector are markets, access, and issues of land rights. Markets suffer from high transaction costs associated partly with paying unofficial fees along main transportation routes. Access, restricted by lack of irrigation and roads as well as the high cost of energy, is a limiting factor in expansion. Finally, there is underinvestment as a result of undefined land rights that arise out of the misuse of land concessions and the insecurity of land tenure (Ear 2005).

Overview of Project Sites

Kampot Province

Kampot is located in the coastal region of southern Cambodia, near the Mekong delta bordering Vietnam. Rice is the primary agricultural commodity produced and Kampot province has had the most severe problems with HPAI of any province in Cambodia. In addition to three confirmed outbreaks in poultry, there have been four human deaths from HPAI since 2004. Poultry outbreaks have occurred in chickens, ducks, and wild birds in several districts. In each case, outbreaks in birds have occurred in March or April, likely related to duck and rice production cycles. All of the human deaths also occurred within a 4 month period from February–May, 2005. The one village in Cambodia with confirmed poultry and human cases of HPAI is located in Banteay Meas district, Kampot province.

During the HPAI outbreaks, income from poultry decreased among smallholders of all income levels. However, case studies in Kampot villages suggest that since the HPAI outbreak the amount of income smallholders derived from poultry production has actually increased (Tables 19.1 and 19.2).

Siem Reap Province

Siem Reap province is located in the Tonle Sap region of northwestern Cambodia. There are three primary zones in Siem Reap: urban, suburban, and terrace. The living

Table 19.1 Confirmed HPAI cases in poultry in Kampot province, 2004–2007

Year	Month	District	Bird type	No. affected
2004	March	Ta Khmau	Chickens/ducks/wild birds	533
2005	March	Benteay Meas	Backyard chickens	28
2006	April	Kampong Bay	Ducks	247

Source: OIE (2008)

Table 19.2 Confirmed HPAI cases in humans in Kampot province, 2004–2007

Year	Month	Gender	Age	Death
2005	February	Female	25	Yes
2005	March	Male	28	Yes
2005	April	Female	8	Yes
2005	May	Female	20	Yes

Source: WHO (2008), http://www.who.int/csr/disease/avian_influenza/country/en/

Table 19.3 HPAI outbreaks in poultry in Siem Reap province, 2004–2007

Year	Month	District	Bird type	No. affected
2004	March	Siem Reap	Chickens/ducks/wild birds	500+
2004	March	Siem Reap	Chickens/ducks/wild birds	500+

Source: OIE (2008)

and farming conditions of each zone are distinct. The urban zone is characterized by main roads and high population density. Agriculture is less important in this zone. The suburban zone tends to lie toward the flood plain of Tonle Sap Lake and is characterized by zones that were formerly primarily agricultural but have recently become more developed as the area urbanized. The terrace zone is removed from urban centers and characterized by a terrace ecosystem in the hills.

Siem Reap province experienced two HPAI outbreaks in poultry, both in 2004, and no confirmed human cases. Prior to 2004 most people in Siem Reap had never heard of HPAI. Since the outbreaks began, most people are aware of its existence; however, many people continue to consume sick or dead birds (AED 2005; IPC 2007). In fact, most household learned about HPAI on television and were unaware of the two outbreaks in the province. The exception is better-off households in urban/suburban areas (Ly et al. 2007). These households are aware of local outbreaks and are now hesitant to handle live poultry and more likely to purchased slaughtered birds and take meat safety into their purchasing decisions. Other households continue to eat sick or dead birds and do not believe this poses a threat (Seng et al. 2008) (Table 19.3).

Overview of Data

Four separate groups were targeted for surveys: consumers, farmers, traders, and vendors. The purpose of the Consumer Survey is to better understand poultry purchasing habits of households that acquire their poultry products through markets. A detailed survey was carried out revolving around household tastes, price sensitivity, breed preference, and other aspects of shopping habits.

The purposes of the farmer surveys were to better understand farmer cost structure, resource utilization, and assess the adjustment of poultry producers in response to HPAI control measures. The surveys assessed these issues by focusing questions on evaluating farmer inputs and outputs, production cost structure, access

Table 19.4 Household weekly food expenditure (1,000 riels/week)

	Kampot (urban)		Kampot (semi-urban)		Siem Reap		Total	
Poultry consumed	Mean	SD	Mean	SD	Mean	SD	Mean	SD
Eating out	29.6	28.6	25.1	23.9	53.7	35.8	38.9	33.5
Eating in	105.8	48.8	94.8	30.1	110.5	63.6	110.9	78.0
All meats	80.4	48.0	73.7	58.0	67.7	68.4	74.7	72.4
Chicken meat	21.7	11.9	23.8	8.8	18.9	21.4	20.8	16.6
Chicken eggs	6.4	4.3	3.1	1.3	4.7	5.2	4.9	4.5
Duck eggs	11.9	4.6	13.3	5.5	12.8	15.4	12.5	9.4
Duck eggs	6.6	3.5	6.1	2.4	5.7	4.8	6.1	4.1

1 USD ~ 4,000 riels

to markets, trading relationships, barriers to expansion, and HPAI experience. From these data collected, we are better able to estimate the cost to producers from shifts in policy or structural changes in the Cambodian poultry sector. The purpose of the aggregator survey was to improve our understanding of the marketing network and trading relationships as well as to collect information on pricing, disease considerations, resource utilization, and operation costs. The survey aimed to include traders of all poultry products including chicken and duck eggs/meat, and chicks/ducklings for sourcing production. The market vendor survey aims to better understand the sources of chicken sold in urban markets as well as to collect price and breed data.

Findings

Household/Consumer Survey Household surveys revealed that respondents in Kampot visited the market more often than respondents in Siem Reap. However,85% of overall respondents visited the market at least one time per day. Respondents in Siem Reap had significantly higher weekly expenditures on eating outside the home than their counterparts in Kampot; however, expenditure on groceries for home preparation was similar across all groups at around 100,000 riels/week (~$25/ week). Chicken meat was the highest poultry expenditure, averaging 20,800 riels/ week (~$5.20/week) followed by duck meat (12,500 riels/week ~ $3.10/week) (Table 19.4).

While the level of average expenditure was low, duck eggs were the most commonly purchased poultry product. One third of respondents reported purchasing duck eggs every day, while more than 80% reported purchasing duck eggs at least once per week. On average, respondents purchased about ten duck eggs per market visit. Approximately, half respondents purchase chicken meat at least once per week purchasing slightly over 1 kg per visit. Fifty percent of respondents reported never purchasing duck meat.

Duck eggs were significantly more popular in semi-urban regions than in urban regions. Semi-urban respondents in Kampot were most likely to purchase live birds; however, the practice was also common in Kampot's urban areas. However, in Siem Reap, the vast majority of consumer purchases slaughtered birds (>80%) while less than 5% of purchased birds were slaughtered in the market. The trend toward selling slaughtered meat in markets appears to have been accelerated in Siem Reap by the HPAI outbreaks, but not in Kampot. This is likely due in part to the nature of the cities; Siem Reap is urban and relies heavily on tourism while Kampot is still relatively rural and the local economy is based primarily on agriculture (Table 19.5).

Prices for poultry products in Cambodia vary greatly depending on the season. Survey respondents were asked about the prices they paid in high season (months of high season) and low season (months of low season). Average prices were highest for all products in urban Kampot during the high season. However, prices in urban Kampot and urban Siem Reap were very similar, with the exception of chicken eggs, which were 20% cheaper in Siem Reap. Nearly all shopping (more than 90%) occurred at traditional wet markets (Table 19.6).

In Kampot, consumers continue to purchase live birds, determining quality of the bird by its live appearance. However, in Siem Reap there has been a trend toward purchasing slaughtered meat, and thus judging quality by meat appearance. The trend of purchasing slaughtered birds in Siem Reap was accelerated by the HPAI outbreaks, which caused more of a reaction in Siem Reap due to the tourism industry. In addition, consumers were asked whether they felt the safety of the poultry products that they purchase could be improved. More than 8 in 10 respondents in Kampot felt that safety could be improved compared with 6 in 10 respondents in Siem Reap suggesting that consumers may feel safer purchasing slaughtered meat. However, despite these concerns for safety, very few people have been enticed to switch to packaged meat with safety guarantees provided by large farms. Instead most consumers continue to show preference for traditional breads of chicken raised by traditional methods (Table 19.7).

Producer Survey

Small-Scale Chicken Producers

The majority of smallholder chicken raising respondents were females in both provinces. This finding supports findings in earlier studies (Seng 2007, Seng et al. 2008) that women often control the income from poultry production. Like most Cambodians, the majority of survey respondents produced crops as their main economics activity. In Kampot, rice was the most frequent crop produced (94.3% respondents) while in Siem Reap vegetables were most common. In addition, nearly 90% of Kampot respondents and 68% of Siem Reap respondents also reared other livestock, most commonly pig and cattle. Off farmer employment was reported in 20% of Kampot households and 10% of Siem Reap households (Tables 19.8 and 19.9).

Table 19.5 Form of poultry purchased (percent) in selected locations

	Kampot (urban)			Kampot (semi-urban)			Siem Reap			Total		
	Live	Slaughter in market[a]	Slaughter	Live	Slaughter in market	Slaughter	Live	Slaughter in market	Slaughter	Live	Slaughter in market	Slaughter
Percent of birds purchased	47.2	39.9	12.9	62.5	6.1	31.4	13.7	4.5	81.8	32.7	17.0	50.3

[a] Slaughtered in market = birds are selected live then slaughtered and prepared by vendor

Table 19.6 Average price paid for poultry by breed and form purchased (riels/kg; riels/egg)

	Kampot (urban)		Kampot (semi-urban)		Siem Reap		Total	
Reason	High season	Low season	High season	Low season	High season	Low season	High season	Low season
Chicken meat	15,800	13,800	14,200	12,200	15,700	13,900	15,500	13,500
Chicken eggs	520	440	500	410	440	380	500	430
Duck meat	10,300	8,600	9,600	7,800	8,700	7,600	9,900	8,200
Duck eggs	530	460	450	340	520	450	520	440

Table 19.7 Methods for determining quality of chicken and duck meat (percent respondents)

	Kampot (urban)		Kampot (semi-urban)		Siem Reap		Total	
Reason	Freq.	Perc.	Freq.	Perc.	Freq.	Perc.	Freq.	Perc.
Live appearance	272	79.1	113	75.8	89	17.7	474	47.6
Meat appearance	26	7.6	49	32.9	392	78.1	467	46.9
Relationship with seller	168	48.8	21	14.1	94	18.7	283	28.4
Knowledge of source	75	21.8	7	4.7	12	2.4	94	9.5
Do not think about safety	7	2.0	1	0.7	1	0.2	9	0.9
Other	1	0.3	2	1.3	10	2.0	13	1.3

Table 19.8 Other household economic activities

	Kampot		Siem Reap		Total	
Age range in years	Nr	Percent	Nr	Percent	Nr	Percent
Rice production	493	94.3	228	60.3	721	80.0
Vegetables	156	38.6	268	70.9	424	47.1
Fruits and nuts	57	10.9	113	29.9	170	18.9
Raise livestock (other than poultry)	468	89.5	254	67.2	722	80.1
Off-farm employment	103	19.7	38	10.1	141	15.7

Table 19.9 Smallholder chicken flock sizes in selected locations

	Kampot		Siem Reap		Total	
Farm size	Freq.	Percent	Freq.	Percent	Freq.	Percent
1–20 chickens	339	65.8	347	98.0	686	78.9
21–50 chickens	166	32.2	7	2.0	173	19.9
51–75 chickens	10	1.9	0	0	10	1.2
>76 chickens	0	0	0	0	0	0
Average flock size	19.0		5.6		13.5	

Most respondents in both provinces raised less than 20 chickens; however, flock sizes were larger, on average, in Kampot. Farmers were asked what type of feed their birds consumed on a daily basis, whether scavenged or provided. The most common feed was paddy rice in both provinces. However, there was some variation between locations. In Siem Reap, human food scraps were a common source of feed, with more than 70% of respondents reporting using this type of feed.

Table 19.10 Buyer of products in selected locations (percent)

Agreement	Kampot	Siem Reap	Total
Aggregator	50.0	45.8	48.6
Vendor	30.7	39.0	33.4
End-user	15.2	13.4	14.9
Food vendor	1.7	1.1	1.5
Other	2.4	0.3	1.0

Contrarily, in Kampot only 30% of respondents fed human food scraps to their birds.Approximately, half of all respondents reported providing pharmaceuticals to their birds. The most common source of medicine was a Village Animal Health Worker in Kampot (34%) and a pharmacy that sells human medicines in Siem Reap (10.1%). Farmers in Kampot were more likely to use veterinary services.

Most respondents sell some birds, consumer some birds, and provide other as gifts. Prices received were higher for meat products in Siem Reap and for eggs in Kampot, reflecting relative demand levels. Prices for slaughtered birds were higher in both provinces with a premium of 800–1,000 riels charged for slaughtering. Aggregators were the most common purchaser of poultry products among all respondents. Vendors were the next most common outlet. Less than 20% of respondents sold birds to their friends and neighbors (Table 19.10).

While aggregators bought close to half of the birds sold, there were a variety of sales locations reported by farmers. In Kampot, most sales took place at the farm; however, roadside sales and bringing birds to market were also common. In Siem Reap, bringing birds to market was most common with only 15% of sales taking place at the farm. Bringing slaughtered birds to market necessitates having a predetermined arrangement with buyers so that the farmer is assured they will not waste birds. Indeed we found that respondents in Siem Reap that brought birds to market almost always had agreements with vendors beforehand. In fact, nearly half of farmer–vendor traders in Siem Reap took place between people whom also interact regularly outside of the poultry trade. These social connections remain important in facilitating poultry trade.

Respondents in Kampot averaged higher income from poultry; however, standard deviations were high in both provinces. This is a result of the high variation in roles that poultry production plays in household economies. Respondents were also asked where they spent their cash income received from poultry production. The most common response was essential consumption. However, more than 20% of respondents reported using the money for school fees and 15–20% reported saving money for emergencies. These findings underlie the importance of poultry production in low income households (Tables 19.11 and 19.12).

Large-Scale Duck Producers

Unlike with small-scale chicken, duck farmers invest significant resources into production and ducks are one of their major economic activities. In fact, in each province, we found that duck farmers spent at least 7 h per day attending to their flocks (compared with less than 20 min for smallholders).

Table 19.11 Income from poultrty production

Variable	Kampot		Siem Reap		Total	
	Mean	SD	Mean	SD	Mean	SD
HH income from poultry 2008 (10,000 riels)	44.1	56.0	24.0	40.2	38.1	52.5
Percent of total cash income from poultry production	24.2	23.2	8.2	16.4	17.5	22.1

Table 19.12 Use of cash income from poultry production

Barrier	Kampot		Siem Reap		Total	
	Freq.	Percent	Freq.	Percent	Freq.	Percent
Save for emergency	113	21.6	59	15.6	172	19.1
School fees	111	21.2	81	21.4	192	21.3
Essential consumption (food, clothing, shelter)	335	64.1	160	42.3	495	55.0
Nonessential consumption	19	3.7	3	0.8	22	2.4
Invest in other economic activities	13	2.5	11	2.9	24	2.7
Other	19	3.7	55	14.6	76	8.4
Don't know	7	1.3	0	0	14	1.6

There are several distinct production structures depending on whether the farmer produces eggs, meat, or both, and whether they provide an input to one of these products (e.g., raise layers to be sold to layer farmers) or whether they want to produce the output. Average flock size varied greatly both across and within these groups. However, most farmers maintained flocks of a few hundred birds. Layer producers averaged about 400 birds kept at one time.

Paddy Rice is the main source of feed. In addition, 75% of respondents in Kampot, and more than 90% of respondents in Siem Reap, provided additional commercial feed to supplement paddy rice as duck feed.

In Kampot, more than 60% of respondents allowed their duck flocks to graze in their rice fields. Furthermore, nearly three-quarters allowed other peoples ducks to graze in their fields. These activities are complimentary because ducks can extract feed from the rice field and the ducks can fertilize the rice and aerate the soil. However, this practice was highly uncommon in Siem Reap, were there was less rice cultivation and less than 10% of respondents reported duck grazing in their rice fields. Duck production is a major economic activity. Consequently, income from duck production is significant and farmers spend more than 7 h per day attending to their flocks.

Despite the occurrence of at least four reported HPAI outbreaks in chicken, and four reported human deaths from HPAI in Kampot, zero respondents reported having their birds culled. In Siem Reap, none of the respondents reported having birds culled either; however, it appears that farm scale was the main determinant of how HPAI impacted households in Siem Reap. The impact of HPAI was felt mostly by commercial producers (medium and large-scale) who were hurt by the loss of a main source of income during the outbreaks. In fact, since 2003 16 of 20 commercial chicken farms have shut down, which likely explains why were unable to find farmers whose flocks had been culled.

Aggregator Survey

Aggregators (aka traders) provide important marketing functions including assembly and delivery of birds. These actors facilitate many of the trades that occur. Income among aggregators, compared with farmers, was relatively high. Trading activities contributed more than 60% of cash income among respondents. Respondents were most likely to trade chicken for meat or layer ducks for egg production. Interestingly, in Siem Reap nearly all respondents who traded chicken also traded layer ducks. However, in Kampot, the people trading layer ducks chicken for meat tended to be specialized in one or the other. Chicken trading volumes were higher in Kampot (per trader) but duck trading volumes were slightly higher in Siem Reap.

Table 19.13 summarizes the operating structure of the average aggregator interviewed in this study. Aggregators were asked all operating costs as well as values and quantities of inputs and outputs in order that we might better understand how they operate. Aside from covering the costs of the birds for trading, gasoline and road fees (payments to officials) were the primary costs of operation. However, the price of gasoline and number of road fees assessed vary greatly and largely determine the profitability of trading. In addition to weekly variation, the poultry market is highly seasonal depending on Khmer holidays and festivals as well as farming cycles. Figure 19.2 depicts the cyclical nature of trading.

Market Vendor Survey

In all locations, the majority of vendors were found to be women. Duck eggs were the most commonly sold product in both Siem Reap and the semi-urban area in Kampot. This coincides with our findings in the consumer survey that duck eggs are the most commonly poultry products purchased by consumers. The prices for poultry products were higher in Siem Reap than in Kampot for all poultry products except chicken eggs (Table 19.14).

Daily trading volumes varied greatly across vendors. On average vendors reported selling more than 35 kg of chicken meat per day, more than 20 kg of duck meat, and a few hundred eggs (though, as noted earlier, not all vendors sell all products). Vendors were most likely to have agreements with traders who delivered birds to their home, with 35% of respondents reporting a verbal arrangement for this type of purchase (Table 19.15).

The average price of chicken meat in the market was found to be approximately \$4.00/kg during the high season and \$3.20/kg during the low season. Duck meat is significantly cheaper at \$2.50/kg during the high season and only \$2.15/kg during the low season.

Table 19.13 Chicken meat aggregator cost structure – average weekly values in Kampot province

Average inputs					Outputs			
Item		Quantity/week	Unit price	Total cost (riels/week)	Item	Quantity/week	Unit price	Total value
Variable costs	Chickens	140 kg/week	13,640 riels/kg	1,909,600	Chickens	140 kg/week	15,300 riels/kg	2,142,000 riels/week
	Gasoline	4.7 L/week	3,860 riels/L	17,500				
	Poultry feed	4.2 kg/week	2,500 riels/kg	8,400				
	Road fees	15 fees/week	1,000 riels/fee	15,000				

Item		Quantity	Percent traders utilizing input	Approximate cost (in dollars)	Author's calculation of monthly trader profit [a]
Fixed costs	Bicycle	1	23	25	Authors estimate of average income from trading ~ 219.00 USD/month
	Motorbike	1	70	550	Average self-reported income from trading ~ 230.00 USD/month
	Cars	1	3	2,750	

[a] Does not take into account fixed costs

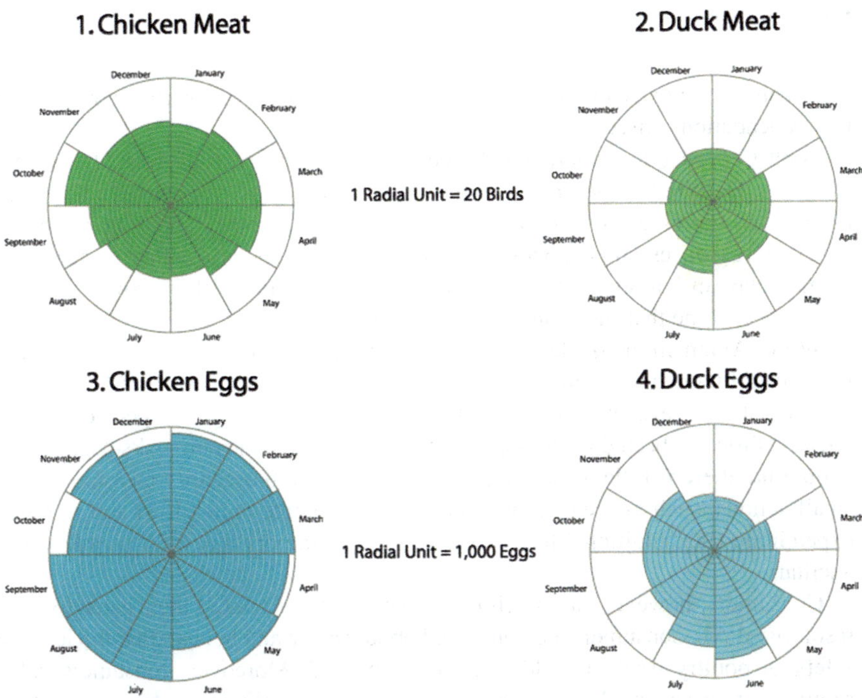

Fig. 19.2 Seasonality of trading (average number of products traded by an aggregator in 1 month)

Table 19.14 Types of poultry products sold (percent of vendors)

	Kampot (urban)		Kampot (semi-urban)		Siem Reap		Total	
Breed	Freq.	Perc.	Freq.	Perc.	Freq.	Perc.	Freq.	Perc.
Chicken meat	14	100	81	60.5	19	43.2	114	58.8
Chicken eggs	2	14.3	12	9.0	20	45.5	32	16.5
Duck meat	8	42.9	39	29.1	2	4.6	47	24.2
Duck eggs	4	28.6	72	53.7	24	54.6	98	50.5

Table 19.15 Average market vendor sale price in selected locations

	Kampot		Siem Reap		Total	
	High season	Low season	High season	Low season	High season	Low season
Chicken meat (riels/kg)	15,400	12,300	16,100	13,200	15,700	12,600
Chicken eggs (riels/egg)	410	410	380	390	410	410
Duck meat (riels/kg)	12,100	8,300	12,500	10,000	10,300	8,600
Duck eggs (riels/egg)	420	410	460	430	480	410

Conclusions

After conducting extensive surveys of Cambodian market chain participants, several recommendations arise.

Poultry production continues to be an important part of rural household activities. Despite the fact that very little time and resources are invested, poultry production serves as a significant source of cash income, allowing increased expenditures on essential goods, schooling, and emergency savings. Moreover, women play an important role in poultry production and are thus provided with a source of income that they can manage from home.

While Avian Influenza has made headlines across the world, farmers appear unaffected by the news. Most respondents had never experienced HPAI, or if they had then it had never been diagnosed because high mortality is expected. Consequently, little has been done to improve biosecurity since the outbreaks began. That being said, there is little incentive for farmers to adopt biosecurity measures. Even small actions, such as creating an enclosure, necessitate more effort and resource expenditure (i.e., additional feed provision) for an activity that most people see as ancillary.

Consumers, however, are much more likely to have altered their behavior as a result of HPAI. Consumers are concerned about safety and believe that the level of safety of poultry products sold could be improved. Moreover, consumers value safety over price and have expressed interest in purchasing safety guaranteed poultry products, something that has not fully been explored in the provinces of Cambodia.

The network that markets poultry products is based on largely informal relationship, often between prior-known acquaintances. An extensive network of farmers, aggregators, and vendors deliver poultry products to urban consumers. Any attempt to address livestock disease or poultry producer livelihoods will need to incorporate all members of the supply chain.

References

AED (2005) "Backyard Poultry Farmers and Avian Flu in Cambodia: A Baseline Survey", carried out by AED on behalf of USAID

Burgos, S., Hinrichs, J., Otte, J., Pfeiffer, D., Roland-Holst, D., Schwabenbauer, K., Thieme, O. (2008) "Poultry, HPAI and Livelihoods in Cambodia- A Review", FAO PPLPI Mekong Team Working Paper No. 3

CelAgrid (2007) "Community Based good practice in chicken raising and AI awareness in three provinces of Cambodia", carried out by CelAgrid on behalf of USAID.

Ear, S. (2005) "The Political Economy of Pro-Poor Livestock Policy in Cambodia", FAO PPLPI Working Paper No. 26

FAO (2008) "Biosecurity for Highly Pathogenic Avian Influenza", FAO Animal Production and Health Paper 68, FAO Rome, 2008

Hickler, B. (2007) "Bridging the Gap Between HPAI 'Awareness' and Practice in Cambodia: Recommendations from an Anthropological Participatory Assessment", Emergency Centre for Transboundary Animal Disease (ECTAD), FAO Regional Office for Asia and the Pacific

Ly S., Van Kerkhove M.D., Holl D., Froehlich Y., Vong S. (2007) "Interaction between humans and poultry, rural Cambodia", Emerging Infectious Diseases, January 2007, Available from http://www.cdc.gov/ncidod/EID/13/1/130.htm

IPC (2007), Evaluating poultry handling behavior among backyard poultry owners, their families and poultry market merchants, KAP survey conducted by IPC for UNICEF (April 2007)

Ministry of Agriculture, Forestry, and Fisheries (2007) "Provincial Websites" available online at http://www.maff.gov.kh/eng/provinces/

National Institute of Statistics (2004) "Socio-Economics Survey" (a.k.a. SES), available online at http://statsnis.org/SURVEYS/CSES2003-04.htm

OIE (2008) "Disease Information", monthly disease reports available online at http://www.oie.int/eng/info/hebdo/a_info.htm

Otte, J., Hinrichs, J., Rushton, J., Roland-Holst, D., and D. Zilberman (2008) "Review: Impacts of avian influenza virus on animal production in developing countries" in CAB Reviews: Perspectives in Agriculture, Veterinary Science, Nutrition and Natural Resources 2008 3, No. 080

Seng, S. (2007) "Gender and socio-economic impacts and its controls: Rural livelihood and bio-security of smallholder poultry producers and poultry value chain in Cambodia", submitted to FAO Cambodia by CEDAC

Seng, S., Samnol, Y., Sok, L. Khemrin, K., Thol, U. (2008) "Gender and socio-economic impacts of HPAI and its control: Rural Livelihood and Bio-Security of Smallholder Poultry Producers and Poultry Value Chain" in Siem Reap Province, Cambodia. Submitted to FAO Cambodia by CENTDOR

Vétérinaires Sans Frontières [VSF] (2004) "Review of the poultry production and assessment of the socio-economic impact of the highly pathogenic avian influenza epidemic in Cambodia", prepared for FAO's TCP/RAS/3010 "Emergency Regional Support for Post Avian Influenza Rehabilitation"

WHO (2008) "Monthly AI Reports", Available online at http://www.who.int/csr/disease/avian_influenza/country/en/

Chapter 20
Avian Influenza in China: Consumer Perception of AI and their Willingness to Pay for Traceability Labeling

Yanhong Jin and Jianhong Mu

Introduction

Highly pathogenic avian influenza (HPAI) or "bird flu" is a contagious animal disease caused by avian influenza viruses (CDC 2007). The HPAI virus spreads rapidly. In particular, the H5N1 subtype of the influenza virus causes a high mortality rate among infected birds (up to 90–100% within 48 h) (CDC 2007; OIE 2008).

The H5N1 subtype was initially detected in poultry on a farm of Scotland, UK, in 1959 (Fang et al. 2008). From then until 1990, there were nine H5N1 outbreaks recorded in Europe, North American, and Australia, contained by stamping out infected flocks (Alexander 2000). From 1990 till 2002, another ten H5N1 outbreaks were confirmed (Peiris et al. 2007). From 2003, H5N1 outbreaks accelerated in scale and geographic distribution. As shown in Fig. 20.1, from 2003 till August 2010, a total of 6,731 H1N5 outbreaks were reported to OIE (World Organization for Animal Health), which took a huge toll on the poultry industry worldwide and directly or indirectly affected both economics and social wellbeing. Peiris et al. (2007) attribute the high frequency of H1N5 outbreaks to the fast expanding, intensive poultry husbandry world wide as well as to the greater mobility of live poultry and poultry products associated with the commercialized large-scale poultry industry.

Even worse, the H5N1 virus appears to have gained ability to cross the species barrier and induces severe disease and even death in humans. The first human disease caused by H5N1 was reported in Hong Kong in 1997, which caused deaths

Y. Jin (✉)
Department of Agricultural, Food and Resource Economics,
Rutgers University, New Brunswick, NJ 08901, USA
e-mail: yjin@aesop.rutgers.edu

J. Mu
Department of Agricultural Economics, Texas A&M University,
College Station, TX 77843, USA
e-mail: mujh1024@gmail.com

D. Zilberman et al. (eds.), *Health and Animal Agriculture in Developing Countries*, Natural Resource Management and Policy 36, DOI 10.1007/978-1-4419-7077-0_20,

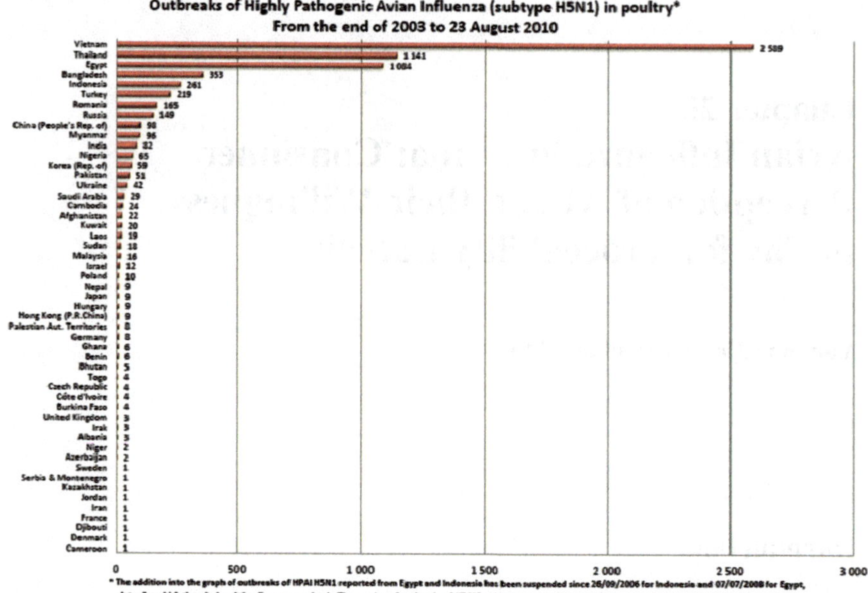

Fig. 20.1 Outbreaks of H5N1 HPAI in poultry (2003–2010).
Source: Available at the OIE website at: http://www.oie.int/downld/AVIAN%20INFLUENZA/Graph%20HPAI/graphs%20HPAI%2023_08_2010.pdf. Last accessed on October 18, 2010

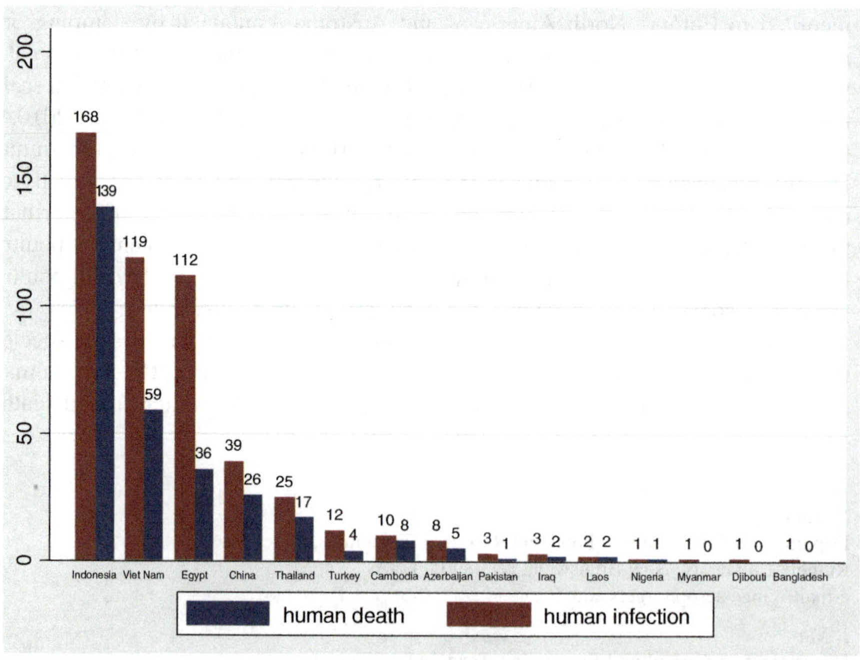

Fig. 20.2 Cumulative number of confirmed human cases of H5N1 HPAI reported to World Health Organization (2003–2010). Last accessed on October 18, 2010.
Source: Complied based on data from World Health Organization. Data are available at http://www.who.int/csr/disease/avian_influenza/country/cases_table_2010_08_31/en/index.html

of six out of 18 infected persons (Chen et al. 2004; Cheung et al. 2006; Subbarao et al. 1998). The source of human infections appeared to be in live-poultry markets where chickens, ducks, and geese were sold for human consumption. Since 2003, the disease spread widely, initially through East and Southeast Asia in 2003–2004 and then into Mongolia, southern Russia, the Middle East and to Europe, Africa, and South Asia in 2005–2006, with outbreaks recurring in various countries in 2007 (FAO 2008). As shown in Fig. 20.2, from 2003 till August 2010, a total of 505 human infections in 15 countries in Asia, Africa, the Pacific and Europe were reported to World Health Organization (WHO), of which 300 died.

China faces a high risk of HPAI due to high population density, long borders, frequent domestic and international human communications, and existence of several cross-border migration routes of wild birds through the Chinese territory, and a significant number of poultry (chicken and turkey) and water fowls (duck gooses). The first H5N1 HPAI was detected in Hong Kong live bird markets in 1997 and resulted in a depopulation of over one million birds and death of six people among 18 infected persons (Chen et al. 2004; Cheung et al. 2006; Subbarao et al. 1998). In February 2003, H5N1 reappeared when two human cases were detected in Hong Kong from travelers returning from Southern China, suggesting that H5N1 was still circulating at least among domestic poultry during the prior year (Elvander 2006). Official reports suggest there were 111 HPAI outbreaks in China from 2004 to 2009, resulting in a total of 1,347,895 birds infected and 35,647,650 depopulated (Table 20.1). Meanwhile, 39 confirmed HPAI human cases have been reported in China, with 26 fatalities (see Fig. 20.2).

In this study, we use direct survey experiments to investigate public perceptions of HPAI and willingness to pay (WTP) for foods safety relating to HPAI.

Public Perceptions and Willingness to Pay for Traceability Labeling of Poultry Products

We employ a contingent valuation method (CVM) using payment cards to solicit consumers' preference and WTP for traceability labeling of poultry meat products through face-to-face interviews at six top grocery stores between August and October 2006 in Beijing. No payment was made to respondents for their participation to avoid a possible endowment effect (Wang et al. 2007).

The survey questionnaire consists of three parts. The first section includes questions about shopping experience, preference and consumption of poultry products, and knowledge and perception of avian influenza and traceability labeling. In the second section of the questionnaire, respondents are asked to choose an amount that best represents their maximum WTP from a series of values listed in a payment card. The WTP question was phrased below: "The current chicken price is RMB9.4/kg. Up to how much more are you willing to pay for chicken with traceability labeling?" The WTP amounts were listed in an increasing order on the payment card. The lowest and highest WTP for traceability labeling in the payment card were set at zero and more than RMB5/kg, respectively. The difference between any two

Table 20.1 H5N1 outbreaks in poultry in China (2004–2009)

	2004			2005			2006			2007			2008			2009		
Province	O	BI	BD	O	BI	BD	O	BI	BD	O	BI	BD	O	BI	BD	O	BI	BD
Hunan	5	5.96	659.70	2	1.00	136.80	1	1.81	217	1	20.8	9.63						
Hubei	10	18.35	747.57	3	8.84	37.80												
Guangxi	2	0.88	754.01															
Guangdong	9	6.88	613.56							1	32.63	153.00	3	326.18	318.43			
Anhui	5	26.46	569.71	2	1.35	294.00	1	0.01	0.2									
Shanghai	1	1.50	365.00															
Yunan	7	65.09	3305.00	1	2.50	53.00												
Zhejiang	1	0.55	67.79															
Jiangxi	3	10.30	1370.75	1	3.10	332.50												
Jiangsu													2	377.00	277.00			
Tibet	1	0.43	36.19	1	0.13	78.80	1*			1	7.67	6.99	3	24.23	22.83	1	3.18	1.68
Henan	1	1.5	20.81															
Xinjiang	1	4.06	15.61	11	7.95	1365.60	2	3.25	357.41	1	24.23	29.38				1	13.74	13.22
Gansu	1	0.40	91.54															
Shanxi	2	2.30	138.55															
Tianjin	1	0.24	288.24															
Jilin	1	0.05	0.94															
Qinghai				1		16.00	1*									2	23.69	23.69
Liaoning				4	125.01	19958.50												
I. Mongolia				3	3.02	118.10	1	0.99	8.99									
Shanxi				1	8.10	67.80	2	17.60	1657.75									
Ningxia				1	0.29	99.40	2	51.30	653.90									
Sichuan				1	1.80	12.90												
Guizhou							1	16.00	42.00				1	242.36	238.36			
Total	50	144.93	9044.96	32	16.31	22571.20	12	90.95	2937.28	4	85.33	199.00	9	969.76	856.62	4	40.61	38.59

Note: "O," "BI" and "BD" indicates "outbreaks," "birds infected" and "birds destroyed," respectively. Both the number of bird infected and destroyed are in thousands. *Asterisk* (*) means the associated outbreak was in wild birds. Data are obtained from OIE (World Organization for Animal Health) at http://www.oie.int/wahis/public.php

Table 20.2 Perceptions of avian influenza among 650 useable survey samples

Percentage of respondents indicating their knowledge of HPAI outbreaks in China	99.13
Percentage of respondents with the level of knowledge about AI	
Proficient	14.83
Somewhat	73.84
Not	11.33
Percentage of respondents perceiving adverse impact of HPAI on humans	96.80

neighboring WTP amounts is designed to be noticeable and increases proportionally according to Weber's Law (see details of Weber's Law in Rowe et al. 1996). In the last section of the questionnaire, the respondent were asked to complete questions concerning their demographic and socio-economic characteristics including gender, age, education, income, employment status, marital status, residence, and whether having a pregnant woman and/or children in their family.

Since respondents are presented with a list of bids that may not necessarily represent actual alternatives, such hypothetical responses may not accurately represent "real-world" behavior, i.e., hypothetical bias to the estimated WTP (Little and Berrens 2004; List and Gallet 2001; Murphy et al. 2005). Additional methods are used to calibrate the hypothetical bias. Some studies add cheap talk into payment cards (Cummings and Taylor 1999; List 2001; Lusk 2003). Following the definition in Cummings and Taylor (1999), "cheap talk" refers to the costless of transmission of signals and information (i.e., cheap talk does not directly affect the payoffs of players in a game and it also refers nonbinding communication of actions by two or more players in an experiment prior to their hypothetical commitment). Lusk (2003) explores the effect of cheap talk in a mass mail survey using a conventional value elicitation technique. He finds that cheap talk was effective in reducing WTP for less informative survey participants, but it did not reduce WTP for knowledgeable respondents. Aadland and Caplan (2003) and Aadland et al. (2007) present a theoretical framework to model the relationship between anchoring bias and cheap talk in contingent valuation surveys. Their results show that adding cheap talk is more effective for relatively high referendum prices than under a simple hypothetical choice scenario. In this study, half of the respondents were presented a cheap talk script and then were asked to answer the WTP questions again.

The total number of participants is 702 for the survey without the cheap talk script, 346 of which were presented the cheap talk script. The total number of usable respondent is 650 without the cheap talk script and 313 with the cheap talk script, respectively. As shown in Table 20.2, among the total 650 usable samples, a greater proportion of participants considered themselves to be knowledgeable about HPAI outbreaks in China (99.13%), and about HPAI at various degrees (14.83% at the proficient level and 73.84% indicating some knowledge of AI). Furthermore, 96.80% of the respondents believed that HPAI virus can adversely affect human health. The significant knowledge about HPAI and the impact on human health among the respondents are echoed by other studies. For example, Xiang et al. (2010) find that 69% of urban residents in Shenzhen and 56% of rural residents in Xiuning considered themselves to be knowledgeable about human infections with AI.

Table 20.3 Willingness to pay for traceability labeling of poultry products

	Without cheap talk	With cheap talk
Summary statistics of WTP		
Mean	1.10	0.99
Median	1.13	0.84
Variance	0.93	0.93
Comparison of WTP between with and without cheap talk		
p-Value of Kolmogorov-Simirnov tests	0.05	
Wilcoxon-Mann-Whitney test for equal mean	2.69^{***} (0.01)	
Pearson's χ^2 test for equal median	7.36^{***} (0.01)	
Levene test for equal variance tests	0.53 (0.46)	

Figures in parentheses are p-values and figures above parenthesis are test statistic. Asterisks (*, **, and ***) indicate 10, 5, and 1% significance levels, respectively

The WTP for traceability of poultry product is highly skewed – the skewness coefficients are 2.25 and 1.95 for the survey with or without cheap talk. Thus, the traditional student t-test for equal mean is not appropriate as it is highly sensitive to the normality assumptions. The nonparametric Wilcoxon-Mann-Whitney tests are undertaken. As shown in Table 20.3, the mean and median WTP without the cheap talk script is statistically higher, but the variance of the WTP between two samples is not statistically different from each other.

We also employ a multivariate regression analysis. The underlying WTP for traceability of poultry products which is denoted by $\overline{\text{WTP}}_k$, is not completely observable to researchers, especially of those who may have a negative perception of traceability and those who have selected the WTP category that is higher than the specified WTP in the payment cards. Thus, a double-censored Tobit model is applied. Let WTP^*_{ik} and WTP_{ik} denote the underlying and stated WTP for each individual i under each survey. We assume the underlying WTP is a linear function of explanatory variables,

$$\text{WTP}^*_{ik} = \beta X_i + \alpha PC_{ik} + \varepsilon_{ik},$$

where PC is a dummy indicating the presence of cheap talk script and α measures the impact of the cheap talk. The other explanatory variables denoted by X are grouped into socio-demographic factors, behavioral characteristics, and perception and knowledge of HPAI and traceability labeling. ε_{ik} are normally distributed error terms. Since we posit that cheap talk will systematically affect the WTP amount, we assume that error terms are correlated among those respondents who were presented cheap talk or among those who were not presented cheap talk. The stated WTP is

$$\text{WTP}_{ik} = \begin{cases} \overline{\text{WTP}} & \text{if } \text{WTP}^*_{ik} \geq \overline{\text{WTP}}, \\ \text{WTP}^*_{ik} & \text{if } \overline{\text{WTP}} > \text{WTP}^*_{ik} > 0, \\ 0 & \text{otherwise.} \end{cases}$$

Table 20.4 Estimated coefficients and marginal effects on the WTP for traceability

	Estimated coef. (std. error)	Marginal effects
With a cheap talk script	−0.136* (0.072)	−0.115** (0.061)
Knowledge, perception, and behavior characteristics		
Family grocery shopper	−0.024 (0.078)	−0.021 (0.070)
Frequent chicken eater	−0.150** (0.071)	−0.128** (0.060)
Knowledge of AI	0.189 (0.127)	0.164 (0.112)
Knowledge of traceability	0.266*** (0.085)	0.230*** (0.074)
Perceived importance of traceability	0.172 (0.159)	0.143 (0.130)
Demographic and socio-economic characteristics		
Age	−0.012** (0.006)	−0.010** (0.005)
Male	−0.122* (0.072)	−0.104* (0.062)
Single	−0.078 (0.112)	−0.067 (0.096)
Education status		
Above high school but less than college	0.118 (0.111)	0.101 (0.095)
College and above	0.041 (0.117)	0.035 (0.100)
Employment status (base = full/part time job)		
Retired	−0.048 (0.152)	−0.041 (0.128)
Unemployed	−0.040 (0.113)	−0.033 (0.096)
Students	−0.091 (0.103)	−0.077 (0.086)
Income (1,000 yuan)	0.006** (0.002)	0.005** (0.002)
Urban resident	0.138 (0.104)	0.116 (0.086)
Family with a pregnant woman	0.123 (0.184)	0.106 (0.161)
Family with a child younger than 12	−0.118 (0.078)	−0.010 (0.065)
Constant	1.109*** (0.268)	
Sigma	0.951*** (0.055)	
R-squared	0.03	
No. of observations	813	

Figures in parentheses are standard error of the estimated coefficients. Asterisks (*, **, and ***) indicate 10, 5, and 1% significance levels, respectively

Table 20.4 presents the estimated coefficients and the marginal effects of the double-censored Tobit regression. Controlling for knowledge and perception of HPAI and traceability, demographic factors, and socio-economics characteristics, the cheap talk script reduces the WTP for traceability by RMB0.11. Furthermore, there was no statistically significant difference of WTP between respondents who are knowledgeable of AI. Respondents who are knowledgeable of traceability labeling pay RMB0.23 more at the 1% significance level. Household annual income has a highly significant, positive impact on WTP. However, the marginal effect of income is minimal – an increase of RMB1,000 increases the WTP amount by RMB0.005 on average. An age effect is detected but the marginal effect is minimal: a one year increase in age decreases WTP by RMB0.01 at the 5% significance level. Females are willing to pay RMB0.10 more than male respondents. Frequently consumers of poultry products have a lower WTP by RMB0.13 at the 5% significance level. Among others, no significant effect is detected for employment status, education levels, whether there is a pregnant woman or a child younger than 12 in the household. Overall, the stated average WTP for traceability labeling in Beijing due to HPAI risks is 11.5%; a cheap talk script reduces the WTP by five percent points. The amount of WTP for traceability is consistent with previous studies in the context of meat traceability in other countries (Dickinson and Bailey 2002, 2005; Hobbs et al. 2005; Hobbs 2003).

Conclusions

China faces a high risk of HPAI outbreaks nad human infections, which poses significant economic consequences not only in the domestic poultry sectors, but also in other sectors in the Chinese economy and China's trading partners. This chapter investigates consumers' perception of HPAI and their WTP for traceability labeling of poultry products in China. On average, Chinese consumers are willing to pay approximately 10–11 percent points more for poultry products with traceability label.

References

Aadland, D., and Caplan, A. J. 2003. "Willingness to Pay for Curbside Recycling with Detection and Mitigation of Hypothetical Bias." *American Journal of Agricultural Economics* 85(2), 492–502.

Aadland,D, A.J. Caplan and Owen Phillips. 2007. "A Bayesian Examination of the Interaction between Anchoring and Cheap Talk in Contingent Valuation", Working paper, University of Wyoming.

Alexander, D.J. 2000. "A review of avian influenza in different bird species." *Veterinary Microbiology* 74:3–13.

CDC (Center for Disease Control and Prevention). 2007. "Key facts about avian influenza (bird flu) and avian influenza A (H5N1) virus." Accessed at http://www.cdc.gov/flu/avian/gen-info/facts.htm

Chen, H., G. Deng, Z. Li, G. Tian, Y. Li, P. Jiao, L. Zhang, Z. Liu, R. G. Webster, and K. Yu. 2004. "The evolution of H5N1 influenza viruses in ducks in southern China." *Proc. Natl. Acad. Sci. USA* 101:10452–10457.

Chen, H.L. 2009. "H5N1 HPAI influenza in China." *Science in China Series: Life Sciences* 52 (5):419–427.

Cheung, C.L., J.M. Rayner, G.J. Simth, P. Wang, T.S. Naipospos, J. Zhang, K.Y. Yuen, R.G. Webster, J.S. Peiris, Y. Guan, and H. Chen. 2006. "Distribution of amantadin-resistant H5N1 avian influenza variants in Asian." *Journal of Infectious Disease* 193:1626–1629.

Cummings, R.G. and L.O. Taylor. 1999. "Unbiased value estimates for environmental goods: a cheap talk design for the contingent valuation method." *American Economic Review* 89:649–665.

Dickinson, D.L., & Bailey, D. 2005. "Experimental evidence on willingness-to-pay for red meat traceability in the United States, Canada, the United Kingdom, and Japan." *Journal of Agricultural and Applied Economics* 37: 537–548.

Dickinson, D.L., & Bailey, D. 2002. "Meat traceability: are U.S. consumers willing to pay for it?" *Journal of Agricultural and Resource Economics* 27: 348–364.

Elvander E. 2006. "Avian Influenza and China." Congressional-Executive Commission on China.

Fang, L-Q., S.J. de Vlas, S. Liang, CWN Looman, P. Gong, B. Xu, L. Yan, H. Yang, J.H. Richardus, WC. Cao. 2008. "Environmental factors contributing to the spread of H5NQ avian influenza in mainland China." *PLoS ONE* 3(5):e2268 1–6.

FAO. 2008. ECTAD HPAI Situation Update. Issue No. 306.

Hobbs, J.E., D. Bailey, D.L Dickinson, and M. Haghiri. 2005. "Traceability in the Canadian red meat sector: do consumers care?" *Canadian Journal of Agricultural Economics* 53:47–65.

Hobbs, J.E., 2003. "Traceability in meat supply chains." Current Agriculture, Food and Resource Issues 4: 36-49. [http://cafri.usask.ca/j_pdfs/hobbs4-1.pdf]

List, J.A. and C.G. Gallet. 2001. "What experimental protocol influence disparities between actual and hypothetical stated values? evidence from a meta-analysis." *Environmental and Resource Economics* 20: 241–254.

List, J.A. 2001. "Do explicit warming eliminate the hypothetical bias in elicitation procedures? evidence from field auctions for sports cards." *American Economic Review* 91:1498–1507.

Lusk, J.L. 2003. "Effects of cheap talk on consumer willingness-to-pay for golden rice." *American Journal of Agricultural Economics* 85:840-856.

Little, L. and R. Berrens. 2004. "Explaining disparities between actual and hypothetical stated values: further investigation using meta-analysis." *Economics Bulletin* 3:1–13.

Murphy, J.J, P.G. Allen, T.H. Stevens, and D. Weatherhead. 2005. "A meta-analysis of hypothetical bias in stated preference valuation." *Environmental and Resource Economics* 30:313–325.

OIE. 2008. "Avian influenza: manual of diagnostic tests and vaccines for terrestrial animals (chapter 2.3.4)", accessed at http://www.oie.int/Eng/normes/mmanual/2008/pdf/2.03.04_AI.pdf

Peiris, J.S.M., M.D. de Jong and Y. Guan. 2007. "Avian influenza virus (H5N1): a threat to human health." *Clinical Microbiology Review* 20: 243–267.

Rowe, R.D., Schulze, W.D. & Breffle, W.S. 1996. "A test for payment card biases". *Journal of Environmental Economics and Management* 31:178–185

Subbarao, K., A. Klimov, J. Katz, H. Regnery, W. Lim, H. Hall, M. Perdue, D. Swayne, C. Bender, J. Huang, M. Hemphill, T. Fowe, M. Shaw, X. Xu, K. Fukuda, and N. Cox. 1998. "Characterization of an avian influenza A (H5N1) virus isolated from a child with a fatal respiratory illness." *Science* 279:393–396.

Wang, T., R. Venkatesh and R. Chatterjee. 2007. "Reservation price as a range: An Incentive Compatible Measurement Approach." *Journal of Marketing Research* XLIV:200-213.

Xiang, N., Y. Shi, J Wu, S. Zhang, M Ye, Z. Peng, L. Zhou, H. Zhou, Q. Liao, Y. Huai, L Li, Z. Yu, X. Cheng, W. Su, Xi. Wu, H. Na, J. Lu, J. McFarland, H. Yu. 2010. "Knowledge, attitudes and practices (KAP) relating to avian influenza in urban and rural areas of China." *BMC Infectious Diseases* 10:34.

Part V
Conclusion

Chapter 21
Conclusion

David Zilberman, Joachim Otte, David Roland-Holst, and Dirk Pfeiffer

The influenza virus and zoonotic diseases are perennial companions of human society, posing substantial direct threats to human lives and livelihoods as well as to animal populations. Zoonotic diseases coevolve with human society, animal husbandry, and technology, and this book presents multidisciplinary frameworks to assess zoonotic-disease impacts and to control them. This research is applied to one of today's most important pandemic threats, Avian Flu (HPAI type H5N1), but it has lessons of relevance to most zoonotic-disease risks—past, present, and future.

It is important to understand zoonotic-disease control measures and their impacts in the wider context of the major processes shaping agriculture's evolution in the new millennium (Zilberman and Lipper 2005). These processes include globalization of goods and capital markets, more dynamic and widespread economic growth, private and public environmentalism, consumerism, population growth, industrialization of agriculture, and parallel emergence of the knowledge sector and the bioeconomy. These processes strongly affect the evolution of animal diseases and provide new tools to address them.

It has been shown here and elsewhere that population growth combined with rapid emerging market growth, especially in Asia, are major contributors to accelerating meat demand and concomitant expansion of the livestock sector (von Braun 2007). At the same time, consumerism and the increased willingness to pay for quality and other food characteristics has stimulated the emergence of

D. Zilberman (✉) • D. Roland-Holst
Department of Agricultural and Resource Economics,
University of California, Berkeley, CA, USA
e-mail: zilber11@berkeley.edu

J. Otte
Food and Agriculture Organization of the United Nations, Rome, Italy

D. Pfeiffer
Veterinary Epidemiology and Public Health Group, Department of Veterinary Clinical Sciences,
Royal Veterinary College, University of London, Hartfield, Hertfordshine, UK

D. Zilberman et al. (eds.), *Health and Animal Agriculture in Developing Countries*,
Natural Resource Management and Policy 36, DOI 10.1007/978-1-4419-7077-0_21,

large food-processing sectors and proliferated industrialized animal products (processed, precut, etc.) even in the developing countries. Heiman, McWilliams, and Zilberman (Chapter 9) (and the literature that they survey) clearly reveal that consumers are ever more willing to pay for time-saving features and safety assurance in food items. Significantly, this research identifies social norms that increase the value and selection of food items that are synergetic with traditional practices. This finding is consistent with the emergence of the slow food markets that promote the production of traditional foods by modern means. Ifft (2011) shows that there is significant willingness to pay for traditional poultry above and beyond the price of varieties used to produce industrial chicken. Ifft further shows that these premia actually increase for traditional poultry products certified to be safe. Jin and Mu (Chapter 20) also find that consumers are concerned about zoonotic disease and are willing to pay for credible assurances of food safety. These findings suggest that there is a demand for diverse meat products that may vary in taste and texture and, at the same time, there is a premium for the ease of use and convenience. Thus, different livestock production systems can apparently coexist, even by different sectors, as long as the final product is certified to be both safe and convenient. In other words, consumers can value product diversity and can sustain heterogeneous production systems as long as they meet acceptable standards of quality and safety.

Globalization and the frequent movement of agrofood-related goods, services, and livestock has been accelerating diffusion of zoonotic-disease risk, and environmental considerations have led to policies that will counter some of the excesses of uncontrolled trade. Wang, Fenichel, and Perrings (Chapter 7) developed inspection strategies and incentives that would reduce transboundary disease risk. Although they argue that the use of complete bans on trade and livestock to control disease can serve as trade barriers, there should be a place for intensive inspection to permit trade even when zoonotic diseases exist. The ability to control animal movement as well as movement of disease vectors is crucial for fully effective disease control (Chapter 13). The ability of operators to move livestock in spite of regulations is a major hindrance to disease-control efforts (Chapters 2 and 19). Environmentalism led to the support of regulations and international cooperation in pursuit of global and environmental problems and has given rise to some of the collaborative efforts that are being introduced to address some global zoonotic-disease risks. Sproul et al. (Chapter 8) suggest that differences in income and willingness to pay to prevent health risks may lead to contributions to global effort to control zoonotic diseases at the source where some of the developing countries in which diseases are endemic are supported by wealthier countries that may suffer the consequences to fight the spread of disease. The design of policies to control disease should combine financial incentives as well as direct control but, to a large extent, depends on technology and quantitative modeling of biological and economic systems, which can be challenging given the dynamic considerations and multiple feedbacks (Chapters 6 and 10).

In spite of significant progress in medical technologies, real-world capacity to deal with zoonotic diseases is constrained by practical limitations and costs of these technologies (Chapter 16). For example, in the case of HPAI H5N1, the effectiveness of mass vaccination is limited because costs are high while the range of problems that it protects against is narrow and protection is short term (Chapter 12). High costs, combined with extended diagnostic response time, undermine the capacity to monitor the evolution and spread of disease in a timely manner. This, in turn, necessitates more generic strategies, such as culling, food controls, etc., as second-order disease-control strategies. Famers' own perceptions and attitude toward risk may exacerbate the cost of implementing damage-control policies (Chapter 15). It is unclear, for example, why it would be rational for them to be internalizing social costs of animal-disease risk (especially global risks). Science is challenged to develop more precise and cheaper diagnostics as well as vaccination technology to reduce the cost and increase the effectiveness of zoonotic-disease control. Whenever new technologies are introduced, farmer training in the use of the technology and a regulatory setup to assess the impact also need to be introduced, combined with incentive-compatible polices that foster appropriate adoption behavior.

Agricultural industrialization provides new mechanisms to control the spread of disease and to improve product quality. Yet, Hennessy and Wang (Chapter 5) suggest that these features may lead to larger-than-optimal operations and change the structure of agriculture, adversely affecting smaller farms and the rural-poor majorities, who inhabit them in many developing countries. Smallholders can continue to operate in the livestock sector, especially if they concentrate on safe, traditional products that may fetch an extra premium because of quality and taste and if they are part of a supply chain that holds down the cost of marketing and distribution (Chapters 17–19). There is, for example, potential to encourage contractual arrangements where farmers are producing poultry and where intermediaries purchase the product and market it to various retail outlets while monitoring and ensuring product safety.

Any strategy to control diseases must include efforts to improve technology and develop appropriate monitoring, infrastructure, and incentives. Yet, as Olmstead and Rhode (Chapter 2) suggest, the key in the control of complex zoonotic diseases is political commitment and consistency. Technology may advance slowly, and there may be incentives to sell and smuggle sick animals and to conduct other counterproductive activities. Furthermore, political cooperation may be lacking. Only long-term commitment by governments and international organizations to monitor agrofood systems, learn from mistakes, and take advantage of opportunities to reduce the cost of disease control will contain zoonotic-disease risk in a socially effective manner.

Salient Lessons from HPAI

Much of the research leading to this book was conducted by teams working in countries of the Mekong region relating to various aspects of HPAI and HPAI control. From the core HPAI research project, 10 main findings have been distilled for policy makers:

1. HPAI (and LPAI/AIV) control in domestic poultry poses an unprecedented challenge for the veterinary profession because of the genetic versatility of these viruses; their invasion of a large and geographically dispersed region; high-turnover, domestic-poultry populations; and the possibility of asymptomatic persistence in domestic ducks and, possibly, other animal reservoirs.
2. Because of the scale of economic risk in their own countries, the Organisation for Economic Cooperation and Development (OECD) economies has given primary global policy and financial impetus to HPAI risk reduction. However, empirical evidence on actual and potential domestic damages suggests that OECD, as well as China and India, should still make much larger investments in overseas risk reduction.
3. HPAIV H5N1 now appears to be endemic in parts of the Greater Mekong Subregion (GMS). We anticipate that it will be difficult to obtain the level of domestic and (especially) external public resources needed to sustain control measures at previous levels.
4. The role of wild birds in the spread of HPAI H5N1 in the GMS is negligible. Indeed, it is apparent that an important source of risk, the covert transboundary trade that pervades the region, has been misattributed to wild birds.
5. Publicly funded, blanket vaccination campaigns are costly and appear to be ineffective against HPAI in areas with a high prevalence of small-scale poultry keepers. Targeted vaccination of specific high-risk groups can achieve comparable risk reduction at a fraction of the cost. The same is true of radial approaches to culling birds and the destruction of smallholder poultry infrastructure, which are very costly to communities but appear to contribute little to risk reduction. Culling should be limited to infected flocks. Infrastructure can be disinfected but should not be destroyed.
6. Attempting to improve the biosecurity of millions of backyard producers is an ineffective use of scarce resources in the GMS countries, especially public funds in countries with many high-priority development objectives.
7. Domestically, effective public- and animal-health policies must arise from and be sustained by sound policy institutions with adequate capacity and coordination at the national, regional, and local levels. Unfortunately, governmental institutions in the GMS are very diverse in all of these aspects and HPAI risk management has, in some cases, been seriously compromised by institutional weakness.
8. Although they comprise the vast majority of poultry keepers in the GMS, smallholders do not presently have a voice in the design of short- and long-term HPAI control and mitigation policies. Omitting this stakeholder group is a mistake that seriously compromises policy effectiveness and legitimacy.
9. Despite the fact that their individual risk is statistically much lower than that of large operations, because of their geographic dispersion and majority status, smallholders play a crucial role in HPAI risk management. It is essential to recognize the rural poor as part of a solution (effective disease defense) rather than a problem (infection risk), enlisting them with socially effective

policies that recognize and reward their contribution to the national and global commons of disease resistance.

10. Market-oriented polices offer vital opportunities for private cost sharing and self-directed poverty reduction. For example, certification and other product quality/safety initiatives can be self-financed and incentive compatible—a socially effective substitute for open-ended fiscal commitments to public disease monitoring and geographically extensive control measures.

References

Ifft, Jennifer, "The Economics of Animal Health: The Case of Avian Influenza in Vietnam." University of California at Berkeley, Ph.D. Dissertation, Department of Agricultural and Resource Economics, 2011.

von Braun, Joachim, *The World Food Situation: New Driving Forces and Required Actions.* Washington, D.C.: International Food Policy Institute, 2007.

Zilberman, David, and Leslie Lipper. "Major Processes Shaping the Evolution of Agriculture, Biotechnology, and Biodiversity." In Joseph Cooper, Leslie Marie Lipper, and David Zilberman (eds.) *Agricultural Biodiversity and Biotechnology in Economic Development.* New York: Springer, 2005.

Index

D. Zilberman et al. (eds.), *Health and Animal Agriculture in Developing Countries*, Natural Resource Management and Policy 36, DOI 10.1007/978-1-4419-7077-0,

B